Biological Engineering

Biological Engineering

Edited by **Suzy Hill**

New York

Published by Callisto Reference,
106 Park Avenue, Suite 200,
New York, NY 10016, USA
www.callistoreference.com

Biological Engineering
Edited by Suzy Hill

International Standard Book Number: 978-1-63239-099-8 (Hardback)

Printed in the United States of America.

Contents

Permissions

List of Contributors

Preface

Biological engineering is a field of engineering in which life and life-sustaining systems are highlighted. The most significant trend in biological engineering is the dynamic advancement at which biotechnology is now able to integrate with biological processes and an explosion in micro/nanoscale technology is permitting the manufacture of nanoparticles for drug delivery into cells, miniaturized implantable microsensors for medical diagnostics, and micro-engineered robots for on-board tissue repairs. This book intends to offer an updated sketch of the latest advances in biological engineering from varied aspects and several applications grouped under two sections: Biochemical Engineering Methods and Applications, and E-Health and Educational Aspects of Bioengineering.

After months of intensive research and writing, this book is the end result of all who devoted their time and efforts in the initiation and progress of this book. It will surely be a source of reference in enhancing the required knowledge of the new developments in the area. During the course of developing this book, certain measures such as accuracy, authenticity and research focused analytical studies were given preference in order to produce a comprehensive book in the area of study.

This book would not have been possible without the efforts of the authors and the publisher. I extend my sincere thanks to them. Secondly, I express my gratitude to my family and well-wishers. And most importantly, I thank my students for constantly expressing their willingness and curiosity in enhancing their knowledge in the field, which encourages me to take up further research projects for the advancement of the area.

Editor

Part 1

Biochemical Engineering Methods and Applications

In Vitro Blood Flow Behaviour in Microchannels with Simple and Complex Geometries

Valdemar Garcia[1], Ricardo Dias[1,2] and Rui Lima[1,2]
[1]Polytechnic Institute of Bragança, ESTiG/IPB, C. Sta. Apolonia, Bragança,
[2]CEFT, Faculty of Engineering of the University of Porto (FEUP), R. Dr. Roberto Frias, Porto,
Portugal

1. Introduction

Over the years, various experimental methods have been applied in an effort to understand the blood flow behaviour in microcirculation. The development of optical experimental techniques has contributed to obtain possible explanations on the way the blood flows through microvessels. In recent years, due to advances in computers, optics, and digital image processing techniques, it has become possible to combine a conventional particle image velocimetry (PIV) system with an inverted microscope and consequently improve both spatial and temporal resolution. The present review outlines our most relevant studies on the flow properties of blood at a microscale level by using current micro-PIV and confocal micro-PIV techniques.

Blood flow in both microvessels and microchannels has been measured by several measurements techniques such as: double-slit photometric (Nash & Meiselman, 1983), laser-Doppler anemometer (Uijttewaal et al., 1994), video-based methods (Parthasarathi et al., 1999). Although the past research findings have been encouraging, detailed studies on the way blood flow behaves at a microscopic level have been limited by several factors such as poor spatial resolution, difficulty to obtain accurate measurements at such small scales, optical errors arisen from walls of the microvessels, high concentration of blood cells, and difficulty in visualization of results due to insufficient computing power and absence of reliable image analysis techniques. In recent years, due to advances in computers, optics, high-speed imaging and image processing techniques, it has become possible to combine a conventional particle image velocimetry (PIV) system with an inverted microscope and consequently improve both spatial and temporal resolution (Santiago et al., 1998; Koutsiaris et al., 1999). This system, known as micro-PIV (see Fig.1), has been applied to study the flow behaviour in several research fields in science and engineering. In the biomedical field, Sugii and his co-workers, by using a conventional micro-PIV system, they have used red blood cells (RBCs) as tracer markers to measure their velocities in straight (Sugii et al., 2002) and they found that the velocity profiles were markedly blunt in the central region. However, later they measured both tracer particles and RBCs through a 100 μm glass capillary and they reported that by using in vitro blood with about 20% Hct the velocity profiles were parabolic (Sugii et al., 2005). By using a microchannel close to a rectangular shape, Bitsch and his co-workers (Bitsch et al., 2005) have reported blunt profiles. More recently, by using liposomes tracer particles the blood-plasma velocity was measured in the beating heart of a chicken embryo (Vennemann et al., 2006). Kim and Lee (2006) have analysed the flow behaviour of blood through a circular opaque microchannel by using an X-ray PIV technique. Their measurements have shown typical non-Newtonian flow characteristics of blood such as yield stress and shear-thinning effects. In addition, Chiu et al. (2003) have also applied the conventional micro-PIV system to analyse the effect of flow in monocyte adhesion to endothelial cells cultured in a vertical step flow chamber. Although, micro-PIV systems are gaining widespread among the biomicrofluidics community due to its high spatial and temporal resolution, the employment of conventional microscope leads to the entire illumination of the flow field resulting in high levels of background noise and consequently errors on the flow measurements (Meinhart et al., 2000). These errors can be partially removed by using a spinning disk confocal microscope (SDCM), i. e., by combining a SDCM with a laser, the emitted light intensity is significantly improved and as a result, it is possible to obtain adequate signal to noise ratio to detect the motion of the RBCs in both diluted and concentrated suspensions (Tanaami et al., 2002; Park et al., 2004, 2006; Lima et al., 2006, 2007, 2008). Moreover, in contrast to the conventional microscope where the entire flow region is illuminated, the confocal systems have the ability to obtain in-focus images with optical thickness less than 1 μm (optical sectioning effect). As a result, confocal systems due to its spatial filtering technique and multiple point light illumination system, confocal micro-PIV has become accepted as a reliable method for measuring in vitro blood flow through microchannels.

In this chapter, our recent studies about in vitro blood flow behaviour in microchannels both in straight and with complex geometries are presented. In straight microchannels we present some phenomena such as Fahraeus effect and Fahraeus-Lindqvist effect, the flow of particles and red blood cells (RBCs) in diluted suspensions, the flow of RBCs in concentrated suspensions, the cell-free layer and sedimentations effects. The most recent studies in blood flow through complex geometries such as bifurcations, confluences and stenosis are also reviewed. By using a chromatographic method, the flow of RBCs through a network of microcapillaries is presented.

2. Conventional micro-PIV and confocal micro-PIV/PTV

The main components of a conventional micro-PIV system consists of a microscope, a high resolution objective lens, optical filters, a high power light source for flow illumination and a high speed camera. Briefly, the light enters the inverted microscope and is reflected by a dichromatic mirror to be transmitted through the objective lens which illuminates the entire flow volume. The light emitted from fluorescent trace particles travels back to the dichromatic mirror and filters out all reflected light only allowing to pass the emitted light from the particles. Finally, the signals from the trace particles are recorded by a high speed camera and then by using a PIV cross-correlation method it is possible to obtain velocity fields of the working fluid (see schematic illustration of a conventional micro-PIV in Figure 1). The resolution of a micro-PIV system is influenced by many factors such as out-of-focus particle images from volume illumination, density and size of the tracer particles, size and optical characteristics of the microchannel and image quality (Lima, 2007).

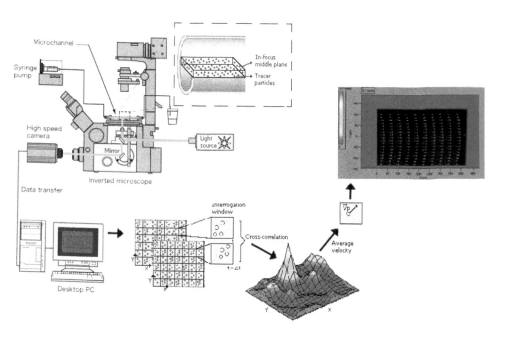

Fig. 1. Experimental setup of a typical conventional micro-PIV system and PIV cross-correlation method.

For the case of a confocal micro-PIV system, the main components of consists of a microscope combined with a confocal scanning unit (CSU), a high resolution objective lens, a high power light source for flow illumination and a high speed camera. In brief, the light emitted by the laser enters the CSU and then is conducted to the microscope to illuminate the microchannel from below the microscope stage. The light emitted from fluorescent particles travels back into the CSU, where a dichromatic mirror reflects it onto a high-speed camera to record the confocal images and by using a PIV cross-correlation method to obtain the velocity fields of the flowing trace particles (see Figure 2).

The main advantages of using a confocal spinning disk (CSD) are: the ability to obtain thin in-focus images, improve image definition and contrast of the trace particles (see Figure 3). As a result confocal micro-PIV systems have potential to obtain three-dimensional information about the fluid flow and also to obtain accurate flow-field measurements. In this way it is possible to study complex blood flow phenomena that often occur in two-phase flows (Lima, 2007).

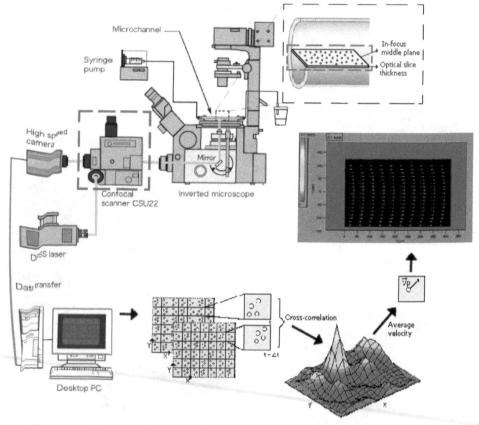

Fig. 2. Experimental setup of a confocal micro-PIV system and PIV cross-correlation method.

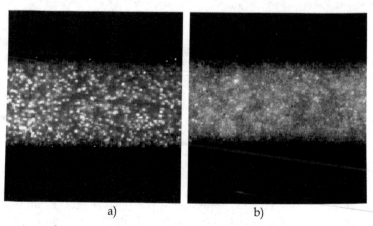

a)   b)

Fig. 3. Comparison of trace particles images from both confocal (a) and conventional micro-PIV (b) for pure water.

The density of particles in the recorded images determines the most adequate PIV methodology to obtain the velocity fields. When the concentration of particles is high enough that every interrogation window contains at least three particles, this method is called high-image-density PIV mode (Adrian, 1991). However, for the case of physiological fluids with high concentrations of cells, the amount of tracer particles captured within the fluid is often very low. Hence, if the number of particles within the interrogation area is small, it is recommended to measure the displacements by tracking individual particles in a Lagrangian way. This low-image-density PIV method is often referred as particle tracking velocimetry (PTV) (Adrian, 1991). The main advantage of PTV method is the ability to obtain detailed quantitative information on the motion of particles and cells flowing within the working fluid. This method is becoming essential in several biomedical fields such as cell biology and microcirculation. The present review will show several examples using a micro-PTV method to investigate in vitro blood flow behaviour in microchannels with both simple and complex geometries. A schematic illustration of a confocal micro-PTV is shown in Figure 4.

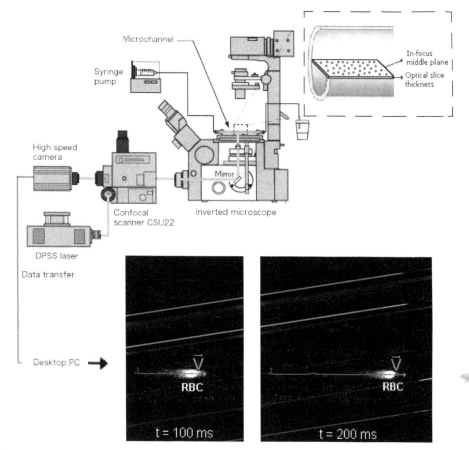

Fig. 4. Experimental setup of a confocal micro-PTV system with a labelled RBC trajectory and correspondent velocity at different times obtained by means of a particle tracking method.

3. Blood composition

In microcirculation, which comprises the smallest arteries and veins, the flow behaviour of individual blood cells and their interactions provide the microrheological basis of flow properties of blood at a macroscopic level. As a result, in microcirculation it is fundamental to study the flow behaviour of blood at cellular level. Thus, blood is not a homogeneous fluid, but one composed of a suspension of cells, proteins and ions in plasma. In normal blood, three types of cells comprise about 46% of its volume. These cells are the red blood cells (RBCs) (also known as erythrocytes) representing 45% of volume, white blood cells (WBCs) (also known as leukocytes) and platelets (also known as thrombocytes) (see Figure 5).

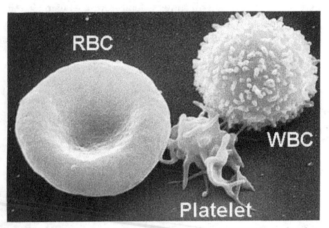

Fig. Scanning electron micrograph of a white blood cells (WBC), a platelet and a red blood cell (RBC) (adpated from NCI-Frederick, 2005).

In vitro blood flow behaviour in microchannels is strongly influenced by the RBCs, since they occupy almost half of whole blood volume. RBCs are formed in the bone marrow and during maturation they lose their nuclei before entering the circulatory system. When suspended in an isotonic medium RBCs has a biconcave discoid shape, with a major diameter of about 8 µm. The density of a RBC is about 1.08×10^3 kg.m^{-3} and its major cellular components are the cytoplasm and a thin membrane composed of lipid bilayer and protein molecules. There is experimental evidence that healthy RBCs are extremely deformable into a variety of shapes in flowing blood in response to hydrodynamic stresses acting on them. Figure 6 shows a RBC deformation in a capillary (Lima et al., 2012).

The WBCs are nucleated cells that represent the major defence mechanism against infections. Generally, their shape is roughly spherical but their surface is not normally smooth (see Figure 5). The diameter of WBCs ranges from about 7 up to 22 µm, depending on its type. Healthy blood contains normally less than 1% of WBCs of the total volume of blood cells (Lima et al., 2012). Little is known about the effect of the WBCs on the blood flow behaviour in microcirculation. The blood flow under pathological conditions may increase amount of WBCs within the flow and consequently they may disturb the flow behaviour in microvessels (see Figure 7).

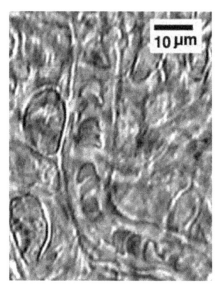

Fig. 6. RBCs deformation *in vivo* capillary (Minamiyama, 2000).

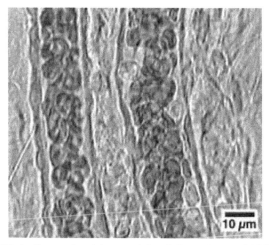

Fig. 7. Rolling of WBCs (yellow colour) in arterioles (Minamiyama, 2000).

Platelets are cells with no nuclei, round or oval discoid shape, in general, and with diameters from about 1 to 2 μm (see Figure 5). The number of platelets is usually less than the WBCs and they may have little effect on the blood flow behaviour. Although platelets play an important role in blood coagulation and thrombus formation, this topic is beyond the scope of the present review. Plasma is a yellowish fluid which contains 90% of water by volume and 10% of proteins, inorganic substances, vitamins, dissolved gases, etc. The proteins within the plasma flow, due to their large molecular size, usually do not pass through the capillary wall, thus generating an osmotic pressure. In *in vitro* experiments the osmotic pressure is an important parameter that needs special attention (Lima et al., 2012).

4. In vitro blood flow behaviour in straight microchannels

4.1 Fahraeus effect and Fahraeus-Lindqvist effect

In large arteries, where the diameter of the blood vessels is large enough compared to individual cells, it has been proved adequate to consider blood as a single-phase fluid (Caro et al., 1978). Accordingly, blood in large arteries may be treated as a homogeneous fluid where its particulate nature is ignored. Moreover, due to the large Reynolds number (Re) in arteries, blood flow is governed by inertial forces. However, arteries divide into successive smaller arteries and consequently the cross-sectional area of the vascular bed increases. As a result both pressure and velocity decrease as the blood flows into the smaller vessels. When the blood reaches the arterioles and capillaries the Re became less than 1, where viscous force dominates over inertial forces. At this microscale it is fundamental to take into account the effects of the multiphase properties of the blood on its flow behaviour (Caro et al., 1978). A clear example of the multiphase nature of the blood is the formation of a plasma layer at microvessels less than 300 μm, known as Fahraeus-Lindqvist effect (Fahraeus & Lindqvist, 1931).

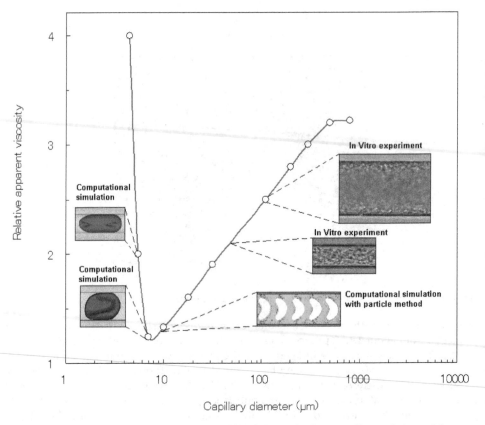

Fig. 8. Relative apparent viscosity of *in vitro* blood through glass capillaries (adapted from Pries et al., 1992; Wada & Kobayashi, 2002; Tsubota et al., 2006).

In the classical work of Robin Fahraeus, he observed that blood flow behaviour and its hematocrit are strongly affected by microtubes diameters less than 300 μm. The Fahraeus effect indicates that the Hct in the glass capillaries (< 300 μm) is lower than the feed Hct, which suggests that the Hct decreases as the blood proceeds through narrower microvessels. This phenomenon results from the axial migration of the RBCs to the centre of the microtube and consequent faster motion of the cells when compared with the suspending medium, such as plasma or dextran (Fahraeus & Lindqvist, 1931). The Fahraeus-Lindqvist effect is somehow related to the above phenomenon. Again for microtubes diameters less than 300 μm, Fahraeus and Lindqvist observed that the apparent blood viscosity decreases as the microtube diameter became smaller (Fahraeus & Lindqvist, 1931). After them, several works have extended their experiment down to diameters of about 3 μm and they have observed that the decrease of the apparent viscosity continues down to diameters of about 10 μm. However, the Fahraeus-Lindqvist effect is reversed at diameters 5 to 7 μm (see Figure 8) (Pries et al., 1992).

4.2 Particles and RBCs in diluted suspensions

In Poiseuille flow, the behaviour of suspended particles depends on several factors such as shear rate, particle deformability, size and shape. Generally, at low shear rates and diluted suspensions, rigid spherical particles and hardened RBCs (HRBCs) tend to move axially without any radial migration. On the other hand, deformable bodies, such as healthy RBCs tend to migrate towards the tube axis due to a radial hydrodynamic force. For higher (>1) particle Reynolds number (Re_p) where the inertial forces become important, both deformable bodies and rigid spheres have axial migration, however the spheres not always migrate toward the centre. The spheres near the wall moves towards the centre whereas the ones in the centre moves towards the wall. At the end they reach an equilibrium radial position of 0.6R, where R is the tube radius. This effect is known as tubular pinch effect (see Figure 9) (Caro et al., 1978).

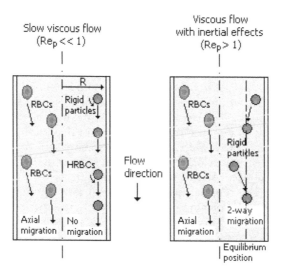

Fig. 9. Schematic representation of migration differences of rigid, healthy RBCs and hardened RBCs (HRBCs) in the centre of a capillary. Re_p corresponds to the particle Reynolds number (adapted from Goldsmith, 1971a, 1971b).

4.3 RBCs in concentrated suspensions

Although the flow properties of RBCs in diluted suspensions were extensively studied for many the years, such is not the case when the RBCs flow within a crowded environment. The reason is mainly related to technical limitations of optical systems to obtain reliable measurements at Hct bigger than 10%. However, Goldsmith and coworkers (Goldsmith & Turitto, 1986) have overcome this technical difficulty by using transparent RBCs (known as ghost cells) as the suspension medium. By using ghost cells they were able to study the behaviour of individual RBC flowing in concentrated suspension of RBC ghosts. The motion of RBCs in concentrated suspensions is appreciably different from those observed in very diluted suspensions (Hct < 1 %). At concentrated suspensions the motion of RBCs is disturbed not only by the collisions with neighbouring cells but also by the plasma layer near the wall. In this way, the cell paths exhibit continuous erratic displacements with the largest ones occurring in the region between 0.5 and 0.8 of the tube radius from the axis. At a given microtube, the magnitude of radial displacements tends to increase with the concentration of RBC ghost cells. However, at concentrations bigger than 50%, the displacement decreases. At Hct > 50%, although the crowded environment leads to an increase of the cell deformation, it also limits the magnitude of the RBC radial dispersion (Goldsmith & Turitto, 1986).

Recently, Lima et al. demonstrated the ability of confocal micro-PIV not only to measure both pure water and suspensions of RBCs through a square glass microchannel (Lima et al., 2006) but also to measure the velocity profiles of both physiological saline (PS) and in vitro blood (20% Hct) in a rectangular polydimethysiloxane (PDMS) microchannel (Lima et al., 2008). Good agreement between the measured velocity profiles of pure water and an established analytical solution was obtained for both studies. Further work was also performed by the Lima et al. but this time to measure both ensemble and instantaneous velocity profiles for *in vitro* blood with Hcts up to 17% (Lima et al., 2007). Although the ensemble velocity profiles were markedly parabolic, some fluctuations in the instantaneous velocity profiles were found to be closely related to the Hct increase. Hence, those results have suggested that the presence of RBCs within the plasma flow influences the measurements of the instantaneous velocity fields (see Figure 10).

Lima and his colleagues have also observed that by using a confocal micro-PIV system (see Figure 2) it was possible to measure with good accuracy blood plasma containing trace particles with Hct up to 9%. For Hct bigger than 9%, the light absorbed by the RBCs has contributed significantly to diminish the concentration of fluorescent tracer particles in the acquired confocal images. This low density images become evident for Hct bigger than 20 %, which generates errors in the velocity fields. Hence, Lima et al. (Lima et al., 2008, 2009) have developed a new approach to track individual tracer cells at high concentration suspensions of RBCs. The confocal micro-PTV system (see Figure 4) was employed, for the first time, in an effort to obtain detailed quantitative measurements on the motion of blood cells at both diluted and high suspensions of RBCs in simple straight microchannels. Lima et al. have successfully labelled both RBCs and WBCs and have measured their motions through microchannels. The ability of the confocal system to generate thin in-focus planes has allowed measurements in flowing blood at concentrated suspensions of: cell-cell hydrodynamic interaction, RBC orientation and RBC radial dispersion at different depths (Lima et al., 2008, 2009). To analyse the ability of the confocal micro-PTV system to track RBCs, the motions of labelled RBCs were followed at several Hcts (3% to 37%).

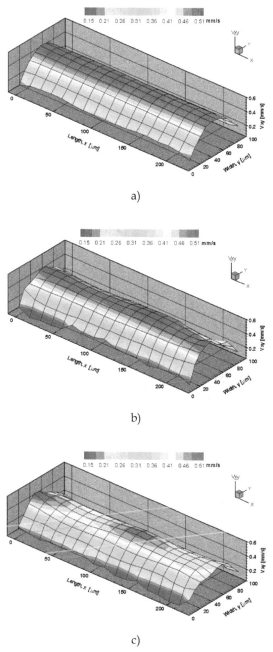

Fig. 10. Time series of the instantaneous velocity profiles of (a) pure water (b) *in vitro* blood with a 9% Hct and (c) *in vitro* blood with a 17% Hct, in the central plane of the microchannel with a Δt = 10 ms (adapted from Lima et al., 2007).

For the calculation of the radial dispersion coefficient (D_{yy}), measurements were performed at middle plane of the microchannels. However to investigate complex microrhelogical events in flowing blood (such as interaction and orientation of blood cells) all measurements were performed near the wall of the microchannel ($z = 20$ μm) with Hct ~ 20% and Re ~ 0.007. Figure 11 shows the trajectories of two-RBC interactions near the wall and within the plasma layer. This figure shows the radial disturbance effect enhanced by the collision of a neighbouring RBC. The hemodynamic interaction effect of WBC on the motion of RBCs was also possible to be measured. Figure 12a shows a RBC interacting with a WBC. In Figure 12a it is possible to observe that the transversal displacement increases due to the collision with a flowing WBC. An interesting measurement of both RBC translational and rotational motion was also possible by adjusting the image contrast (see Figure 12b). The translational motion was measured at the centre of the RBC whereas the rotational was measured along the RBC membrane.

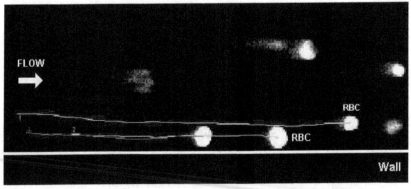

Fig. 11. Two-RBC interactions in a straight microchannel.

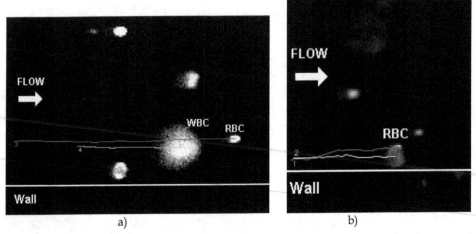

Fig. 12. (a) RBC-WBC interaction in a straight microchannel; (b) Translational and rotational motion of a RBC.

Further research was carried by using in vitro blood with several Hcts and low Reynolds numbers (Re ~ 0.005). The paths of hundreds labeled RBCs were measured in the centre plane of both straight glass and PDMS circular microchannel (Lima et al., 2008, 2009, 2009). The RBC dispersion coefficient (D_{yy}) for two different diameters (75μm and 100μm) and for several Hcts is shown in Figure 13. The results show that RBC D_{yy} rises with the increase of the Hct and that RBC D_{yy} tends to decrease with the diameter.

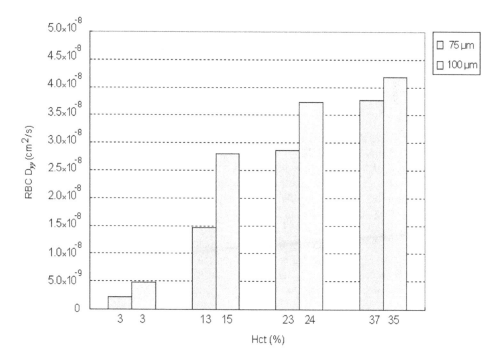

Fig. 13. Effect of the Hct on the RBC D_{yy} at 75μm PDMS microchannel and 100 μm glass capillary (adapted from Lima et al., 2008, 2009).

4.4 Cell-free layer (CFL)

In microcirculation the cell-free layer is believed to reduce the friction between red blood cells (RBCs) and endothelial cells and consequently reduce blood flow resistance. However, the complex formation of the cell-free layer has not yet been convincingly described mainly due to multi-physical and hemorheological factors that affect this phenomenon. Recent studies have measured the effect of hematocrit (Hct) on the thickness of the cell-free layer in straight circular polydimethylsiloxane (PDMS) microchannels. The formation of a cell-free layer is clearly visible in the images captured (see Figure 14) and by using a combination of image analysis techniques we are able to detect an increase in the cell-free layer thickness as Hct decreases.

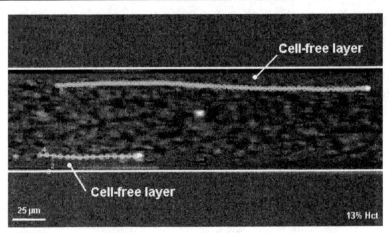

Fig. 14. Labelled RBCs flowing around the boundary region of the cell-free layer (Lima et al., 2009).

Labelled RBCs flowing around the boundary region of the cell-free layer were able to track manually by using the MtrackJ plugin from Image J. Figure 14 shows an example of the trajectories of two labelled RBCs flowing close to the boundary of the cell-free layer. The radial position of the tracked RBCs is then determined and the corresponding thickness of the cell-free layer is calculated. Finally, the data is time-averaged. When the number of labelled RBCs flowing close to the boundary region is not significant, additional analysis is performed by manually measuring the distance from the boundary region to the wall for several points along the microchannel (Lima et al., 2009; Lima et al., 2009). Examination of Figure 15 shows an overall reduction of the thickness of the cell-free layer as Hct is increased. In particular, the thickness decreases almost four fold as Hct is increased from 3% to 37% (Lima et al., 2009; Lima et al., 2009).

4.5 Sedimentation effects

Recently Garcia and his colleagues (Garcia et al., 2010, 2011) have investigated the flow behaviour of two different physiological fluids frequently used in biomedical microdevices. The working fluids used in this study were physiological saline (PS) and dextran 40 (Dx40) containing about 6% of sheep red blood cells (RBCs), respectively. By using a syringe pump and a video camera it was possible to measure qualitatively the flow behaviour within a horizontal capillary. To analyze the dynamic sedimentation of PS and Dx40 containing RBCs they decided to use flow rates close to the one observed in vivo, i.e., 10 µl/min. During the experiment we made flow qualitative visualizations measurements in glass capillaries with diameters of about 1,2 mm. The visualizations were captured by a camera for about 15 minutes. Figure 16 shows the flow qualitative measurements for 0 minutes and 15 minutes. This image shows clearly that for a period of 15 minutes the RBC tend to settle down in the fluid with PS whereas using Dx40 we did not observe any RBC sedimentation. Although not shown in Figure 16, for the case of PS fluid we did not observe any RBC sedimentation for the first 10 minutes. According to our visualization the RBC tend to settle down for period of time superior to 10 minutes.

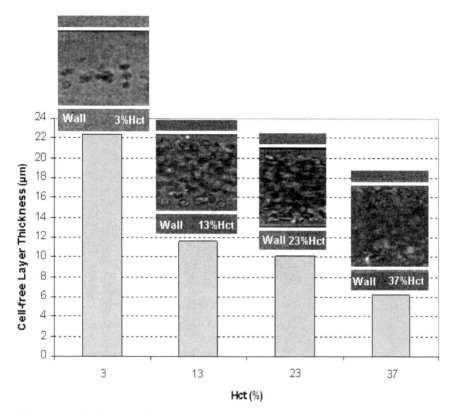

Fig. 15. Average thickness of the cell-free layer at several Hcts layer (Lima et al., 2009; Lima et al., 2009).

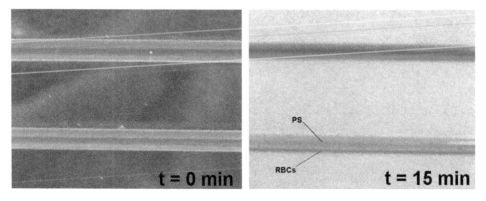

Fig. 16. Dynamic sedimentation measurements for two different time periods of PS and Dx40 containing RBCs, with a flow rate of 10 µl/min (Garcia et al., 2010, 2011).

Additionally, flow visualization measurements were also performed in glass microchannels (see Fig.17) and compared with *in vivo* blood flow (see Fig.17c). Figure 17 shows that for the case of Dx40 there is a clear formation of cell-free layer adjacent to the walls of microchannels. However, in the fluid with PS the RBCs do not exhibit a clear tendency to migrate into the microtube axis. The *in vivo* visualization measurements (Fig. 17c) have shown a clear tendency for the formation of a plasma layer in microvessels (Kim et al., 2006; Minamiyama, 2000). These results indicate that *in vitro* blood containing Dx40 has a flow behaviour closer to the one observed *in vivo* microvessels.

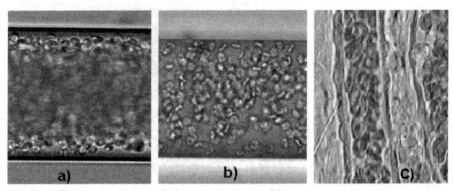

Fig. 17. *In vitro* flow visualization in glass microchannels for a period time bigger than 10 minutes a) Dx40 containing RBCs; b) PS containing RBCs. c) *In vivo* flow visualization in a microvessel (Lima et al., 2009a, 2009b; Minamiyama, 2000).

5. In vitro blood flow behaviour in microchannels with complex geometries

5.1 Bifurcation and confluence

By using a soft lithographic technique it is possible to fabricate polydimethysiloxane (PDMS) microchannels with complex geometries similar to human blood arterioles and capillary networks (Lima et al., 2011). In this section we show the application of a confocal micro-PTV system to track RBCs through a rectangular polydimethysiloxane (PDMS) microchannel with a diverging bifurcation and a confluence (see Figure 18).

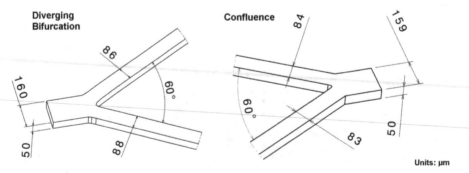

Fig. 18. Dimensions of the a) diverging and b) confluence used in this study (Leble et al., 2011a, 2011b; Lima et al., 2011).

By using a confocal system, we have measured the effect of both bifurcation and confluence on the flow behaviour of both fluorescent particles suspended in pure water (PW) and RBCs in concentrated suspensions. For the case of trace particles in PW we have observed that the trajectories were almost symmetric and do not present so many fluctuations for both geometries. These results are consistent with the Stokes flow regime. In contrast, for the case of labelled RBCs the trajectories are more asymmetric when compared with PW trajectories. Additionally, we can also observe several fluctuations on their trajectories possibly due to cell interactions enhanced by the high local Hct originated at this region (Lima et al., 2011a, 2011b).

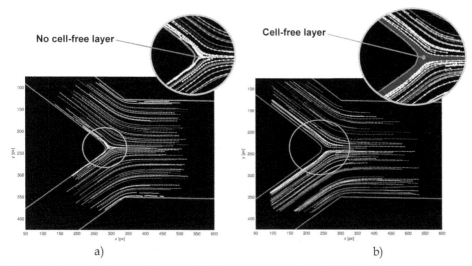

a)   b)

Fig. 19. Trajectories in a confluence of a) fluorescent particles in PW and b) labelled RBCs in Dx40 (Leble et al., 2011a, 2011b; Lima et al., 2011).

In the confluence, it is possible to observe that the trace particles tend to flow very close to the inner walls and as a result they tend to flow in the centre of the microchannel, just downstream of the confluence apex (see Fig.19a). However, for the case of labelled RBCs we could not measure any trajectory passing in this centre region (see Fig.19b). This is due to the existence of a cell-free layer (CFL) in both inner walls and a consequent formation of a triangular CFL in the region of the confluence apex (see Fig.19b) (Leble et al., 2011a, 2011b; Lima et al., 2011). As this triangular CFL seems to play an important role on the *in vitro* blood flow characteristics, a detailed quantitative study, to clarify the CFL effect in the velocity profiles, is currently under way and it will be published in due time (Leble et al., 2011).

5.2 Stenosis

The behaviour of RBCs in a microchannel with stenosis was also investigated using a confocal micro-PTV system. Individual trajectories of RBCs in a concentrated suspension of 10% hematocrit (Hct) were measured successfully. Results indicated that the trajectories of healthy RBCs became asymmetric before and after the stenosis, while the trajectories of tracer particles in pure water were almost symmetric. The effect of deformability of RBCs on the cell-free layer thickness, by hardening RBCs using a glutaraldehyde treatment, was also investigated.

Fujiwara et al. have found that deformability has a considerable effect on the asymmetry of the cell-free layer thickness and as result they concluded that the motions of RBCs are influenced strongly by the Hct, the deformability, and the channel geometry (Fujiwara et al., 2009).

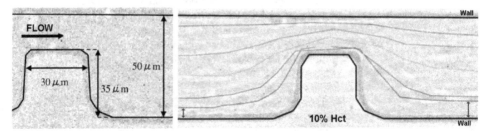

Fig. 20. Geometry of the stenosis and Flow of blood with. trajectories of labelled RBCs with 10% Hct (Fujiwara et al., 2008, 2009).

5.3 Capillary networks

The flow of red blood cells through a column packed with soda lime glass spheres was recently studied by Couto and her colleagues (Couto et al., 2011). Since the diameter of the glass spheres was 337.5 µm and the packing porosity 0.4 it was easy to determine (Dias et al., 2006, 2007, 2008; Dias, 2007) that the average capillary diameter of the porous network was 150 µm, this diameter being within the range of the typical dimensions observed in human blood arterioles.

The samples used in the referred study (Couto et al., 2011) were suspensions of sheep RBCs (with major diameter close to 6.5 µm) in physiological saline (PS) containing 0.1 to 1% Hct and the experimental apparatus was that shown in Fig. 21 and contains: a pump (1); injector (2); a chromatographic column packed with the above mentioned glass spheres (3); refractive index detector (4) and data acquisition system (5a and 5b). The mobile phase was PS and the flow rates varied between 0.2 ml/min and 1 ml/min.

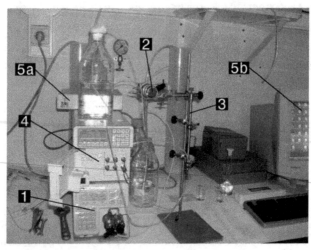

Fig. 21. Chromatographic apparatus (Couto et al., 2011).

An infinite small sized marker (sucrose) dissolved in PS was used in order to know the velocity of the carrier fluid (PS). The different experiments (different flow rates and Hcts) have shown that the RBCs migrate faster through the capillary network than the fluid (PS) that suspends the RBCs, as can be seen in Fig. 22. The authors (Couto et al., 2011) suggested that this behaviour can be explained by the theory that supports the hydrodynamic fractionation (Tijssen et al., 1986; Stegeman et al., 1993; Dias et al., 2008; Dias, 2007, 2008) and by the formation of a cell-free layer adjacent to the walls of the capillaries.

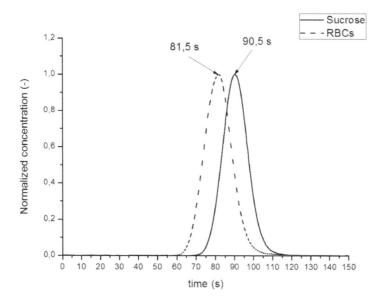

Fig. 22. Chromatogram for 1ml/min.

6. Conclusions

Advances in computers, image analysis, and optics have made it possible to combine both particle image velocimetry (PIV) and particle tracking velocimetry (PTV) system with a spinning disk confocal microscope. This revision summarise the most relevant studies of *in vitro* blood flow, through simple and complex microchannels, by means of a confocal micro-PIV/PTV system.

In vitro blood studies showed that the confocal micro-PIV system is able to measure with good accuracy blood plasma with Hct up to 9%, in a 100-µm square microchannel. However, for Hct bigger than 9%, the light absorbed by the RBCs contributes to diminish the concentration fluorescent particles in the acquired confocal images. This low density images become more evident for Hct bigger than 20%, which generates spurious errors in the velocity profiles. Hence, a Langragian approach was proposed to the confocal system in order to track individual blood cells at high Hcts. Owing to its optical sectioning ability and consequent improvement of the image contrast and definition, the proposed confocal micro-PTV system eliminates several problems and concerns of the microvisualization systems used in the past and provides additional detailed description on the RBC motion. Hence, the proposed confocal system has shown detailed RBC trajectories of RBC-RBC interaction at moderate and high Htcs (Hcts up to 35%) never obtainable by other conventional methods. As a result, by measuring hundreds of RBC trajectories, it was possible to conclude that RBC paths are strongly dependent on the Hct and as a result the RBC dispersion coefficient tends to increase with the Hct. Another unique measurement was the possibility to obtain both translational and rotational motion of RBCs flowing in highly concentrated suspensions. Hence, it was possible to compare in detail the motion of a RBC without any kind of interaction (rolling on the wall of the microchannel) with a RBC interacting with neighbouring blood cells. These remarkable results, have shown for the first time, that RBC rolling on the wall and without interaction tends to rotate in regular and periodically way whereas RBCs that interact with neighbouring cells tend to rotate in a rather erratic way. Very recently, our confocal system have shown also for the first time, a very astonishing blood flow phenomenon happening downstream of a confluence. Our results show the formation of a triangular cell-free layer (CFL) in the region of the confluence apex which consequently promotes a development of thin CFL in the middle of the microchannel. This finding suggests that in our microcirculatory system we may have CFLs not only on the walls but also in the middle of the microvessels. Additionally, the confocal system have also proved to be powerful tool to obtain further insight onto the flow behaviour of blood through other kinds of complex geometries such as diverging bifurcations, and stenosis.

The net of capillaries obtained by packing chromatographic columns with glass spheres may be used as a simple model in order to further understand the blood flow in complex microvascular networks.

7. Future directions

Past *in vitro* research performed with classical glass microchannels has revealed several amazing phenomena happening in microcirculation. However, recently several studies on the blood flow behaviour have shown quantitative difference between the results in microvessels and in glass microchannels. The reason for these differences still remains unclear mainly because glass microchannels differ from microvessels in several aspects, such as elasticity and geometry and biological effect by the endothelial inner surface. These limitations encountered in glass microchannels can be overcome by using PDMS microchannels. Hence, in the near future we are planning to develop a biochip to mimic the *in vivo* environment. By using an innovative cellular micropatterning technique based on electrochemical method, we expect to culture living cells on their surfaces and consequently to provide experimental evidence on several microcirculation mysteries, such as the role of the glycocalyx on the flow behaviour of blood and the thrombogenesis process.

Computational modelling is a powerful way to obtain detailed information on the mechanical behaviours of multiple RBCs. Recently, several numerical methods have been proposed to analyse microscale blood flows in microvessels such as the particle method and the elastic RBC flow model. We expect in the near future to compare the reported experimental results with the numerical methods mentioned above. This collaborative work between experimental and computational methods will contribute not only to improve the modelling of cell behaviour but also to provide better understanding on several microcirculation phenomena

Another area for further research is the improvement of the current particle tracking methods. In this work a manual tracking method was used due to the inability of the automatic methods to track RBCs when they experienced collisions. Hence, a more reliable automatic tracking method needs to be developed to reduce the time-consuming and also to avoid the tedious work performed during the post-processing procedure.

8. Acknowledgment

The authors acknowledge the financial support provided by: PTDC/SAU-BEB/108728/2008, PTDC/SAU-BEB/105650/2008 and PTDC/EME-MFE/099109/2008 from the FCT (Science and Technology Foundation) and COMPETE, Portugal. The authors would like also to thank all the colleagues and students for their valuable comments, suggestions and technical assistance throughout this research work.

9. References

Adrian, R. (1991). Particle-imaging techniques for experimental fluid mechanics. *Annual Review of Fluid Mechanics.*, 23, pp. 261-304

Bitsch, L., Olesen, L., Westergaard, C., Bruus, H., Klank, H., & Kutter, J. (2005). Micro particle-image velocimetry of bead suspensions and blood flows. *Experiments in Fluids*, 39 (3), pp. 505-511.

Caro, C., Pedley, T., Schroter, R., & Seed, W. (1978). *The mechanics of the circulation*, Oxford University Press, Oxford, England.

Chiu, J., Chen, C., Lee, P., Yang, C., Chuang, H., Chien, S., & Usami, S. (2003). Analysis of the effect of distributed flow on monocytic adhesion to endothelial cells. *Journal of Biomechanics*, 36, pp. 1883-1895.

Couto, A., Teixeira, L., Leble, V., Lima, R., Ribeiro, A., & Dias, R. (2011). Flow of red blood cells in capillary networks ", in T Yamaguchi, Y. Imai, MSN Oliveira, R Lima (ed.), *Japan-Portugal Nano-BME Symposium: proceedings of the 2011 conference*, Porto/Braganca, Portugal, pp. 35-38, 2011.

Dias, R. (2007). Transport phenomena in polydisperse porous media (in Portuguese). P.hD. Dissertation, Minho University, Braga, Portugal.

Dias, R. (2008). Size fractionation by slalom chromatography and hydrodynamic chromatography, *Recent Patents on Engineering*, 2, pp. 95-103

Dias, R., Teixeira, J., Mota, M., & Yelshin, A. (2006). Tortuosity variation in low density binary particulate bed. *Separation and Purification Technology*, 51, pp. 180-184

Dias, R., Fernandes, C., Teixeira, J., Mota, M., & Yelshin, A. (2007). Permeability and effective thermal conductivity of bisized porous media. *International Journal of Heat and Mass Transfer*, 50, pp. 1295-1301

Dias, R., Fernandes, C., Teixeira, J., Mota, M., & Yelshin, A. (2008). Starch analysis using hydrodynamic chromatography with a mixed-bed particle column, *Carbohydrate Polymers*, 74, pp. 852-857

Dias, R., Fernandes, C., Teixeira, J., Mota, M., & Yelshin, A. (2008). Permeability analysis in bisized porous media: Wall effect between particles of different size, *Journal of Hydrology* 349 470-474

Electron Microscopy Facility at The National Cancer Institute at Frederick (NCI-Frederick). A three-dimensional ultra structural image analysis of a T-lymphocyte (right), a platelet (center) and a red blood cell (left), using a Hitachi S-570 scanning electron microscope (SEM) equipped with a GW Backscatter Detector. 2005. Available from: http://en.wikipedia.org/wiki/File:Red_White_Blood_cells.jpg

Fahraeus, R., & Lindqvist, T. (1931). The viscosity of the blood in narrow capillary tubes. *American Journal of Physiology.*,Vol. 96, n°3, pp. 562-568

Fujiwara, H., Ishikawa, T., Lima, R., Marsuki, N., Imai, Y., Kaji, H., Nishizawa, M., & Yamaguchi, T. (2008). Motion of Red Blood Cells and Cell Free Layer Distribution in a Stenosed Microchannel, *Proceedings of 16th Congress of the European Society of Biomechanics*, pp. 14-17, Lucerne, Switzerland, May 17-22

Fujiwara, H., Ishikawa T., Lima, R., Marsuki, N., Imai, Y., Kaji, H., Nishizawa, M., & Yamaguchi, T. (2009). Red blood cell motions in a high hematocrit blood flowing through a stenosed micro-channel. *Journal of Biomechanics*, 42, pp. 838-843

Garcia, V., Correia, T., Dias, R., & Lima, R. (2010). Flow of physiological fluids in microchannels: the sedimentation effect, *Proceedings of 6th World Congress of Biomechanics*, pp. 1071-1074, Singapore, Singapore, August 1-6, 2010

Garcia, V., Correia T., Dias R., & Lima R. (2011). Dynamic Sedimentation Measurements of Physiological Fluids in Biomedical Devices", in T Yamaguchi, Y. Imai, MSN Oliveira, R Lima (ed.), *Japan-Portugal Nano-BME Symposium: proceedings of the 2011 conference*, Porto/Bragança, Portugal, pp.47-48, 2011

Goldsmith, H. (1971a). Red cell motions and wall interactions in tube flow, *Federation Proceedings 30* (5), 1578-1588

Goldsmith, H. (1971b). Deformation of human red cells in tube flow. *Biorheology*, 7, pp. 235-242.

Goldsmith, H. & Turitto, V. (1986). Rheological aspects of thrombosis and haemostasis: basic principles and applications. ICTH-Report-Subcommittee on Rheology of the International Committee on Thrombosis and Haemostasis. *Thrombosis and Haemostasis.*, Vol.55, No.3, pp. 415–435

Kim, S., Kong, R., Popel, A., Intaglietta, M., & Johnson, P. (2006). A computer-based method for determination of the cell-free layer with in microcirculation. *Microcirculation*, 13 (3), pp. 199-207

Kim, G., & Lee, S. (2006). X-ray PIV measurements of blood flows without tracer particles. *Experiments in Fluids*, 41, pp. 195-200

Koutsiaris, A., Mathioulakis, D., & Tsangaris, S. (1999). Microscope PIV for velocity-field measurement of particle suspensions flowing inside glass capillaries. *Measurement Science and Technology*, 10, pp. 1037-1046

Leble, V., Dias, R., Lima, R., Fernandes, C., Ishikawa, T., Imai, Y., & Yamaguchi, T., (2011) "Motions of trace particles and red blood cells in a PDMS microchannel with a converging bifurcation", in T Yamaguchi, Y. Imai, MSN Oliveira, R Lima (ed.), *Japan-Portugal Nano-BME Symposium: proceedings of the 2011 conference*, Porto/Braganca, Portugal, pp. 29-30

Leble, V., Lima, R., Fernandes, C., Dias, R. (2011). Flow of red blood cells through a microchannel with a confluence", *Proceedings of the Congresso de Métodos Numéricos em Engenharia 2011*, CD-ROM paper ID267.pdf, Coimbra, Portugal, 2011.

Leble, V., Lima, R., Ricardo, D., Fernandes, C., Ishikawa, T., Imai, Y., &Yamaguchi, T. (2011). Asymmetry of red blood cell motions in a microchannel with a diverging and converging bifurcation ", *Biomicrofluidics*, 5 (4), 044120 (15 pages).

Lima, R., Wada, S., Tsubota, K., & Yamaguchi, T. (2006). Confocal micro-PIV measurements of three dimensional profiles of cell suspension flow in a square microchannel. *Measurement Science and Technology*, Vol. 17, n° 4, pp. 797-808

Lima, R. (2007). Analysis of the blood flow behavior through microchannels by a confocal micro-PIV/PTV system, PhD (Eng.), Bioengineering and Robotics Department, Tohoku University, Sendai, Japan, 2007.

Lima, R., Wada, S., Takeda, M., Tsubota, K., & Yamaguchi, T. (2007). In vitro confocal micro-PIV measurements of blood flow in a square microchannel: the effect of the haematocrit on instantaneous velocity profiles. *Journal of Biomechanics*, 40, pp. 2752-2757

Lima, R., Ishikawa, T., Imai, Y., Takeda, M., Wada, S., & Yamaguchi, T. (2008). Radial dispersion of red blood cells in blood flowing through glass capillaries: role of heamatocrit and geometry, *Journal of Biomechanics*, 41, pp. 2188-2196

Lima, R., Wada, S., Tanaka, S., Takeda, M., Ishikawa T., Tsubota, K., Imai, Y., & Yamaguchi, T. (2008). In vitro blood flow in a rectangular PDMS microchannel: experimental observations using a confocal micro-PIV system. *Biomedical Microdevices*, Vol. 10, n° 2, pp. 153-167

Lima, R., Nakamura, M., Omori, Y., Ishikawa T., Wada, S., & Yamaguchi T. (2009). Microscale flow dynamics of red blood cells in microchannels: an experimental and numerical analysis, In: Tavares and Jorge (Eds), *Advances in Computational Vision and Medical Image Processing: Methods and Applications*, Springer, Vol.13, 203-220

Lima, R., Ishikawa, T., Imai, Y., Takeda, M., Wada, S., & Yamaguchi, T. (2009). Measurement of individual red blood cell motions under high hematocrit conditions using a confocal micro-PTV system. *Annals of Biomedical Engineering*, Vol.37, n° 8, pp. 1546-1559

Lima R., Oliveira, M., Ishikawa, T., Kaji, H., Tanaka, S., Nishizawa, M., & Yamaguchi, T. (2009). Axisymmetric PDMS microchannels for in vitro haemodynamics studies. *Biofabrication*, Vol.1, n° 3, pp. 035005.

Lima, R., Oliveira, M., Cerdeira, T., Monteiro, F., Ishikawa, T., Imai,Y., & Yamaguchi, T. (2009). Determination of the cell-free layer in circular PDMS microchannels, *ECCOMAS Thematic Conference on Computational Vision and Medical Image Processing*, Porto, Portugal.

Lima, R., Fernandes, C., Dias, R., Ishikawa, T., Imai, Y., Yamaguchi, T. (2011). Microscale flow dynamics of red blood cells in microchannels: an experimental and numerical analysis, In: Tavares and Jorge (Eds), *Computational Vision and Medical Image Processing: Recent Trends*, Springer, Vol.19, pp. 297-309.

Lima, R., Dias, R., Leble, V., Fernandes, C., Ishikawa, T., Imai, Y., Yamaguchi, T. (2012). Flow visualization of trace particles and red blood cells in a microchannel with a diverging and converging bifurcation, *ECCOMAS Thematic Conference on Computational Vision and Medical Image Processing*, Olhão, Portugal, pp. 209-211.

Lima, R., Ishikawa, T., Imai, Y., & Yamaguchi, T. (2012). Blood flow behavior in microchannels: advances and future trends, In: Dias et al. (Eds), *Single and two-Phase Flows on Chemical and Biomedical Engineering*, Bentham Science Publishers Springer, (in press).

Meinhart, C., Wereley, S., & Gray, H. (2000). Volume illumination for two-dimensional particle image velocimetry. *Measurement Science and Technology*, 11, pp. 809-814

Minamiyama, M. (2000). In Vivo Microcirculatory Studies: In Vivo Video Images of Microcirculation Part 2, Available from:
http://www.ne.jp/asahi/minamiya/medicine/

Nash, G. & Meiselman, H. (1983). Red cell and ghost viscoelasticity. Effects of hemoglobin concentration and in vivo aging. *Biophysical Journal*, 43. pp. 63-73

Park, J., Choi, C., & Kihm, K. (2004). Optically sliced micro-PIV using confocal laser scanning microscopy (CLSM). *Experiments in Fluids*, 37, pp. 105-119.

Park, J., & Kihm, K. (2006). Use of confocal laser scanning microscopy (CLSM) for depthwise resolved microscale-particle image velocimetry (μ-PIV). Optics and Lasers in Engineering, 44, pp. 208-223.

Parthasarathi, A., Japee, S. & Pittman, R. (1999). Determination of red blood cell velocity by video shuttering and image analysis. *Annals of Biomedical Engineering*, 27, pp. 313-325

Pries, A., Neuhaus, D. & Gaehtgens, P. (1992). Blood viscosity in tube flow: dependence on diameter and hematocrit. *Am. J. Physiol.*, Vol. 263, n° 6, pp. 1770-1778

Santiago, J., Wereley, S., Meinhart C., Beebe, D., & Adrian, R. (1998). A particle image velocimetry system for microfluidics. *Experiments in Fluids*, 25, pp. 316-319.

Stegeman, G., Kraak, J., & Poppe, H. (1993). Hydrodynamic chromatography of polymers in packed columns. *Journal of Chromatography A*, 657(2), pp. 283-303

Sugii, Y., Nishio, S., & Okamoto, K. (2002). In vivo PIV measurement of red blood cell velocity field in microvessels considering mesentery motion. *Physiological Measurement*, 23(2), pp. 403-416

Sugii, Y., Okuda, R., Okamoto, K., & Madarame, H. (2005). Velocity measurement of both red blood cells and plasma of in vitro blood flow using high-speed micro PIV technique. *Measurement Science and Technology*, 16, pp. 1126-1130

Tanaani, T., Otsuki, S., Tomosada, N., Kosugi, Y., Shimizu, M., & Ishida, H. (2002). High-speed 1-frame/ms scanning confocal microscope with a microlens and Nipkow disks. *Applied Optics*, 41 (22), pp. 4704-4708

Tijssen, R., Bos, J., & Kreveld, M. (1986). Hydrodynamic Chromatography in Open Microcapillary Tubes. *Anal. Chem.*, 58, pp 3036-3034

Tsubota, K., Wada, S., & Yamaguchi, T. (2006). Simulation study on effects of hematocrit on blood flow properties using particle method. *Journal of Biomechanical Science and Enginneering*, 1, pp. 159-170

Uijttewaal, W., Nijhof, E. & Heethaar, R. (1994). Lateral migration of blood cells and microspheres in two-dimensional Poiseuille flow: a laser Doppler study. *Journal of Biomechanics*, 27, pp. 35-42

Vennemann, P., Kiger, K., Lindken, R., Groenendijk, B., Stekelenburg-de Vos, S., Hagen, T., Ursem, N., Poelmann, R., Westerweel, J., & Hierk, B. (2006). In vivo micro particle image velocimetry measurements of blood-plasma in the embryonic avian heart. *Journal of Biomechanics*, 39, pp. 1191-1200

Wada, S., & Kobayashi, R. (2002). Simulation of the shape change of a red blood cell at the entrance of a capillary, *Proceedings of the 4th World Congress of Biomechanics*, Calgary, Canada.

Physiological Analysis of Yeast Cell by Intelligent Signal Processing

Andrei Doncescu[1], Sebastien Regis[1], Katsumi Inoue[2] and Nathalie Goma[3]
[1]*University of Toulouse, LAAS CNRS UPR 8001*
[2]*National Institute of Informatics*
[3]*IPBS CNRS*
[1,3]*France*
[2]*Japan*

1. Introduction

Before getting into the details of this chapter, let us make some light upon the abiotic/biotic debate. The issue could be summarized as two main differences between biotic and abiotic: the first one is the internal regulation of biotic systems and what is generally called by some researchers "cellular intelligence" related to the possibility of communication between cells. More precisely, in both types of systems it is possible to identify mass and energy exchanges (thermodynamic laws control), but within the biotic systems only one can find a certain project capability, using an ascending informational system (genome towards metabolic networks and environment adaptation systems) and a descending one (inverse).

The living organisms science gave birth to two main research areas: biomimetics and systemic biology. Biomimetics is a new discipline based not on what could be extracted from nature but on what could be learned from it, through the biologic systems or biosystems. Many attempts on defining systems in general exist. The classic one is the definition given by Bertalanffy: the system is an "organized complex", delimited by the existence of "strong interactions or non trivial interactions", *i.e.* non linear interactions. A biologic system is supposed to replicate itself and develop a system of reactions to exterior perturbations. This replication is done on the basis of a non-systemic, not organized or less organized exploitation of the surrounding environment. As the biologic system "works" as an algorithm, it is quite normal that since the first days of the molecular biology (1959) the engineers intensified and diversified their references to biology, even before the general acceptance in the scientific community of the term nanotechnology. The convergence between biotechnology and nanotechnology is due to the conceptual statement "bio is nano". Many examples of biomimetics achievements may be recalled: the artificial pancreas, the artificial retina, biomaterials. The latter is very interesting because biomaterials are not homogeneous. Biosystems are also used to develop innovating methods allowing measuring physiologic parameters, to find diagnostics for diseases, and to evaluate the effectiveness of new therapeutic compounds. Encouraging activities take place in the biomedical field: the cancer, brain/heart vascular diseases, infectious diseases.

The biologic systems, as a result of billions of years of evolution, are complex and degenerated systems and this double characteristic makes their understanding extremely difficult. In general, for a biologic system, the cyclic issues are related to the measure redundancy, to variables pertinence and to the significant correlation between the parameters and the type of the model that has to be used. The difficulty of these models is the intrinsic nature of some of the constituting elements and their phenomenological reductionism. By giving a few examples their limited character will better understood: the phenomenological modeling does not take into account the metabolic capabilities of the system, the stoechiometrical modeling does not take into account the dynamics of the system, thus a "time-space" analysis is not possible, etc. Such observations drive us to envisage a new approach of biologic problems: passing from the analytic paradigm to the complexity paradigm, via the so called approach of biology of systems or biosystemics or systemic biology.

The systemic biology is, as a simple definition, the integration of mathematics, biology, physics and computer science for creating new models for the biologic systems. Kitano, one of the fathers of Biology of Systems defines the biosystemic strategy (13) as:

1. Defining and analysing the structure of these systems;
2. Studying their behaviour and characteristics under different conditions;
3. Studying the regulation processes through which the systems control their equilibrium states and manage their variations;
4. Identifying the processes that allow building systems that are adapted to a given function

From a purely biologic standpoint, there exist complex groups of interacting proteins, performing:

1. the metabolism;
2. the synthesis of DNA;
3. information processing;

These interactions are network-like organized. The purpose of this biochemical diagram where each node represents a specific protein, which regulates the biochemical conversions, is to explain the cellular physiology starting from the dynamics of the net, in a cells population context. We can now understand that, in this context, the biologic systems are uncertain and subordered to viability constraints. From a qualitative dynamics standpoint, the viability constraints induct two states: the homeostasis and the chaos. The homeostasis defined by C. Bernard and W. Cannon is the capability of a living system to preserve its vital functions, by maintaining, in a certain structural stability range, its parameters and, in a certain viability range, its internal variables. The longevity of this living is a function of this stability. The chaos, in its common definition, is more opposed the concept of order than to those of stability and viability. It is possible to continuously pass from order to chaos, by increasing the complexity of the studied system, and this can be done by simple variations of some of its bifurcation parameters, variations that might provoke a change in the nature of the systemÕs dynamics attractors by overriding critical bifurcation values. The two notions are derived from the notion of attractor of a dynamic system in the context of a wide regulation network. We can give as an example the nature of biologic regulation: direct, indirect, or causal, governing the response of these systems. As the evolution of these systems is still rather unknown due to the continuous or discrete way of evolution we can suppose the existence of response thresholds. Biology inspires an analysis of the non-linear dynamics in terms of feedback loops (positive or negative). A positive feedback loop within the interaction graph of a differential system is a necessary condition of multistationarity which becomes, in biology, the cellular differentiation. A negative feedback loop within the interaction graph is a necessary condition for a periodic stable behaviour, very important in biology. It is quite possible that the oscillators coupling and multiple feedback loops interaction could lead to better oscillations coherence. The biologic regulation networks allow, among other things, to model genes interactions within a cell. Consequence to genome sequency, including the humain genome sequence, the postgenomics field has the goal to characterize the genes, the functions and the intercations beetween genes. The network of regulation takes an important and difficult role to assume the analysis at different scales, which means: "setup of new models using the biologic data and predicts the behavior of biologic systems from information extracted from genome sequence". But how Monod had enhanced: "everything which exists in universe is the result of hazard and necessity which it is not in contradiction with Occam principle to do not introduce supplementary clauses when it is not necessary. It is important to mention that the hazard presented in biology is not stochastic which means "to drive" the hazard but tychastic which describes some phenomenaÕs which escape to all statistic regularity. Therefore it is impossible to speak about the predictive capacity of the model which is the final goal of theoretical biology. From a realist view point we have made the hypothesis that the DNA could not include the finest maps of the organism but only some information about the iterative process of bifurcation and growth. If the approaches proposed before are in-silico, in the case of the in vitro the extraction of correct information is primordial. Basically, the information obtained is realized by direct observation and also by the interpretation of the response of the system. To be more accurate in this explication this approach is based on the perturbation of the analyzed system, supposed in a steady state, by a pulse of metabolite which belongs to the non linear differential model which described the biological system. Using the response of the system it is possible to deduce the relations between this variable and some variables of the model. For the complex models the analysis of the picks and the slope of the response directly link to the concentration of the metabolite it is not possible. The extraction of the information from responses supposes the cooperation of two parties :

1. the development of a model able to observe the dynamic of the system with a lot of precision

2. using of the mathematical tools allowing to tune the model to experimental observations

which means to solve a regression problem. If the model is linear the regression is linear therefore without any theoretical interest. Comparing with linear models the non-linear models offer infinity of possibility without the difficulties enclosed in the resolution of this kind of approach. One strategy is the Lotka-Voltera modelling. This method has been applied in ecology and focuses on the interaction between two species of type predator Ð prey. The inconvenient of this method in study of metabolic pathway or fermentation is that he metabolite depends of many compounds. The deduction of a non linear model from experimental data is an inverse problem which could be solved by a regression method or genetic algorithms which minimize the error between the model and the data. In practice the local minima which stop the convergence of the algorithms. One smart solution is the utilisation of Bayesian Methods or Simulated Annealing when the systems are small size, no noise and we use a PC cluster. Another approach is the non linear estimation by NARMAX but these methods did not work with strong non linearties. The challenge in the case of the modeling of biological systems is the elucidation of the optimal evolution of the equations system when initial conditions are incomplete or missing. In this case, the analysis of the trajectories of the system is the central point to make the difference between regular trajectories (periodical or quasi periodic) and chaotic paths in the space of phases. The change of trajectories is directly related to the parameters of the model. Periodic trajectories will be identified by Poincaré sections and by using the Harmonic Wavelet Transform we will be able to make the difference between quasi-periodical theoretical and the chaos. The interest of the analysis by wavelet is the linearization of the model to find the steady states stationary of the biological system excited by a non-stationary process. The original system will be decomposed into linear subsystems, each having the response in a frequency band of well defined. The final response is calculated by the addition of each subsystem response. This type of analysis allows the studying of the influence of modes of regulation on the time of relaxation of the cell and to find out the stationary states of the cellular cycle.

Today, the pace of progress in fermentation is fast and furious, particularly since the advent of genetic engineering and the recent advances in computer sciences and process control. The high cost associated with many fermentation processes makes optimization of bioreactor performance trough command control very desirable. Clearly, control of fermentation is recognized as a vital component in the operation and successful production of many industries. Despite the complexity of biotechnological processes, biotechnologists are capable of identifying normal and abnormal situations, undertaking suitable actions accordingly. Process experts are able to draw such conclusions by analysing a set of measured signals collected from the plant. The inexistence of satisfactory mathematical models impedes model-based approaches to be used in supervisory tasks, thus involving other strategies to be applied. That suggests that, despite the lack of process models, measured data can be used instead in the supervisory system development. The advances in measurement, data acquisition and handling technologies provide a wealth of new data which can be used to improve existing models.

In general, for a biologic system, the cyclic issues are related to the measure redundancy, to variables pertinence and to the significant correlation between the parameters and the type of the model that has to be used. The difficulty of these models is the intrinsic nature of some of the constituting elements and their phenomenological reductionism. Moreover, the dynamic nature and the inherent non-linearity of bio-processes make system identification difficult. The majority of kinetic models in biology are described by coupled differential equations and simulators implement the appropriate methods to solve these systems. Particularly, analysis of states occuring during experiences is a key point for optimization and control of these bioprocesses. Thus, model-based approaches using differential equations (26), expert system (31), fuzzy sets and systems (23), (9), neural networks (8) have been developed. However, although model-based approaches give more and more accurate results close to real outputs (10), these methods using simulation techniques can lead to wrong conclusions, because of lack of description parameters or during an unexpected situation. Non-model-based methods have an increasing success and are based on the analysis of the process biochemical signals. The detection and the characterization of the physiological states of the bioprocess are based on signal processing and statistical analysis of signals. For example, methods based on covariance (21) and moving averages (4) have been proposed, but they do not take account of the changes occurring in the signals. Wavelet transform is a powerful tool for non-stationary signal analysis due to its good localization in time and frequency domains. Wavelets are thus sensitive to changes in signals. Bakshi and Stephanopoulos (3), then Jiang et al. (12) have successfully used wavelets to analyze and detect states during bioprocesses.

One of the main contribution of Artificial Intelligence to biological or chemical processes turns out to be the classification of an increasing amount of data. Can we do more than that and can an AI program contribute to help in discovery of hidden rules in some such complex process. In fact, even if we can predict, for instance, mutagenicity of a given molecule or the secondary structure of proteins, with high degree of accuracy, this is not sufficient to give a deep insight of the observed behavior. In this paper we present a method using Maximum of Modulus of Wavelets Transform, Hölder exponent evaluation and correlation product for the detection and the characterization of physiological states during a fermentation fed-batch bioprocess. Therefore, we consider the estimation of nonoscillating and isolated Lipschitz singularities of a signal.

2. Yeast biotechnology

The main process we are concerned is a bio-reaction, namely the dynamical behavior of yeast during chemostat cultivation. Starting from the observation of a set of evolutive parameters, our final aim is to extract logical rules to infer the physiological state of the yeast. Doing so, we obtain not only a better understanding of the system's evolution but also the possibility to integrate the inferred rules in a full on-line control process. The first thing we have to do is to capture and analyze the parameters given by the sensors. These signals must be treated to be finally given to the logic machine. Thus, two things have to be done : first, to denoise the signals, secondly to compute the local maximum values of the given curves. In fact, we are more interested in the variations of the signals than in their pure instantaneous values. We use a method issued from wavelets theory (1) and which tends to replace classical Fourier analysis. At the end of this purely analytic treatment, we dispose of a set of clean values for each critical parameter. Now, our idea is to apply Inductive Logic Programming to exhibit, starting from a finite sample set of numerical observations, a number of logical formulae which organize the knowledge using causal relationships. Inductive logic programming is a sub-field of machine learning based upon a first-order logic framework. So instead of giving a mathematical formula (for instance a differential equation) or a statistical prediction involving the different parameters, we provide a set of implicative logical formulae. A part of these formulae can generally be inferred by a human expert, so it a a way to partially validate the mechanism. But its remains some new formulae which express an unknown causality relation : in that sense, this is a kind of knowledge discovery. As far as we know, one of the novelties of our work is the introduction of a time dimension to simulate the dynamic process. In logic, this time variable is in general not considered except with some specific modal logics. So, we modelize the time with an integer-valued variable.

The methodology has been applied to a biotechnological process. Saccharomyces Cerevisiae is studied under oxidative regime (i.e., no ethanol production) to produce yeast under a laboratory environment in a bioreactor. Two different procedures are applied: a batch procedure that is followed by a continuous procedure. The batch procedure is composed by a sequence of biological stages. This phase can be thought as a start-up procedure. Biotechnologists state that the behaviour in the batch procedure influences later in induced phenomena in the continuous phase. So complete knowledge of the batch phase is of great importance for the biotechnologist. The traditional way to get acquainted of such knowledge is at present carried out through offline measurements and analysis which most of the time produce results when the batch procedure has ended, thus lacking of real time performance. Instead, the proposed methodology allows for real time implementation. This example deals with the batch procedure. Among the set of available on-line signals the expert chooses the subset of signals which, according to the expert knowledge contain the most relevant information to determine the physiological state:

1. DOT : partial oxygen pressure in the medium.

2. O2 : oxygen percent in the output gas

3. CO2 : carbon dioxide percent in the output gas

4. pH.

5. OH- ion consumption : derived from control action of the pH regulator and the index of reflectivity.

The consumption of negative OH ions is evaluated from the control signal of the pH regulator. The actuator is a pump, switched by an hysteresis relay, that inoculates a basic solution (NaOH). The reflectivity, which is measured by the luminance, seems to follow the biomass density. Nevertheless its calibration is not constant and depends on the run. Yeasts are a very well-studied micro-organisms and today, such micro-organism like *Saccharomyces cerevisiae* which make the object of this study, are largely used in various sectors of the biomedical and biotechnology industrial processes. So, this is a critical point to control such processes. Two directions have been explored:

1. the *on-line* analysis : it does not allow to identify in an instantaneous manner and with certainty the physiological state of the yeast.

2. the *off-line* analysis : it allows to soundly characterize the current state, but generally too late to take into account this information and to adjust the process on the fly by actions of regulators allowing to adjust some critical parameters such that pH, temperature (addition of basis, heats, cooling).

To remedy these drawbacks, computer scientists in collaboration with micro-biologists develop tools for supervised control of the bioprocess. They use the totality of informations provided by the sensors during a set of sample processes to infer some general rules to which the biological process obeys. These rules can be used to control the next processes. This is exactly the problem we tackle in this paper. To sum up, our application focus on the evolutive behavior of a *bio-reactor* (namely yeast fermentation) that is to say an evolutive biological system whose interaction with physical world, described with pH, pressure, temperature, etc..., generates an observable reaction. This reaction is studied by the way of a set of sensors providing a large amount of (generally) numerical data, but, thanks to the logical framework, symbolic data could also be integrated in the future. For an approach based upon classification and fuzzy logic, one can see (24) : this work is devoted to discover the different states of the bio-reactor but not to predict its behavior.

In a yeast culture, measures result of biology phenomena and physical mechanisms. That is why to bring the culture, it is always decisional between biology and physico-chemical. The biological reaction is function of the environment and an environmental modification will improve two types of biological responses. The first one is a quasi steady-state response, the micro-organism is in equilibrium with the environment. The biological translation of this state is kinetics of consummation, production and this phenomenon is immediate. The second biological response is a metabolic one, which can be an oxidative or fermentative mode, or a secondary metabolism. The characteristic of this response is that the time constants are relatively long. For cultures, in term of production, the essential parameters are metabolism control and performance (productivity and substrate conversion in biomass yield). With this goal, the process must be conducted by a permanent intervention in order to bring the culture to an initial point to a final point. This control can be done from acquired measures on process, which are generally gases. Indirect measures show the environmental dynamic, which is shown by gas balance, with respiratory quotient (RQ) and pH corrector liquid (see figure 1).

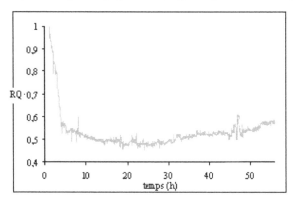

Fig. 1. An example of respiratory quotient evolution during a culture. x-axis is the time of the experience, y-axis is the amplitude of the signal.

Then, there are physical phenomenon, which are associated to real reactors. These mechanisms can be decomposed in many categories : transfer phenomenon (mass, thermal and movement quantity), regulation (realised by an operator), introduction of products, and mixing. These mechanisms interfere with biology and it is significant to notice that relaxation times of these phenomena are of size order of response time of biological response. With all these phenomena, a variable can be described by the following equation (see (26)) :

$$\frac{dV}{dt} = \Delta.\left(\frac{V_{equilibrium} - V(t)}{\tau_{physical}}\right) + r_{V(t)} + \Phi_{V(t)} \tag{1}$$

where:

- $\frac{dV}{dt}$ corresponds to the dynamic of the system.

- Δ.$\left(\frac{V_{equilibrium}-V(t)}{\tau_{physical}}\right)$ is variable variation between biological and physical parameters. $\tau_{physical}$ is the time constant of physical phenomena; this constant can not be characterised because it depends on reaction progress.

- $r_{V(t)}$ is the volumic density of reaction of the variable V, it is a biological term.

- $\Phi_{V(t)}$ corresponds to an external intervention which results of a voluntary action.

Moreover, it is essential to observe that there is a regulation loop between biology and physic (see figure 2). The problematic is, from measures, to isolate or eliminate perturbations. These responses depend on physical phenomena or human interventions (process regulation). It is to quantify biological kinetics and by this way to optimise biological kinetics and control that is to say identify modifications of the biological behaviour. For example, in the case of yeast production, it is important to maintain an oxidative metabolism by the control of glucose residual concentration, fermentative metabolism is prejudicial to the yield. The aim is to maintain an optimal production to avoid the diminution of substrate conversion yield, that is to say to remark the biological change between oxidative and fermentative metabolism.

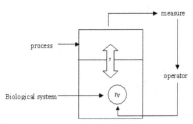

Fig. 2. Interactions between the biological system, the process and the operator.

3. Knowledge based methodology

In learning there is a constant interaction between the creation and the recognition of concepts. The goal of the methodology is to obtain a model of the process, which can be used in a supervisory system for condition monitoring. The complexity of this model imposes the co-operation of data mining techniques along with the expert knowledge. When only expert knowledge is used to identify process situations or states, any of these situations can arise: Ø the expert can express only a partial knowledge from process, Ø he does know the existence of several states but he ignores how to recognise them from on-line data, or/and Ø he doesn't have a clear idea on which states to recognise. For example, in the yeast production batch phase, biotechnologists apply expert rules when recognising some of the physiological states from on-line data. Nevertheless those rules usually don't take into account other phenomena that can change the evolution of signals without any influence in the physiological state. This leads to wrong conclusions. It is mainly due to the fact that the expert is not able to draw conclusions from the analysis of multiple signals between which there exist true relationships. Nevertheless, a classification tool copes well with this drawback. This proves the need of an iterative methodology to identify the biological states, which refines the expert knowledge with the analysis of past data sets.

Initialize : $E' = E$ (initial set of examples)
$\qquad\quad H = \emptyset$ (initial hypothesis)
While $E' \neq \emptyset$ do
$\qquad$ Choose $e \in E'$
$\qquad$ Compute a covering clause C for e
$\qquad H = H \cup \{C\}$
$\qquad$ Compute $Cov = \{e' \mid e' \in E, B \cup H \models e'\}$ $\qquad\qquad E' = E' \setminus Cov$ End while

Fig. 3. General Progol scheme

3.1 Standard ILP task

We stay within the pure setting i.e. where programs do not involve negation. In that case, the meaning of a logic program is just its least Herbrand model, which is a subset of the Herbrand universe i.e. the full set of ground atoms. In that setting, a concept C is just a subset of the Herbrand base. As shortly explained in our introduction, an ILP machine takes as input :

- a finite proper subset $E \ =< \ E^+, E^- >$ (the training set in Instance Based Learning terminology) where E^+ can be considered as the positive examples i.e. the things known as being true and is a subset of C, E^- as the negative examples and is a subset of $\overline{C}$.

- a logic program usually denoted B (as background knowledge) representing a basic knowledge we have concerning the concept to approximate. This knowledge satisfies two natural conditions : it does not explained the positive examples : $B \not\models E^+$ and it does not contradict the negative ones : $B \cup E^- \not\models \perp$

So the ILP task consists in finding a program H such that $H \cup B \models C$. One of the most popular method is to find H such that $H \cup B \models E^+$ and $H \cup B \cup E^- \not\models \perp$. In the field of classification, it is known that this approach, minimizing the error rate over the sample set (here we have zero default on the sample set) does not always guaranty the best result for the whole concept C.

Nevertheless, as far as we know, no alternative induction principle is used for ILP. Of course, as explained in the previous section, an ILP machine could behave as a classifier. Back to the introduction, the sample set $S = \{(x_1, y_1), \dots, (x_i, y_i), \dots, (x_n, y_n)\}$ is represented as a finite set of Prolog facts $class(x_i, y_i)$ constituting the set E^+. The ILP machine will provide an hypothesis H. Consulting H with a Prolog interpreter, for a given element x, we get the class y of x by giving the query $class(x, Y)$? to the interpreter.

3.2 Progol machinery

Back to the standard ILP process, instead of searching for consequences, we search for premises : it is thus rather natural to reverse standard deductive inference mechanisms. That is te case for Progol which uses the so-called inverse entailment mechanism ((20)). Progol is a rather complex machine and we only try to give a simplified algorithm schematizing its behavior in figure 3. One can read the tutorial introduction of CProgol4.4[1] from which we take our inspiration.

The main point we want to update is the choice of the relevant clause C for a given training example e. Let us precise here how this clause is chosen.

---

[1] available on **http://www.cs.york.ac.uk/mlg/progol.html** where a full and clear description is given.

3.3 The choice of the covering clause

It is clear that there is an infinite number of clauses covering e, and so Progol need to restrict the search in this set. The idea is thus to compute a clause C_e such that if C covers e, then necessarily $C \models C_e$. Since, in theory, C_e could have an infinite cardinality, Progol restricts the construction of C_e using mode declarations and some other settings (like number of resolution inferences allowed, etc...). Mode declarations imply that some variables are considered as input variables and other ones as output variables : this is a standard way to restrict the search tree for a Prolog interpreter.

At last, when we have a suitable C_e, it suffices to search for clauses C which θ-subsume C_e since this is a particular case which validates $C \models C_e$. Thus, Progol begins to build a finite set of θ-subsuming clauses, $C_1, \ldots, C_n$. For each of these clauses, Progol computes a natural number $f(C_i)$ which expresses the *quality* of C_i : this number measures in some sense how well the clause explains the examples and is combined with some compression requirement. Given a clause C_i extracted to cover e, we have :

$$f(C_i) = p(C_i) - (c(C_i) + h(C_i) + n(C_i))$$

where :

- $p(C_i) = \#(\{e \mid e \in E, B \cup \{C_i\} \models e\})$ i.e. the number of covered examples
- $n(C_i) = \#(\{e \mid e \in E, B \cup \{C_i\} \cup \{e\} \models \perp\})$ i.e. the number of incorrectly covered examples
- $c(C_i)$ is the length of the body of the clause C_i
- $h(C_i)$ is the minimal number of atoms of the body of C_e we have to add to the body of C_i to insure output variables have been instantiated.

The evaluation of $h(C_i)$ is done by static analysis of C_e. Then, Progol chooses a clause $C = C_{i_0} \equiv \arg \max_{C_i} f(C_i)$ (i.e. such that $f(C_{i_0}) = max\{f(C_j) \mid j \in [1, n]\}$). We may notice that, in the formula computing the number $f(C_i)$ for a given clause C_i covering e, there is no distinction between the covered positive examples. So $p(C_i)$ is just the number of covered positive examples. The same computation is valuable for the computation of $n(C_i)$ and so success and failure could be considered as equally weighted.

To abbreviate, we shall denote $Progol(B, E, f)$ the output program P currently given by the Progol machine with input B as background knowledge, E as sample set and using function f to chose the relevant clauses. In the next section, we shall explain how we introduce weights to distinguish between examples.

4. A boosting-like mechanism for Progol

As explained in our introduction, a Progol machine is a consistent learner i.e. it renders only hypothesis with no error on the training set : so the sample error at the end of a learning loop, $\epsilon^t = \Sigma_{\{i \mid x_i \text{ misclassified}\}} w_t(i)$, is 0 since each example is necessarily correctly classified. So we cannot base our solution over the computation of such an error since the nullity of this error is a halting condition for a standard boosting algorithm. So we introduce a new way to adjust the weights. Given an example e_i, since it is covered we have $B \cup H_t \vdash e$. Given an other problem instance e, we claim that the longer the proof for $B \cup H_t \vdash e$, the riskier the prediction for e.

5. Inductive logic programming : basic concepts

Mathematical logic has always been a powerful representation tool for declarative knowledge and Logic Programming is a way to consider mathematical logic as a programming language. A set of first order formulae restricted to a clausal form, constitutes a logic program and as such, becomes executable by using standard mechanisms of theorem proving field, namely unification and resolution. Prolog is the most widely distributed language of this class. In this context, the data and their properties, i.e. the observations, are represented as a finite set of logical facts E. E could generally been discomposed into the positive examples E^+ and the negative ones E^-. In case of background knowledge, it is described as a set of Horn clauses B. This background knowledge is supposed to be insufficient to explain the positive observations and the logical translation of this fact is : $B \not\models E^+$ but there is no contradiction with the negative knowledge: $B \cup E^- \not\models \bot$. So an ILP machinery ((20)), with input E and B, will output a program H such that $B \cup H \models E$. So H constitutes a kind of explanation of our observations E. Expressed as a set of logical implications (Horn clauses) $c \rightarrow o$, c becomes a possible cause for the observation $o \in E$. We give here a simple scheme giving a functional view of an ILP machine.

ILP machine

Fig. 4. ILP machine functional scheme

It is important to note that a logic program is inherently non deterministic since a predicate is generally defined with a set of distinct clauses. To come back to our introductive notation, we can have two clauses of the form (using Prolog syntax) $o \leftarrow c$ and $o \leftarrow c'$: this means that c and c' are potential explanations for o. The dual situation $o \leftarrow c$ and $o' \leftarrow c$ where the same cause produces distinct effects is also logically consistent. So this is a way to deal with uncertainty and to combine some features of fuzzy logic. The main difference is that we have a sound and complete operational inference system and this is not the case for fuzzy logic.

6. Formalization of our problem

We have 4 potential states for the bio-reactor : we shall denote e_1, e_2, e_3 and e_4 these states to avoid unusefull technical words. e_4 will be considered as a terminal state where the bio-reactor is stable because of the complete combustion of the available ethanol. We add a specific state e_5 corresponding to a stationary situation where the process is going on without perturbation. The transition between two states is a critical section where the chosen observable parameters (bio-mass, pH and O_2 rate) give rise to great variations.

The predicate to learn with our ILP machine is :

$$\textbf{to-state}(E_i, E_t, P_1, P_2, P_3, T)$$

meaning that the bio-reactor is going into state E_t knowing that at step T, the current bio-reactor parameters are the P_i's and the current state E_i. It is thus clear that we can deal with as many parameters as we want, but we restrict to 3 in this experiment. As explained in our introduction, we introduce the variable T to simulate the dynamic behavior of our process. As far as we know, previous experiments using inductive logic programming generally compute causal relationship between parameters which do not involve time. So we have to learn a transition system where we do not know what are the basic actions activating a transition. Our informations about the system are given by sensors providing numerical signals for p_1 (bio-mass), p_2 (pH value) and p_3 (O_2 rate),. These signals are analyzed using a wawelet-based system, we visualize the curve of the different functions and we extract the values of the differential for each given function. These values constitutes the input of our learning system. So, we want to obtain a causal relationship between the transitions of the system and the values of the differentials of the curve describing the evolution of our parameters.

So, we add a predicate **derive**$(P, T, P1)$ which expresses the fact that, for the curve of the parameter P,

at time T, the value of the differential is $P1$. It is thus easy to describe what is a pike for the curve describing P : this is included in our background knowledge. These pikes correspond to

local minima/maxima for the given parameter. So, we are also interested in the sign of

the derivative and we include specific predicates (**positive/2, negative/2**)

to compute and test this sign.

As background knowledge (corresponding to the B input of our scheme 4), we have the definitions of predicates **derive/3**,

positive/2, negative/2, pike/2. Here is an overview of the mode declaration

to describe the potential causes of a state transition.

```
:- modeb(*,pike(+parameter,-float))?
:- modeb(3,pike(-parameter,+float))?
:- modeb(1,positive(+parameter,+float))?
:- modeb(1,negative(+parameter,+float))?
:- modeb(1,between(+float,+float,+float))?
:- modeb(1,between(-float,+float,+float))?
% a (very little) part of

% our background knowledge.
pike(P,T) :- derive(P,T, P1),P2 is P1, between(P2,-0.001,0.001).
```

The results in this paper are obtained using the last implementation of Progol, namely **CProgol4.4** freely available on the following site **http://www.cs.york.ac.uk/mlg/progol.html**. Of course, we get a lot of rules, depending on the quantity of introduced examples. Some of them are not really interesting : they only generalize one or two examples. But we get, for instance, the next one (among the simplest ones to explain) :

```
to_state(E,E,A,B,C,T)  :- derive(p1,A,T),
        derive(p2,B,T), derive(p3,C,T),
        positive(p1,T), positive(p2,T),
        positive(p3,T).
```

This rule indicates that there is no evolution of the metabolism state (the bio-reactor remains in the same state) when the parameters have an increasing slope but that we do not encounter maxima or minima. In general, the obtained rules are long except those ones generalizing only one or two examples. Nevertheless, there are some observations where this rule could be overcame : this means that we need (at least) an other parameter $p4$ to better understand the behaviour of the machinery.

7. Detection and characterization of physiological states

In microbiology, a physiological state (or more simply, a state) is, qualitatively, the set of potential functionalities of a micro-organism, and, quantitatively, the level of expression of theses functionalities. The environment has a strong influence on the activity of the micro-organism due, on one hand, to its chemical composition (nature of substrate, pH...) and, on the other hand to its physical properties (temperature, pression...). Yeast can react on the availability of substrates such as carbon and nitrogen sources, or oxygen, by a flexible choice of different metabolic pathway. It is possible to analyze the global metabolism by genetical analysis, biochemical or biophysical analysis but the complexity of the biological system requires a simplification of the characterization by the analysis of some functionalities of some known mechanisms. The quantification of materials and energy interactions flows between the micro-organism and the environment enables to have a macroscopic characterization of several intrinsic metabolism of yeast population which, by correlation, enables to differentiate several physiological states even if the biological characterization is unknown. Thus the detection, as far as we know, is based on the analysis of biochemical signals measured during the bioprocess. A bioprocess is the set up of the fermentors protocol. Fermentors are composed of a number of different components which can be grouped by their functions, i.e. temperature control, speed control, continuous culture accessories. In this context the ultimate aim of bioprocess analysis therefore is a detailed monitoring of biological system, the chemical and physical environment and how these interact. However, no reliable technique exist to carry out real-time measurement of non-volatile substrates and metabolites in the fermentor. Several works using various approaches, lead to the conclusion that the limits of a state are linked to the singularities of biochemical signals: Steyer et al. (31) (using expert system and fuzzy logic), Bakshi and Stephanopoulos (3) (using expert system and wavelets) and Doncescu et al. (6) (using inductive logic) show that the beginning and the end of a state correspond to singularities of the biochemical signals measured during the process. In a fed-batch bioprocess, a physiological state can occur several times during the experience. After the detection of states, it is then necessary to characterize these states. The characterization is often based on the statistical properties of the biochemical signals. Experts in microbiology characterize the states by analysing and comparing the variations and the values of different biochemical signals and by a deductive reasoning using " if-then" rules. These approaches can be linked to mathematical methods based on correlation. Classification methods based on Principal Components Analysis (PCA) (27), adaptive PCA (15), and kernel

PCA (14) enable to distinguish and characterize the different states. However, these methods (except the adaptive PCA) do not take into account the temporal variation of the signals. The adaptive PCA is a PCA applied directly on wavelet coefficients in order to take into account the variations of the biological system. It has been shown that it can characterize the Lipschitz singularities of a signal by following the propagation across scales of the modulus maxima of its continuous wavelet transform. For identifying the boundaries of states, we propose to use the Maximum of Modulus of Wavelets Transform (17)(16) to detect the signals singularities. The singularities are selected according to their Hölder exponent evaluation between -1 and 1. The characterization of the states is based on the correlation product between the signals on intervals whose boundaries are the selected singularities.

8. Detection and selection of singularities by wavelets and Hölder exponent

The singularities of the biochemical signal correspond to the boundaries of the states. These signals are non-stationary and non-symmetrical; they are not chirps and have no infinite oscillations (see figure 5).

Several authors have proposed to use wavelets to detect the singularities of the signals for the detection of states: Bakshi and Stephanopoulos (3) and more recently Jiang et al. (12). Besides singularities correspond to maxima of modulus of wavelets coefficients. Bakshi and Stephanopoulos (3) propose to detect the maxima by analysing the variation of the wavelet coefficients through a multi-scale analysis but they don't explicitly characterize the nature of detected singularities. Jiang et al. (12) propose to select meaningful singularities by using a threshold on the finest scale, but the determination of the threshold remains empirical. After the detection of singularities by the Maxima of Modulus of Wavelet Transform, we propose to use the evaluation of Hölder exponent to characterize the type of singularities and eventually select meaningful singularities.

The wavelets are a powerful mathematical tool of non-stationary signal analysis, signals whose frequencies change with time. Contrarily to the Fourier Transform, Wavelet Transform can provide the time-scale localization. The performance of the Wavelet Transform is better than of the windowed Fourier Transform. Because of these characteristics, Wavelet Transform can be used for analyzing the non-stationary signals such as transient signals. Wavelets Transformation (WT) is a rather simple mechanism used to decompose a function into a set of coefficients depending on scale and location. The definition of the Wavelets Transform is:

$$W_{s,u}f(x) = (f \star \psi_{s,u})(x) = \int f(x)\psi(\frac{x-u}{s})dx \qquad (2)$$

where ψ is the wavelet, f is the signal, $s \in R^{+*}$ is the scale (or resolution) parameter, and $u \in R$ is the translation parameter. The scale plays the role of frequency. The choice of the wavelet ψ is often a complicated task. We assume that we are working with an admissible real-valued wavelet ψ with r vanishing moments ($r \in N^*$).

The wavelet is translated and dilated as in the next relation :

$$\psi_{u,s} = \frac{1}{\sqrt{s}}\psi(\frac{t-u}{s}) \qquad (3)$$

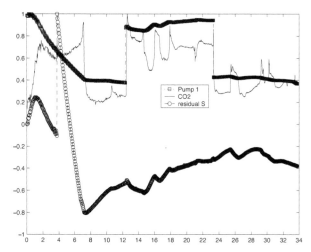

Fig. 5. Example of biochemical signals measured during the bioprocess. Pump 1 is a pump providing substance in the process, CO2 is the measured carbon dioxid and residual S is the residual substrate of the micro-organisms of the bioprocess. The signals have been normalized.

The dilation allows the convolution of the analyzed signal which different sizes of "window" wavelet function. For the detection of the singularities and of the inflexion points of the biochemical signal, we use the Maxima of Modulus of Wavelets Transform (16). The idea is to follow the local maxima at different scales and to propagate from low frequencies to high frequencies. These maxima correspond to singularities, particularly when the wavelet is the derivative of a smooth function:

$$\psi(x) = \frac{d\theta(x)}{dx}$$

$$W_{s,u}f(x) = f * \psi_{s,u} = f(x) * \frac{d\theta(x/s)}{dx}$$

Yuille and Poggio (35) have shown that if the wavelet is derivative of the Gaussian, then the maxima belong to connected curves which are continuous from a scale to another. The detection of the singularities of the signal is thus possible by using the wavelets (see for example figure 6).

The discretization form of Continuous Wavelet Transform is based on the next form of the Mother Wavelet :

$$\psi^{m,n}(t) = a_0^{-m/2}\psi\left(\frac{t - nb_0a_0^m}{a_0^m}\right) \tag{4}$$

By selecting a_0 and b_0 properly, the dilated mother wavelet constitutes an orthonormal basis of $L^2(R)$. For example, the selection of $a_0 = 2$ and $b_0 = 1$ provides a dyadic-orthonormal Wavelet Transform (DWT). The decomposed signals by DWT will have no redundant information thanks to the orthonormal basis.

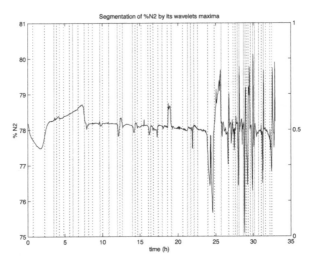

Fig. 6. Segmentation of N2 (nitrogen). Each vertical dotted line correspond to a singularity of the signal detected by wavelets. The wavelet is a DOG (first derivative of Gaussian) and the scales go from 2^0 to 2^3.

Jiang et al. (12) have proposed to select the maxima by using thresholding. Besides, all the singularities are not relevant only some of them are meaningful. However, as stated above, the thresholds proposed by Jiang et al. are chosen empirically. To select the meaningful singularities, we proposed using the Hölder exponent. The Hölder exponent is a mathematical value allowing characterization singularities. The fractal dimension could also be used but only the Hölder exponent can characterize locally each singularity. A singularity in a point x_0 is characterized by the Hölder exponent (also called Hölder coefficient or Lipschitz exponent). This exponent is defined like the most important exponent α allowing to verify the next inequality:

$$|f(x) - P_n(x - x_0)| \leq C|x - x_0|^{\alpha(x_0)} \tag{5}$$

We must remark that $P_n(x - x_0)$ is the Taylor Development and basically $n \leq \alpha(x_0) < n + 1$. Hölder exponent measures the remainder of a Taylor expansion and more of this measures the local differentiability:

1. $\alpha \geq 1$, $f(t)$ is continuous and differentiable.

2. $0 < \alpha < 1$, $f(t)$ is continuous but non-differentiable.

3. $-1 < \alpha \leq 0$, $f(t)$ is discontinuous and non-differentiable.

4. $\alpha \leq -1$, $f(t)$ is not longer locally integrable.

Therefore Hölder exponent could be extended to the distribution. For example the Hölder exponent of a Dirac is equal to -1. A simple computation leads to a very interesting result of the Wavelets Transform (11):

$$|W_{s,u}f(x)| \simeq s^{\alpha(x_0)} \tag{6}$$

This relation is remarkable because it allows to measure the Hölder exponent using the behavior of the Wavelets Transform. Therefore, at a given scale $a = 2^N$ the $W_{a,b}f(x)$ will be maximum in the neighborhood of the signal singularities. The detection of the Hölder coefficient is linked to the vanishing moment of the wavelet: if n is the vanishing moment of the wavelet, then it can detect Hölder coefficients less than n (16). We use a *DOG* wavelet (DOG: first derivative of Gaussian) with a vanishing moment equal to 1; consequently we can only detect Hölder coefficients smaller than 1. This is not a real problem because we are interested (*in this application*[2] by the singularities as step or dirac and the Hölder coefficient of these singularities are smaller than 1. Moreover, the meaningful singularities of the fed-batch bioprocess have Hölder exponents smaller than 1 which correspond to sharp singularities. This type of variations are meaningful for the fed-batch bioprocess fermentation because of many external regulations of the process. Moreover, for Hölder coefficients greater than 1 particularly for integer values, there are difficulties to interpret the Hölder coefficient (see (19) cited in (17)). To evaluate the Hölder coefficient using the wavelets, there are two main ways:

(1) the graphical method which consists in finding the maximum line i.e. the maximum which propagates through the scales, and computes the slopes of this maximum line (often using a log-log representation). The computed slope corresponds to the Hölder coefficient (16).
(2) the minimization method which consists in minimizing a function which has one of the parameters the Hölder coefficient (17). The function is the following:

$$\sum_j \left(ln_2(|s_j|) - ln_2(C) - j - \frac{\alpha(x_0) - 1}{2} ln_2(\sigma^2 + 2^{2j}) \right)^2 \qquad (7)$$

where s_j represents the maximum at scale j, C is a constant depending on the singularity localized in x_0, σ is the standard deviation of a Gaussian approximation (see (17)), and $\alpha(x_0)$ the Hölder exponent.

In (17), a gradient descent algorithm is proposed to solve the minimization, but this technique is very sensitive to local minima. Recently, a minimization using Genetical Algorithms has been proposed (18) and used in bioprocess. More precisely it uses Differential Evolutionary (DE) algorithms. The DE algorithms was introduced by Rainer Storn and Kenneth Price (33).

9. Differential evolution

Differential Evolution (DE)(33) is one of Evolutionary Algorithms (EA) which are a class of stochastic search and optimization methods including Genetic Algorithms (GA), evolutionary programming, evolution strategies, genetic programming and all methods based on genetics and evolution. Through its fast convergence and robustness properties, it seems to be a promising method for optimizing real-valued multi-modal objective functions. Compared to traditional search and optimization methods, the EAs are more robust and straightforward to use in complex problems : they are able to work with minimum assumptions about the objective functions. These methods are slower because due to the generation of the population

---

[2] However it is always possible to use other wavelets with greater vanishing moment for others applications in bioprocesses

and the selection of individuals for crossing. The goal is to obtain the trade-off between accuracy and computing time.

The generation of the vectors containing the parameters of the model is made by applying an independent procedure :

$$X_{i,G} = X_{1,i}.....X_{D,i} \tag{8}$$

with $i = 1...NP$, is the index of one individual of the population; D is the number of parameters which have to be estimated; NP is the number of individuals in one population; G is the index of the current population; i is one individual of the population and $X_{j,i}$ is the parameter j of the individual i in the population G.

As Genetic Algorithms are stochastic processes, the initial population has been chosen randomly, but the initialization of the parameters is based on experts knowledge. Trial parameter vectors are evaluated by the objective function. Several objective functions are tested to produce results on Hölder coefficient detection. For simple GA algorithms the new vectors are the result of the difference between two population vectors and the result is added to a new one. It's a simple crossing operation. The objective function determines if the new vector is more efficient than a candidate population member and replace it if this simple relation is true. In the case of the DE the generation of the new vectors are realized by the difference between the "old vectors" given an weight to each one.

We have tested and compared different schemes of individual generations :

- $DE/rand/1$: For each vector $X_{i,G}$ a perturbed vector $V_{i,G+1}$ is generated according to :

$$V_{i,G+1} = X_{R1,G} + F * (X_{R2,G} - X_{R3,G})$$

$R1, R2, R3 \in [1, NP]$: individuals of population, chosen randomly, $F \in [0,1]$: controls the amplification $(X_{R2,G} - X_{R3,G})$
$X_{R1,G}$: the perturbed vector. There is no relation between $V_{i,G+1}$ and $X_{i,G}$. The objective function must evaluate the quality of this new trial parameter with respect to the old member. If $V_{i,G+1}$ yields a lower objective function value, $V_{i,G+1}$ is set to $X_{i,G+1}$ in the next generation or there is no effect.

- $DE/best/1$: It is like $DE/rand/1$ but is generating $V_{i,G+1}$ by integrating the most performante vector :
$$V_{i,G+1} = X_{best,G} + F * (X_{R1,G} - X_{R2,G})$$

$X_{best,G}$: best vector of population G, $R1, R2 \in [1, NP]$: individuals of population, chosen randomly. As $DE/rand/1$, the objective function compares the quality of $V_{i,G+1}$ and $X_{i,G}$; the smallest of the two is kept in the next population.

- Hydrid Differential evolution algorithms: As DE algorithms, a perturbed vector is generated, but the weight F is a stochastic parameter.

To increase the diversity potential of the population, a crossover operation is introduced. $X_{i,G+1} = (X_{1i,G+1}, X_{2i,G+1}.....X_{Di,G+1})$ becomes :

$$V_{ji,G+1} \begin{cases} j = (n)_D, (n+1)_D, (n+L-1)_D \\ X_{ij,G} \text{ otherwise} \end{cases}$$

$n \in [1, D]$: starting index, chosen randomly
$(n)_D$ = n mod D
$L \in [1, D]$: number of parameters which are going to be exchanged

10. Use of GA for Hölder's coefficients detection

10.1 Implementation

The cost function we have to minimize is the following (17):

$$\sum_j \left(log_2(|a_j|) - log_2(C) - j - \frac{h(x_0) - 1}{2} log_2(\sigma^2 + 2^{2j}) \right)^2 \qquad (9)$$

In the Holder objective function three parameters have to be estimated : $h(x_0)$, C and σ. Thus, one individual X in GA's population is represented by the vector X_h, X_C, X_σ. In our case, the size of population equals 30.

Using the graphical method and the DE, the Hölder coefficient found is quite close to -1, whereas the value computed by gradient descent is not correct. Moreover, if we consider the Hölder coefficient of the Step function, only DE provides quite good results while the graphical method and the values of the gradient descent are too far from the theoretical value. The last median square is not so accurate as DE's. The results obtained indicate that the *DE* can be used for the analyzed data .

For this simulation, the results are summarized in the following table :

Singularity	Dirac	Step 1	Step 2
Theoritical Hölder Coef.	-1	0	0
Hölder Coef. by Graph. Method	-0.5	0.51	0.51
Hölder Coef. by Grad. Descent	0.26	0.89	0.89
Least Median Square	-0.5	0.802301	0.802301
Hölder Coef. by AG	-0.5	-0.03	-0.04

We note that the graphical method is the fastest and used method, but the evaluation of the Hölder coefficient is sometimes imprecise as noted in (34), (22).

On a simple signal (see figure 7), this new method using DE provides better results than those of existing methods as shown in table 1.

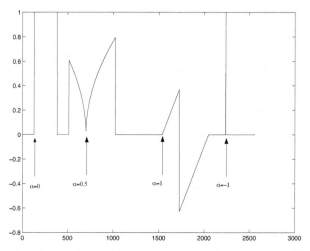

Fig. 7. Simple signal with singularity (step, cups, ramp and dirac) whose Hölder exponents α are known .

Singularity	Dirac	Step	cups	Ramp
theoritical Hölder coef.	-1	0	0,5	1
Hölder exponent by graphical method	-1,13	0,16	0,61	0,84
Hölder exponent by gradient descent	-0,24	0,39	0,74	1,20
Hölder exponent by DE	-1,02	0,02	0,52	1,0007

Table 1. Results of Hölder exponent evaluation by several methods. The wavelet used here is a LOG (second derivative of Gaussian).

11. Characterization by correlation product and classification

Once the states are bounded by the detected and selected singularities using the wavelets, they are characterized by the analysis of the correlations between the biochemical signals. On each interval defined by the singularities, a product of correlation is computed between all pairs of signals. The correlation coefficient (also called Bravais-Pearson coefficient, see (30)) is given by the equation:

$$\frac{\frac{1}{n}\sum_{i=1}^{n}(x_i - \bar{x})(y_i - \bar{y})}{\sigma_x \sigma_y} \tag{10}$$

where x_i represent the values of one parameter (in a given interval), y_i the values of the second parameter (in the same interval), n the number of elements, $\bar{x}$ the average of the elements x_i (of the first biochemical signal), $\bar{y}$ the average of the elements y_i (of the second biochemical signal), et σ_x et σ_y the standard deviation of each of the two signals.

The correlation coefficient is equivalent to the cosine of the scalar product between two biochemical signals projected in the correlation circle of a PCA realized between the two biochemical signals. On each interval, the sign of each correlation coefficient between two signals is kept. Each interval is thus characterized by a set of positive of negative signs. The intervals with the same set of signs are put in the same class as illustrated in the figure 8.

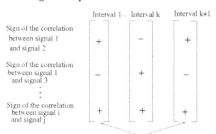

Fig. 8. Principle of the classification method based on wavelets, Hölder exponent and correlation coefficient

Ruiz et al. (27) propose a classification method based on PCA for a neighboring application (wastewater treatment): the data are projected in the space generated by the two first principal components. The method enables to reduce the size of the data space and to take account of the correlation of the signals. However the PCA doesn't take account of the time: the temporal evolution of the process is not taken into account. Ruiz et al. propose to use time analysis window of fixed size. But as the window has a fixed size, it doesn't really take account of the changes occurring during the bioprocess. So the method proposed in this article seems to be more adapted if it is necessary to take account of the variation of the process.

12. Experimental results

Tests have been done on two fed-batch fermentation bioprocesses and the first results have been presented in (25). The two bioprocesses are biotechnological processes using yeast called *Saccharomyces Cerevisiae*. In the first bioprocess we have applied the method to differentiate intrinsic biological phenomena from reactions of micro-organism to extern actions (changes in the environment). In the second bioprocess we directly use the method to detect and classify the states of the bioprocess. For the two fed-batch, the maximum scale is chosen empirically. Mallat and Zhong (17) propose to use as maximal scale $log_2(N) + 1$ where N is the number of measured samples of the signals. However if we use this maximal scale, several singularities would be removed. The empirical value which has been found it is 12. Concerning the Hölder exponent we are interested by the singularities between -1 and 1. For the evaluation of Hölder exponent using Genetical Algorithms, tests have shown that 100 iterations are sufficient for an acurate evaluation (18).

12.1 Differentiation between biophysical and biological phenomena

The first bioprocess is a bioprocess lasting about 25 hours. 12 biochemical signals have been measured during the bioprocess.
In a fed-batch bioprocess, there are two kinds of signals: the signals given by parameters

regulated by an extern action (expert in microbiology or control system) and the signals given by non regulated parameters. An example of regulated parameter is the agitation which is the speed of the rotor of the bioreactor and an example of non regulated parameter is the N2 (nitrogen). The actions on regulated parameters induce modifications of the physiology of the micro-organisms and physical changes in the bioreactor: there are *biophysical* phenomena. On the other hand, during the bioprocess, the micro-organisms have intrinsic physiological behavior: there are *biological* phenomena.

Is it possible to distinguish biophysical phenomena and biological phenomena?

To answer this question, we propose the following steps:

1. search the variations of the regulated signals. These variations are sharp variations which correspond to singularities as Dirac or step.

2. compare the sign of correlation product between regulated signals and non regulated signals before and after each detected singularity of the regulated signals. If the sign is the same before and after, there is no influence: it is a biological phenomenon. If the sign changes before and after, there is an influence: it is a biophysical phenomena.

We must note that:

- only the singularities of the regulated signal are detected and selected,

- to compare the sign of correlation product before and after each singularity, me must choose a reference temporal interval. Besides, the first temporal interval (delimited by the detected singularities) is considered as a biological interval as the bioprocess begins and the initial conditions are considered as biological.

An example of comparison between the agitation and the nitrogen is given in figures 9 and ??.

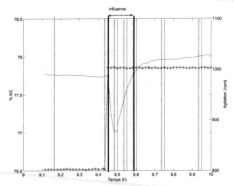

Fig. 9. An example of intervals (horizontal lines are singularities and correspond to boundaries of the temporal interval) with the segmentation given by the detection of regulated signals. This example has a duration of one hour (from 9 hours to 10 hours) taken from the first fed-batch. There are 14 intervals. Signals are agitation (stars) expressed in rotation per minute (rpm) and the percentage of nitrogen N2 (solid line).

Results confirm the observations of the expert. All the intervals considered as biological by the proposed method are considered as biological by the expert. Particularly, the last interval is considered as a biological phenomenon, which is well known by experts, as at the end of a bioprocess, regulated signals are not modified. Another example is given by biological intervals located in the middle of the bioprocess which correspond to spontaneous oscillations.

12.2 Detection and classification of states

We have studied Saccharomyces cerevisiae dynamical behaviour during fed-batch aerated cultivation in oxidative metabolism. The maximal growth rate of this yeast was calculated to 0,45 h-1. The aim of our work was to determine by on line analysis, different physiological states of the yeast behavior only with the available sensors (pH, temperature, oxygen Ě). Off-line metabolites and intracellular carbohydrate reserve analysis help in a first approach to identify the physiological states. State recognition is performed by signal processing technics. The second bioprocess is a bioprocess lasting about 34 hours. 11 biochemical signals have been measued during the bioprocess.

We recall the used method for the characterization of intervals for the classification is given in section 4 and summarised in figure 8. The classification provided by the method gives interesting results shown in the figure 10. Once again, results obtained correspond to the experts observations. Particularly, the most interesting result concerns the detection and the characterization of a state resulting of an external action. Besides, the class number 8 corresponds to the addition of an acid[3] (the acid is not a regulated parameter as in the first example, but is directly introduced by the expert during the experience) in the bioprocess. All apparition of class 8 correspond exactly to an acid addition. These results were confirmed and validated. As far as we know, it is the first time that this kind of non-model-based approach can find and characterize automatically the addition of acid in a fed-bacth process. The results are promising and further analysis of the classification is necessary.

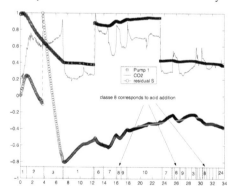

Fig. 10. Classification provided by the method. The wavelet is a DOG and the scales go from 2^0 to 2^{10}.

13. Discussion and conclusion

We apply logical tools to get explanation rules concerning the behavior of a bio-reactor. The ability to incorporate background knowledge and re-use past experiences marks out ILP as a

---

[3] because of industrial confidentiality, we are not allowed to gives more information

very effective solution for our problem. Instead of simply giving classification results, we get some logical rules establishing a causality relationship between different parameters of the bio-machinery. Among these rules, some ones are validated by expert knowledge, but some new ones have been provided. It yet appears that some previous rules have to be removed or modified to fit with new observations.

One of the main interest of this kind of approach is the fact that the resulting theory is easy to understand, even for a non specialist : the first order logic is, from a syntactic viewpoint, close to the natural language.

Intelligibility of resulting explanations is an other argument in favor of the ILP tools. A drawback of standard logic is the difficulty to deal with the time dimension : in some sense, standard logic is static and thus, not well suited to described dynamic process. One could hope that modal logic would be of some help, but it remains to design an inductive machine dealing with the temporal modalities, i.e. a way to reverse temporal logic inference system.

14. References

[1] Arneodo A. and al. Ondelettes, multifractales et turbulence de l'ADN aux croissances cristallines. *DIDEROT EDITEUR*, Paris, 1995.

[2] J. Aguilar-Martin, J. Waissman-Vilanova, R. Sarrate-Estruch, and B. Dahou. Knowledge based measurement fusion in bio-reactors. In *IEEE EMTECH*, May 1999.

[3] Bakshi, B. and Stephanopoulos, G. (1994). Representation of process trends-III. multiscale extraction of trends from process data. *Computer and Chemical Engineering*, 18(4):267–302.

[4] Cao, S. and Rhinehart, R. (1995). An efficient method for on-line identification of steady state. *Journal of Process Control*, 5(6):363–374.

[5] Domingo P. and Pazzani M.. On the optimality of the simple bayesian classifier under zero-one loss. *Machine Learning*, 29,103-130, 1998.

[6] Doncescu, A., Waissman, J., Richard, G., and Roux, G. (2002). Characterization of bio-chemical signals by inductive logic programming. *Knowledge-Based Systems*, 15(1-2):129–137.

[7] Edelman, G. M. and Gally, J. A. (2001). Degeneracy and complexity in biological systems. In *Proc Nat Acad Science USA*.

[8] Gadkar, K., Mehra, S., and Gomes, J. (2005). On-line adaptation of neural networks for bioprocess control. *Computer and Chemical Engineering*, 29:1047–1057.

[9] Guillaume, S. and Charnomordic, B. (2004). Generating an interpretable family of fuzzy partitions from data. *IEEE Transactions on Fuzzy Systems*.

[10] Hvala, N., Strmcnik, S., Sel, D., Milanic, S., and Banko, B. (2005). Influence of model validation on proper selection of process models-an industrial case study. *Computer and Chemical Engineering*.

[11] Jaffard, S. (1997). Multifractal formalism for functions part 1 and 2. *SIAM J. of Math. Analysis*, 28(4):944–998.

[12] Jiang, T., Chen, B., He, X., and Stuart, P. (2003). Application of steady-state detection method based on wavelet transform. *Computer and Chemical Engineering*, 27(4):569–578.

[13] Kitano, H. (2002). Computational systems biology. In *Nature 6912, 206*.

[14] Lee, J.-M., Yoo, C., Lee, I.-B., and Vanrolleghem, P. (2004). Multivariate statistical monitoring of nonlinear biological processes using kernel PCA. In *IFAC CAB'9*, Nancy, France.

[15] Lennox, J. and Rosen, C. (2002). Adaptative multiscale principal components analysis for online monitoring of wastewater treatment. *Water Science and Technology*, 45(4-5):227–235.

[16] Mallat, S. and Hwang, W.-L. (1992). Singularity detection and processing with wavelets. *IEEE Trans. on Information Theory*, 38(2):617–643.

[17] Mallat, S. and Zhong, S. (1992). Characterization of signals from multiscale edges. *IEEE Trans. on PAMI*, 14(7):710–732.

[18] Manyri, L., Regis, S., Doncescu, A., Desachy, J., and Urribelarea, J. (2003). Holder coefficient estimation by differential evolutionary algorithms for saccharomyces cerivisiae physiological states characterisation. In *ICPP-HPSECA*, Kaohsiung, Taiwan.

[19] Meyer, Y. (1990). *Ondelettes et Opérateurs*, volume I. Hermann.

[20] S. Muggleton. Inverse entailement and Progol. New Gen. Comput.,13:245-2,1998.

[21] Narasimhan, S., Mah, R., Tamhane, A., Woodward, J., and Hale, J. (1986). A composite statistical test for detecting changes in steady state. *American Institute of Chemical Engineering Journal*, 32(9):1409–1418.

[22] Nugraha, H. B. and Langi, A. Z. R. (2002). A wavelet-based measurement of fractal dimensions of a 1-d signal. In *IEEE APCCAS*, Bali, Indonesia.

[23] Polit, M., Estaben, M., and Labat, P. (2002). A fuzzy model for an anaerobic digester, comparison with experimental results. *Engineering Applications of Artificial Intelligence*, 15(5):385–390.

[24] Rocca J.. Technical report of Laboratoire d'automatisme et architecture des sytemes. 1998.

[25] Régis, S., Doncescu, A., Faure, L., and Urribelarea, J.-L. (2005). Détection et caractérisation d'un bioprocédé par l'analyse de l'exposant de hölder. In *GRETSI 2005*, Louvain-la-Neuve, Belgium.

[26] Roels, J. (1983). *Energetics and kinetics in biotechnology*. Elsevier Biomedical Press.

[27] Ruiz, G., Castellano, M., Gonzàlez, W., Roca, E., and Lema, J. (2004). Algorithm for steady states detection of multivariate process: application to wastewater anaerobic digestion process. In *AutMoNet 2004*, pages 181–188.

[28] S. Régis. *Segmentation, classification, et fusion d'informations de séries temporelles multi-sources: application à des signaux dans un bioprocédé*. Thèse de Doctorat (PhD thesis), Université des Antilles et de la Guyane, Novembre 2004.

[29] S. Régis, L. Faure, A. Doncescu, J.-L. Uribelarrea, L. Manyri, and J. Aguilar-Martin. Adaptive physiological states classification in fed-batch fermentation process. In *IFAC CAB'9*, Nancy, France, March 2004.

[30] Saporta, G. (1990). *Probabilités, et Analyse des données et Statistique*. Technip.

[31] Steyer, J., Pourciel, J., Simoes, D., and Uribelarrea, J. (1991). Qualitative knowledge modeling used in a real time expert system for biotechnological process control. In *IMACS International Workshop "Decision Support Systems and Qualitative Reasoning"*.

[32] J.P. Steyer. *Sur une approche qualitative des systèmes physiques : aide en temps rèel à la conduite des procédés fermentaires*. Thèse de Doctorat, Université Paul Sabatier, Toulouse France, Décembre 1991.

[33] Storn, R. and Price, K. (1996). Minimizing the real functions of the icec'96 contest by differential evolution. In *Proc. of the 1996 IEEE International Conference on Evolutionary Computation*.

[34] Struzik, Z. R. (1999). *Fractals: Theory and Application in Engineering*, pages 93–112. Springer Verlag.

[35] Yuille, A. and Poggio, T. (1986). Scaling theorems for zero-crossing. *IEEE Transaction for zero-crossing*, 8(1):15–25.

3

Electroporation of *Kluyveromyces marxianus* and β-D-galactosidase Extraction

Airton Ramos[1] and Andrea Lima Schneider[2]
[1]State University of Santa Catarina
[2]University of Joinville
Brazil

1. Introduction

Nowadays electroporation is well established as a method for increasing the permeability of biological membranes aiming to include some kind of molecule inside cells (anticancer drugs, for example) or to extract molecules from cells (enzymes, DNA, etc.). The permeabilized cells show a distribution of hydrophilic pores with diameters of several nanometers in the regions where the induced transmembrane potential rises above a threshold value from 200 mV to 1 V (Hibino *et al*, 1993; Teissie & Rols, 1993). The pore density and their diameters are dependent on the stimulation conditions: field strength, waveform and duration of stimulus (Weaver & Chizmadzhev, 1996; Teissié *et al*, 2005; Miklavčič & Puc, 2006; Chen *et al*, 2006)

Some previous studies have demonstrated experimentally the possibility of enzyme extraction from yeast by electroporation. Galutzov and Ganev (Galutzov & Ganev, 1999) investigated the extraction of glutathione reductase, 3-phosphoglicerate kinase e alcohol dehydrogenase from *Saccharomyces Cerevisiae* by electroporation. They used sequences of electric field pulses with strength 275 kV/m and duration of 1 ms. They observed increase in concentration of enzymes in the supernatant to about 8 hours after exposure to the field. The extraction efficiency was higher compared to other methods with mechanical or chemical lyses. Treatment with dithiothreitol before exposure to the field accelerated the extraction of glutathione reductase and alcohol dehydrogenase. This effect was considered to be related to the increased porosity of the cell wall due to the break of disulfide bonds in the layers of mannoproteins.

Ganev *et al* (Ganev *et al*, 2003) developed and used an electroporation system to stimulate a flow of yeast suspension up to 60 mL/min. They obtained the extraction of hexokinase, 3-phosphoglicerate kinase and 3-glyceraldehyde phosphate dehydrogenase from *Saccharomyces Cerevisiae*. They found that the maximum extraction is obtained after 4 hours with 15 pulses of frequency 6 Hz, but similar results were obtained for different combinations of field strength between 270 and 430 kV/m and pulse duration between 1 and 3 ms, so that stronger fields require shorter pulse to produce the same result in the extraction. The activity of enzymes extracted by electroporation was about double of those extracted by enzymatic and mechanical lysis of cells.

The extraction of β-D-galactosidase has a great interest due to its use in food and pharmaceutical industries. This enzyme is responsible for hydrolysis of lactose, disaccharide of low sweetness present on milk, resulting in its monosaccharide glucose and galactose. Its industrial importance is the application in dairy products for the production of foods with low lactose content, ideal for lactose intolerant consumers by improving the digestibility of milk and dairy. This is an important issue since lactase deficiency is present in about two thirds of the adult population worldwide, mainly in developing countries (Swagerty et al, 2002). In addition, lactase allows getting better technological characteristics and improves sensory and rheological properties of dairy products. Increase the power and sweetness that reduces the addition of sucrose, reducing the calorie content of foods. The formation of monosaccharides in dairy products helps in the metabolism of yeast fermented products. Lactase also reduces the probability of occurrence of the Maillard reaction, as galacto-oligosaccharides obtained did not act as reducing sugars.

The enzyme extraction method based on cell membrane permeabilization by electric field pulses applied to a suspension of yeast has advantages over chemical and mechanical methods due to its simplicity, efficiency in the extraction and preservation of enzyme activity and the cell itself, depending on the stimulation conditions. This study aims to evaluate the relationship between the change in membrane conductance during electrical stimulation of the cells and the enzyme β-D-galactosidase activity released by cells and to determine the stimulation conditions that maximize the extraction of this enzyme from *Kluyveromyces marxianus* yeasts.

The electroporation dynamics (opening and closing pores) is not completely understood. Part of this problem is because the electroporation is evaluated by indirect measurements (Kinosita & Tsong, 1979; He et al, 2008; Saulis et al, 2007). A number of methods have been used in the study of electroporation based on electrical measurements (Kinosita & Tsong, 1979; Huang & Rubinsky, 1999; Kotnik et al, 2003; Pavlin et al, 2005; Pliquett et al, 2004; Ivorra & Rubinsky, 2007; Suzuki et al, 2011). Electrical measurements and modeling were used to evaluate the effectiveness of electroporation in individual cells (Koester et al, 2010; Haque et al, 2009), cell suspensions (Kinosita & Tsong, 1979; Pavlin et al, 2005; Suzuki et al, 2011) and tissues (Grafström et al, 2006; Ivorra & Rubinsky, 2007; Laufer et al, 2010).

The membrane conductance after electroporation is related to electrical conductivity of cell suspension. Models are developed for isolated spherical or spheroidal cells using proposed membrane conductance distributions (Kinosita & Tsong, 1979; Pavlin & Miklavcic, 2003) and are obtained from the static solution to Laplace's equation using the simplest possible structure of a cell with a nonconductive thin membrane filled internally by a homogeneous medium. More complex situations in which the interaction between cells is not neglected or when the membrane conductance is calculated based on dynamic models of electroporation can be solved by numerical methods (Neu & Krassowska, 1999; Ramos et al, 2004; Ramos, 2010).

Conductivity measurement have been used in previous studies in order to determine the variation of the membrane conductance during electroporation (Kinosita & Tsong, 1979; Pavlin et al, 2005; Suzuki et al, 2011). But, caution is needed when using this approach. The impedance of the electrode interface with the suspension has strong dispersion of reactance and resistance at low frequencies, up to about 1 kHz (McAdams et al, 1995). Ionic diffusion inside double layer of cells also results in low-frequency dielectric dispersion. Additionally, interfacial polarization shows strong dispersion with relaxation times of less than 1 μs for cells of a few microns in diameter (Foster & Schwan, 1995). Due to the reactive and dispersive effects, the precise determination of the suspension conductivity cannot be done using instantaneous values of voltage and current with pulsed waveform, since the spectral content of these signals is very wide.

The heating of the electrolyte during the pulse application also affects the relationship of conductivity with the membrane conductance. The variation of conductivity due to power dissipation in an aqueous electrolyte can be estimated by $\Delta\sigma/\sigma=\alpha\sigma_0 E^2\Delta t/\rho c$, where α is the temperature coefficient, ρ and c are respectively the density and specific heat of water, E is the electric field strength applied and Δt is the time length of the pulse. For an aqueous NaCl electrolyte with initial conductivity of 20 mS/m and field of 400 V/m (typical values in our experiments) the variation of conductivity is 1.53% per millisecond. Thus, for a time of 10 ms, we can estimate a contribution of about 15% of variation in the suspension conductivity due only to heating of the sample. The conductivity of the external medium can also vary due to the efflux of ions through the hydrophilic pores created in the cell membrane. The external conductivity increases while the internal conductivity decreases, changing the relationship between suspension conductivity and membrane conductance.

Simple estimates of conductivity by measuring the instantaneous current and voltage during the pulse application are not reliable for a correct evaluation of the membrane conductance. In order to accomplish this goal in this study we measure the electrical impedance of the sample in a wide range of frequencies before and after the pulse application. Using a delay of 60 seconds to measure the impedance after the pulse, the generated heat is allowed to dissipate through the metal electrodes into the air. Once the volume of the suspension is small (about 300 μL), we can predict that thermal effects are negligible after 60 seconds.

The analysis of the impedance spectra enables us to identify and separate the effects of dispersion at low frequencies. In addition, one can adjust the parameters of a dielectric dispersion model for interfacial polarization of cell suspensions to obtain estimates of cell concentration, internal and external conductivities and membrane conductance for both intact cells and electroporated cells.

2. Methods

2.1 Experimental method

Cells of yeast *Kluyveromyces marxianus* CBS 6556 were used in the electroporation experiments. The cells were grown in stirred flasks containing 0.5 L with 150 mL of Bacto peptone (2%), yeast extract (1%) and lactose (2%) at 30 °C for 12 hours and 150 rpm. The colonies in the stationary phase were then washed three times in distilled water and centrifuged each time for six minutes at 12,000 rpm in an Eppendorf centrifuge and finally re-suspended in distilled water.

Samples in the final suspension were observed and photographed under a Olympus CX31 microscope with attached camera Moticam 1000 and software Motic Images Plus 2.0. 120 cell diameter measurements were performed in six different colonies used in the enzyme activity and electrical impedance assays. The average radius obtained was 4.87 µm with a standard deviation of 1.07 µm.

The electroporator used in the experiments consists of a differential power amplifier with a gain of 50 V/V and an arbitrary waveform generator implemented in LabView™ (National Instruments) triggering a PCI 6251 card (National Instruments). Figure 1 shows a schematic representation of the electroporator. A current meter probe using a LTSP 25-NP sensor (LEM Components) and a voltage probe using resistive dividers and an INA 111 amplifier (Burr Brown) was implemented and used to measure current and voltage applied to the sample. The signals from the probes were acquired by the program through the PCI 6251 card using 16-bit resolution and 500 kHz of sampling rate simultaneously with the generation of the applied voltage. The sample holder is made of a cylindrical tube of nylon covering two steel electrodes 0.02 m in diameter and 0.001 m in spacing. The sample of the yeast suspension is injected and removed from the space between electrodes using syringes. The volume stimulated in each pulse application is 311 µL. All experiments were performed at room temperature of 25 °C. For each experiment, the electroporator was configured to generate an output voltage pulse from 100 to 400 Volts with duration from 1 to 10 ms.

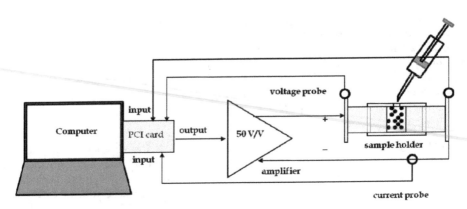

Fig. 1. Schematic of the electroporator

The enzyme activity assays were made 4 hours after pulsation. The cells were centrifugated for six minutes at 12,000 rpm in an Eppendorf centrifuge. To the supernatant was assigned the fraction of enzyme suspended outside the cells; to the pellets were assigned the enzyme associated to the cell walls. The cromogen ONPG, described by Lederberg (Lederberg, 1950) was used as a substrate. The assay mixture contained 780 µL sodium/phosphate buffer with pH 7.6 (containing 47 mM 2-mercaptoethanol and 1mM $MgCl_2$), 110 µL 35 mM ONPG and 110 µL of the enzyme solution. After one minute of incubation in a Thermomixer (Eppendorf) at 30° C, the reaction was stopped by 0.22 mL 1M Na_2CO_3. One unit (U) of β-galactosidase is defined as the amount of enzyme that releases 1 µmol of orto-nitrophenolate per minute under the assay conditions. This can be evaluated by means of measuring the increase of absorbance at 405 nm in a LKB spectrophotometer, using 1 cm optic path glass cuvettes. The molar extinction coefficient used was 3.1 $µmol^{-1}$ cm^2 (Furlan et al, 2000).

The electrical impedance of the sample was measured before and after electroporation pulsation. We used a 4294A impedance analyzer (Agilent Technologies) in the range 40 Hz to 40 MHz. The time interval between the pulse application and the second measurement was kept in 60 seconds for all samples. This time is needed to allow disconnecting the sample holder from electroporator and connecting to the impedance analyzer as well as to allow the impedance reading to stabilize.

2.2 Mathematical method

The dispersion model for interfacial polarization in suspensions of spherical cells is obtained from Maxwell-Wagner theory of dispersion. For spherical particles of complex conductivity γ_c suspended in a medium of complex conductivity γ_o and occupying a volume fraction p, the following relationship applies between these quantities and the complex conductivity of the suspension γ_s (Foster & Schwan, 1995):

$$\frac{\gamma_s - \gamma_o}{\gamma_s + 2\gamma_o} = p\frac{\gamma_c - \gamma_o}{\gamma_c + 2\gamma_o} \tag{1}$$

Where $\gamma_c = \sigma_c + j\omega\varepsilon_c\varepsilon_o$, $\gamma_o = \sigma_o + j\omega\varepsilon_w\varepsilon_o$ and $\gamma_s = \sigma_s + j\omega\varepsilon_s\varepsilon_o$. In these equations ω is the angular frequency, ε_o is the vacuum permittivity, σ_c, σ_o and σ_s are the conductivities of cells, external medium and suspension, respectively, and ε_c, ε_w and ε_s are the dielectric constant of cells, water and suspension, respectively. The complex conductivity of cells can be estimated using the simple model consisting of a homogeneous internal medium with conductivity $\gamma_i = \sigma_i + j\omega\varepsilon_w\varepsilon_o$ surrounded by a spherical membrane of thickness h much smaller than the radius R and with complex conductivity $\gamma_m = h(G_m + j\omega C_m)$, where G_m and C_m are the conductance and capacitance per unit area of membrane, respectively. The result obtained after making approximations that keeps only terms of first order in (h/R) is the following (Foster & Schwan, 1995):

$$\gamma_c \cong \frac{\gamma_i + (2h / R)\gamma_m}{1 + (h / R)(\gamma_i - \gamma_m) / \gamma_m} \tag{2}$$

Substituting (2) in (1), the resulting expression can be separated into first order dispersion equations for the process of interfacial polarization at the cell membrane surface. The general expressions for the conductivity and dielectric constant of the suspension are written below:

$$\sigma_s = \sigma_{so} + \frac{\omega^2\tau_m^2\Delta\sigma_s}{1 + \omega^2\tau_m^2} \tag{3}$$

$$\varepsilon_s = \varepsilon_\infty + \frac{\Delta\varepsilon_s}{1 + \omega^2\tau_m^2} \tag{4}$$

$$\Delta\sigma_s = \frac{\varepsilon_o\Delta\varepsilon_s}{\tau_m} \tag{5}$$

In these equations, σ_{so} and ε_∞ are respectively the low frequency conductivity and high frequency dielectric constant of the suspension. $\Delta\sigma_s$ and $\Delta\varepsilon_s$ are the dispersion amplitudes of conductivity and dielectric constant, respectively, and τ_m is the relaxation time for interfacial polarization. Based on the model described by equations (1) and (2), the dispersion parameters of the first order equations (3) and (4) are presented below (Foster & Schwan, 1995):

$$\Delta\varepsilon_s = \frac{9pRC_m}{4\varepsilon_0\left[1+\dfrac{P}{2}+\dfrac{RG_m}{\sigma_i\sigma_o}\left(2\sigma_o+\sigma_i+p\left(\sigma_o-\sigma_i\right)\right)\right]^2} \tag{6}$$

$$\sigma_{so} = \sigma_o\frac{\left[1-p+\dfrac{RG_m}{\sigma_i\sigma_o}\left(\sigma_o+\dfrac{\sigma_i}{2}+p\left(\sigma_i-\sigma_o\right)\right)\right]}{\left[1+\dfrac{P}{2}+\dfrac{RG_m}{\sigma_i\sigma_o}\left(\sigma_o+\dfrac{\sigma_i}{2}-\dfrac{P}{2}(\sigma_i-\sigma_o)\right)\right]} \tag{7}$$

$$\tau_m = RC_m\frac{\sigma_i+2\sigma_o-p(\sigma_i-\sigma_o)}{2\sigma_i\sigma_o(1+p)+RG_m\left[(\sigma_i+2\sigma_o)-p\left(\sigma_i-\sigma_o\right)\right]} \tag{8}$$

The numerical method is initially applied to the parameterization of the impedance model of the sample. By considering the geometry of the electrodes with parallel faces and with radius much larger than the spacing, the following expression adequately represents the sample impedance:

$$Z_m = \frac{R_{ct}}{1+\left(j\omega\tau_e\right)^\beta} + \frac{d/A}{\sigma_s+j\omega\varepsilon_s\varepsilon_o} \tag{9}$$

Where A and d are the area and spacing of the electrodes, respectively. The first term in the second member represents the impedance of the electrode-electrolyte interface (McAdams *et al*, 1995). The second term is the impedance of the suspension. The parameters of the interface impedance model are: the charge transfer resistance R_{ct}, the surface relaxation time τ_e and the constant β. In the suspension impedance model the parameters to be obtained are shown in equations (3) and (4): $\sigma_{so}, \varepsilon_\infty, \Delta\sigma_s$ or $\Delta\varepsilon_s$ and τ_m.

The parameterization algorithm used is based on successive approximations. Initially, each parameter is assigned a range of search. The parameters are stored in integer variables with 16 bits of resolution. Each interval is divided into $2^{16}-1 = 65,535$ subintervals. The approximation process is performed by calculating all the answers of the equation (9) for a given parameter that varies throughout its subintervals, keeping the other parameters fixed at initial values. The value that minimizes the mean square error between the model and the measured impedance spectrum is selected. The process is repeated for all parameters, while maintaining the selected values of the parameters already adjusted. After several cycles of this procedure, the parameters of the model specified in equation (9) converge to the desired response. The convergence with mean square error less than 1% of the mean square value of the impedance magnitude was achieved on average with 10 cycles of calculation. Figure 2 shows the result of the parameterization for an assay with applied field of 200 kV/m and time length of 10 ms. The obtained parameters are shown in the figure.

Using the values of σ_{so}, $\Delta\sigma_s$, $\Delta\varepsilon_s$ e τ_m obtained from the impedance spectrum measured before electroporation, the volume fraction p and membrane capacitance were calculated. In this case, the membrane conductance is very small, usually between 1 and 10 S/m² (Foster & Schwan, 1995) and the terms containing G_m in equations (6) to (8) can be neglected. Accordingly, these equations can be rewritten in the approximate forms:

$$\Delta\varepsilon_s = \frac{9pRC_m}{\varepsilon_0(2+p)^2} \tag{10}$$

$$\sigma_{so} = 2\sigma_0\frac{(1-p)}{2+p} \tag{11}$$

$$\tau_m = \frac{RC_m}{2\sigma_0}\frac{(1-p)}{(1+p)} \tag{12}$$

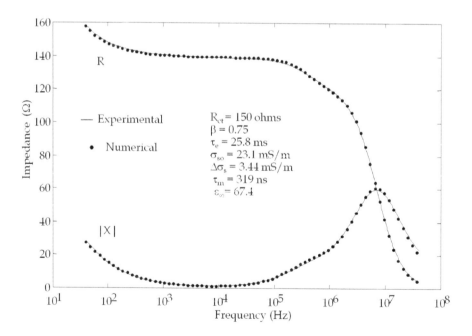

Fig. 2. Experimental and Theoretical impedance spectra to an assay with 200 kV/m and 10 ms. R – resistance; |X| modulus of reactance.

For the equation (12) it was assumed that $\sigma_i \gg \sigma_o$, since the final suspension is obtained by diluting the cells in distilled water. Combining these equations with equation (5), we obtain an equation for calculating the volume fraction from σ_{so} e $\Delta\sigma_s$:

$$\frac{\Delta\sigma_s}{\sigma_{so}} = 9\frac{p(1+p)}{(2+p)(1-p)^2} \tag{13}$$

With the value of p in equation (10) yields the membrane capacitance C_m. Using the values of σ_{so}, $\Delta\sigma_s$, $\Delta\varepsilon_s$ and τ_m obtained from the impedance spectrum measured after electroporation, the membrane conductance and internal and external conductivities were calculated. Newton's method was used for this purpose. Defining the following variables: $x_1=RG_m$, $x_2=1/\sigma_i$ e $x_2=1/\sigma_o$, equations (6), (7) and (8) can be written in the respective forms:

$$F_1 = a_1 + 2a_1x_1x_2 + 2a_1a_2x_1x_3 - 1 = 0 \tag{14}$$

$$F_2 = b_1x_3 + b_1x_1x_2x_3 + b_1b_3x_1x_3^2 - x_1x_2 - b_2x_1x_3 - 1 = 0 \tag{15}$$

$$F_3 = c_1 + c_1c_2x_1x_3 + c_1c_3x_1x_2 - c_2x_3 - c_3x_2 = 0 \tag{16}$$

Where the coefficients are given in the Table 1 below.

$a_1 = \sqrt{\dfrac{\varepsilon_o\Delta\varepsilon_s(2+p)}{9pRC_m}}$	$a_2 = \dfrac{1-p}{2+p}$	———
$b_1 = \sigma_s\dfrac{(2+p)}{2(1-p)}$	$b_2 = \dfrac{(1+2p)}{2(1-p)}$	$b_3 = a_2$
$c_1 = \dfrac{\tau_m}{RC_m}$	$c_2 = \dfrac{(1-p)}{2(1+p)}$	$c_3 = \dfrac{(2+p)}{2(1+p)}$

Table 1. Coefficients of the dispersion equations for applying Newton's method.

Applying Newton's method the convergence was obtained in five steps with $\left\|[F_1F_2F_3]\right\|_\infty < 10^{-10}$ using the initial values: $G_m = 1000$ S/m², $\sigma_i = 500$ mS/m e $\sigma_o = 20$ mS/m.

3. Results and discussion

3.1 Impedance measurement

Each electroporation experiment was performed by applying a single pulse of electric field with strength 100, 200, 300 and 400 kV/m and time length from 1 to 10 ms. Figure 3 shows the typical behavior of the apparent conductance of the sample during electrical stimulation. These curves were obtained as the ratio between the voltage and current measured in the sample during the pulse application. This ratio is named apparent conductance because the sample actually has complex impedance with frequency dependent resistance and reactance. Since the applied pulse has a wide range of spectral components, it is not possible to obtain the conductance of the medium simply by dividing the instantaneous voltage and current in the sample. In any case, the initial apparent conductance increases with the applied field, indicating the occurrence of electroporation. The slope of each curve also indicates that electroporation increases during the pulse. For more intense pulses the conductance shows greater variation. For intense fields, however, the heating of the sample is appreciable and this contributes to increase the sample conductance. Some authors, in previous studies, used instantaneous measurements of voltage and current for conductivity calculation and deduced the membrane conductance from these measurements (Kinosita & Tsong, 1979, Pavlin et al, 2005; Suzuki et al, 2011). It may indeed have a special interest in this approach since it would allow an assessment of the electroporation dynamics (pore opening rate) during application of the pulse. However, without an adequate measurement technique, the results may be adversely affected by other effects, such as the surface impedance of the electrodes and dielectric dispersion due to ion diffusion and accumulation on cell surface.

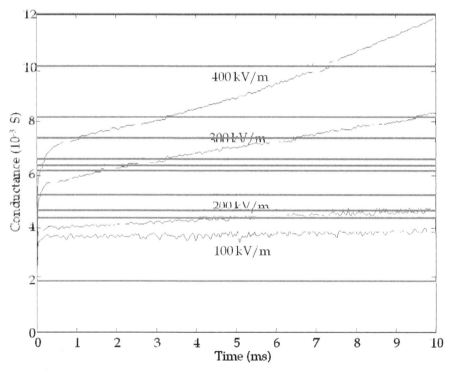

Fig. 3. Apparent sample conductance during the electroporation pulse with strength from 100 to 400 kV/m and time length of 10 ms.

A measurement technique that can be used to compensate for the reactive effects is the technique of small signals. In this case, one can use a small amplitude sinusoidal signal added to the electroporation pulse. This signal can be used to measure the change in conductance of the sample during the pulse. If the small-signal frequency is sufficiently far from the dispersion band of the medium, the sample behaves like a real conductance and the conductivity obtained can be properly related to the variation of the membrane conductance. According to the impedance spectrum shown in Figure 2, the frequency of 10 kHz is adequate for this purpose. Another aspect to consider is the heating that results in increased conductivity of the electrolyte in the suspension. This effect seems to happen in the increase of the curve inclination for 400 kV/m in Figure 3 for pulse length higher than 5 ms. To avoid the interference of heating is necessary to limit the intensity of the applied field or the time length of the pulse.

Figure 4 shows the impedance spectra obtained before and after pulse application with 10 ms and field strengths of 100, 200, 300 and 400 kV/m. There is a strong reduction of the resistance through the frequency range up to 10 MHz and a significant reduction of the reactance at high frequencies for fields greater than 100 kV/m. One can assign both variations to the increase in membrane conductance. However, another effect that decreases the electrical impedance is the efflux of ions from the cytoplasm through the pores created in the cell membrane. This process has little effect on the apparent sample conductance shown in Figure 3, because its time constant is of the order of several seconds (Pavlin et al, 2005), but significantly affects the sample impedance measured after 60 seconds from pulse application. Another process that can have an effect is the cell swelling due to water influx through the pores due to the osmotic pressure difference. This process has time constant of tens of seconds to mouse melanoma cells (Pavlin et al, 2005). In yeast the effect of osmotic pressure is possibly reduced by the presence of the cell wall. Furthermore, in the experiments conducted in this study due to the small volume fraction of the suspension (about 3%) and low conductivity of the external medium (about 20 mS/m), the effects of cell swelling in the impedance are small. The field of 100 kV/m possibly lies just above the threshold of electroporation for yeast. The effects are relatively small as can be seen in Figure 4. There was also a tendency of saturation in the sample impedance for intense fields, suggesting that a maximum permeation can be achieved for fields of the order 400 kV/m.

3.2 Suspension conductivity and membrane conductance

Electroporation assays were repeated three times in all conditions of stimulation with samples prepared using the same procedure described above. The mathematical method used allowed to obtain the volume fraction, the membrane capacitance and conductivity of the external medium before electroporation. The range of values obtained for the set of 72 samples showed small standard deviation. The results obtained are: $p = 0.0332 \pm 0.0011$, $C_m = 3.5 \times 10^{-3} \pm 2.8 \times 10^{-4}$ F/m^2 and $\sigma_o = 24.6 \pm 2.7$ mS/m. The low membrane capacitance over the typically reported value for animal cells (of the order of 10^{-2} F/m^2) probably is due to the effect of the ionic distribution in the cell wall on the outer surface of the cell membrane. The cell wall of yeasts consists of two main types of macromolecules, polysaccharides glucan and mannan-type and several proteins, forming a reticulated structure with a thickness much larger than the cell membrane. The electrostatic interactions between these molecules and ions carried to the membrane by an applied electric field determine the spatial distribution of charge different from that seen in cells devoid of cell wall. The cell wall can also act as a filter for the efflux of macromolecules through the pores of the cell membrane as will be seen in relation to enzyme activity in this study.

The mathematical method applied to electroporated suspensions allowed to determine the change in conductivity of the external medium and membrane conductance. Figure 5 shows the variation of conductivity obtained with the impedance measurement before and after 60 seconds of pulse application as a function of field strength and pulse length. Figure 6 shows the distribution of membrane conductance obtained in the same analysis. The two sets of values are very similar. The correlation between them is mainly due to the fact that both depend on the number and size of pores in the cell membrane created by the applied field. Ions leaving the cells through these pores increase the conductivity of the external medium. This increase is small for 100 kV/m because this field is only slightly above the electroporation threshold. But as the applied field increases, the efflux of ions is increased and also shows a dependence on the pulse length. The conductivity measurement of the external medium can be used as a reliable indicator of the permeation state, provided that the reactive and dispersive effects in the sample are properly compensated.

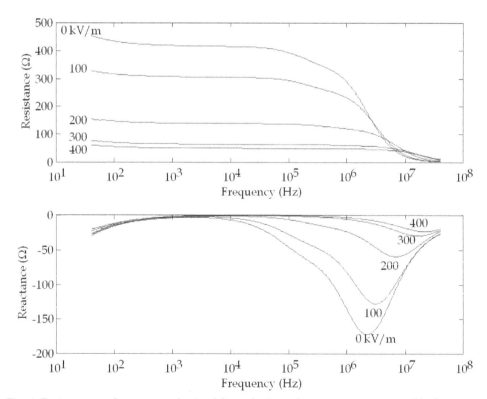

Fig. 4. Resistance and reactance obtained from the impedance spectra measured before (0 kV/m) and after (60 s) electroporation pulse with 10 ms and strength from 100 to 400 kV/m.

In Figure 6 the conductance increases rapidly with increasing applied field. The pulse length has increasing influence on membrane conductance as the applied field increases, but shows saturation for field of 400 kV/m and time length higher than 2 ms. The proposed models of pore creation in electroporation are based on the Boltzmann statistical distribution that depends on the energy stored in the pores of the membrane (Glaser et al, 1988; Krassowska & Neu, 1999; Ramos, 2010). So these models have terms that depend exponentially on the squared transmembrane potential. It is expected therefore that the strength of the applied field has great influence on the change in membrane conductance, as shown in this figure. However, saturation of the conductance for intense fields means that the permeation state does not allow the transmembrane potential to grow indefinitely.

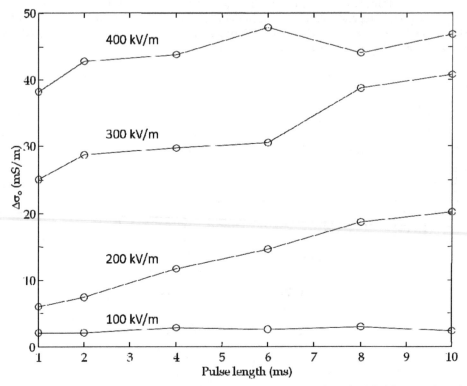

Fig. 5. Increase of external medium conductivity as a function of applied field strength and pulse duration.

Some previous studies with electroporation of animal cells resulted in very different membrane conductance values. Kinosita and Tsong (Kinosita & Tsong, 1979) used human red blood cells suspensions stimulated with single pulses of amplitude between 100 and 600 kV/m and duration up to 80 μs. They calculated the conductivity using instantaneous values and modeled the conductivity from a static solution to the electric potential internal and external to the cells. By fitting the model with the experimental results, they obtained conductance values of 10^5 and 10^6 S/m^2 to a field of 400 kV/m with duration of 2 and 80 μs, respectively. Human erythrocytes and the yeasts used in this study have different shapes but about the same size. However, the difference in the conductance obtained is big compared to the results in Fig. 6. Possibly, there are significant influences of the conductivity measurement procedures and modeling method used. In addition, the cells in question are very different. As the erythrocyte has an oblate spheroid shape with axial ratio 0.28, the yeast is approximately spherical and the cell wall can play a role in reducing the transmembrane potential and membrane conductance. In the study by Pavlin *et al.* (Pavlin *et al*, 2005) mouse melanoma cells were electroporated with 8 pulses of electric field of the same amplitude and duration 100 μs each. The suspension conductivity was calculated using instantaneous values of voltage and current. They estimated the cell conductivity and membrane conductivity from numerical methods Membrane conductivity values obtained with a field of 84 kV/m were 3.5×10^{-5} S/m and 1.4×10^{-5} S/m for two medium conductivity, 1.58 S/m and 127 mS/m, respectively. Considering a membrane thickness of 5 nm, the membrane conductance is 7×10^3 and 2.8×10^3 S/m^2 respectively. The critical field for electroporation is probably lower for melanoma cells than for yeast, since the former are larger and do not have cell wall. The conductance values are within the order of magnitude as were obtained in this study.

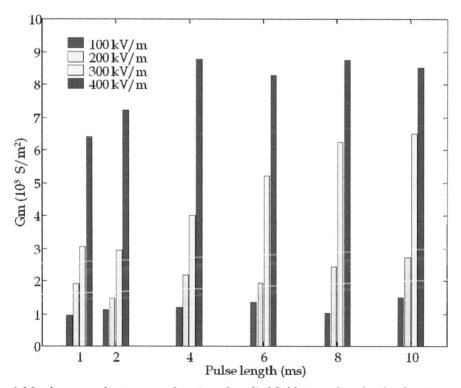

Fig. 6. Membrane conductance as a function of applied field strength and pulse duration.

3.3 Enzyme activity

Figure 7 shows the results of the β-D-galactosidase activity assays in suspensions electroporated with field strength from 100 to 400 kV/m and pulse length from 2 to 10 ms.

The fraction of enzyme activity found in the supernatant was very low in all trials. The values marked supernatant in the figure were obtained with field of 400 kV/m. The highest activity was detected in the cells themselves after centrifugation, indicating that although the enzyme is available outside the cell, it is not diluted in the supernatant. The enzyme is possibly trapped in the molecular network of the cell wall. Note that there is high correlation between the distributions of membrane conductance and enzymatic activity, especially above 200 kV/m. The enzymatic activity in the cell fraction for 400 kV/m is located just above 1 U for all pulse lengths in the same way that the conductance of the membrane shows saturation just above 8,000 S/m² for this field. Since the enzyme molecules must pass through the pores of the membrane to reach the cell wall, it can be predicted that the conditions that maximize the conductance of the membrane also maximize the extraction of the enzyme. The fact of the enzyme to be attached to the cell wall suggests an important application of the technique by means of immobilization. Cell immobilization consists in confinement of cells maintaining their catalytic activities. The enzyme β-D-galactosidase is not excreted naturally by the microorganism *Kluyveromyces marxianus*, but by electroporation it is possible to get it on the cell wall. Combining electroporation and cell immobilization is possible to obtain high catalytic activity in small volumes, provided that the cells remain viable after electrical stimulation and confinement.

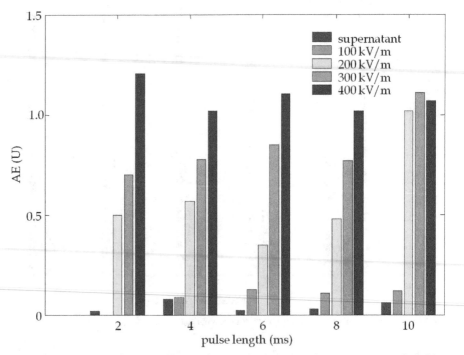

Fig. 7. Enzymatic activity for β-D-galactosidase in electroporated suspensions with fields strength from 100 to 400 kV/m and pulse length from 2 to 10 ms. The activity in the supernatant corresponds to the assay with 400 kV/m.

4. Conclusion

In this work it was studied the influence of the applied field strength and pulse duration on the increase in membrane conductance and activity of the enzyme β-D-galactosidase released by yeasts *Kluyveromyces marxianus* in suspension as a result of electroporation. The numerical technique for analyzing the electrical impedance spectra of the suspension has advantages over other modeling cited in the text because it allows properly compensate the reactive and dispersive effects caused by the impedance surface of the electrodes and ion accumulation on the cell membrane. It was found that membrane conductance and electrolyte conductivity measured after electroporation are strongly correlated, since both depend on the number and size of pores in the membrane. The protocol used with only a single pulse was able to produce large changes in membrane conductance for fields greater than 100 kV/m. The enzyme activity is also correlated to the membrane conductance. The membrane conductance increased between 8,000 and 9,000 S/m² for 400 kV/m. The enzyme activity was slightly greater than 1 U for all pulse duration. Both the membrane conductance and enzyme activity showed saturation for field of 400 kV/m, that is, the result is about independent on pulse length for pulses longer than 2 ms. The enzyme molecules after going through the pores in the cell membrane possibly get stuck in the molecular network of the cell wall. This was concluded based in the verification that the enzyme activity in the supernatant was very low in all assay conditions. Impedance measurement can allow the assessment of the permeation state of the membrane after pulse application, provided that the reactive and dispersive effects are properly modeled. This can be used as a probe for electroporation effectiveness evaluation aiming to control the enzyme extraction.

5. Future directions

Comparative studies with other methods of enzyme extraction, by chemical or mechanical process, should be conducted to determine the efficiency of electroporation compared to traditional techniques. The cell viability after electroporation also should be studied under different stimulation conditions, because it is an important factor to consider in certain applications. Another important study to conduct refers to the obtaining and using of immobilized enzymes on the cell wall of yeasts from electroporation process. The possibility of combining the technique of electroporation with the immobilization of enzymes appears promising but needs to be carefully tested and characterized.

6. References

Chen, C.; Smye, S.W.; Robinson, M.P. & Evans, J.A., (2006) Membrane Electroporation theories: a review, *Med. Biol. Eng. Comput.*, vol. 44, pp. 4-14

Foster, K.R. & Schwan, H.P. (1995) Dielectric properties of tissues, in: Handbook of Biological Effects of Electromagnetic Fields, Polk, C. & Postow, E. (Ed), 2ª Ed., pp. 25-102, CRC, New York

Furlan, S.A.; Schneider, A.L.S.; Merkle, R.; Carvalho, J.M.F. & Jonas R. (2000) Formulation of a lactose-free, low cost culture medium for the production of β–D-galactosidase by *Kluyveromyces marxianus*, *Biotechnology Letters*, vol. 22, pp. 589-593

Ganeva, V. & Galutzov, B. (1999) Electropulsation as an alternative method for protein extraction from yeast, *FEMS Microbiology Letters*, vol. 174, pp. 279-284

Ganeva, V.; Galutzov, B. & Teissie, J. (2003) High yield electroextraction of proteins from yeast by a flow process, *Analytical Biochemistry*, vol. 315, pp. 77-84

Glaser, R.W.; Leikin, S. L.; Chernomordik, L. V.; Pastushenko, V. F. & Sokirko, A. I. (1988) Reversible electrical breakdown of lipid bilayers: formation and evolution of pores, *Biochim. Biophys. Acta*, vol. 940, pp. 275-287

Grafström, G., Engström, P., Salford, L. G. & Persson, B. R. (2006) 99mTc-DTPA uptake and electrical impedance measurements in verification of *in vivo* electropermeabilization efficiency in rat muscle, *Cancer Biother.Radiopharm.*, vol. 21, pp. 623–635

Haque, A., Zuberi, M., Diaz-Rivera, R. E. & Porterfield, D. M. (2009) Electrical characterization of a single cell electroporation biochip with the 2-D scanning vibrating electrode technology, *Biomed Microdevices*, vol. 11, pp. 1239–1250

He, H., Chang, D. C. & Lee, Y. K. (2008) Nonlinear current response of micro electroporation and resealing dynamics for human cancer cells, *Bioelectrochemistry*, vol. 72, pp. 161-168.

Hibino, M.; Itoh, H. & Kinosita, Jr. K. (1993) Time courses of cell electroporation as reveled by submicrosecond imaging of transmembrane potential, *Biophysical Journal*, v. 64, pp. 1798-1800

Huang, Y. & Rubinsky, B. (1999) Micro-electroporation: improving the efficiency and understanding of electrical permeabilization of cells, *Biomed. Microdevices*, vol. 2, pp. 145–150

Ivorra, A. & Rubinsky, B. (2007) In vivo electrical impedance measurements during and after electroporation of rat liver, *Bioelectrochemistry*, v. 70, pp. 287–295

Kinosita, K. & Tsong, T.Y. (1979) Voltage-induced conductance in human erythrocyte membranes, *Bioch. Bioph. Acta*, vol. 554, pp. 479-497

Kotnik, T., Pucihar, G., Rebersek, M., Miklavcic, D. & Mir, L. M. (2003) Role of pulse shape in cell membrane electropermeabilization, *Biochimica et Biophysica Acta*, vol. 1614, pp. 193-200

Koester, P. J., Tautorat, C., Beikirch, H., Gimsa, J. & Baumann, W. (2010) Recording electric potentials from single adherent cells with 3D microelectrode arrays after local electroporation, *Biosensors and Bioelectronics*, vol. 26, pp. 1731-1735

Laufer, S., Ivorra, A., Reuter, V. E., Rubinsky, B. & Solomon, S. B. (2010) Electrical impedance characterization of normal and cancerous human hepatic tissue, *Physiol. Meas.*, vol. 31, pp. 995-1009

Lederberg, J. (1950) The β-D-galactosidase of *Escherichia coli* strain K-12, *J. Bacteriol.* , vol. 60, pp. 381-392

McAdams, E.T.; Lackermeier, A.; McLaughlin, J.A.; Macken, D. & Jossinet, J. (1995) The linear and non-linear electrical properties of the electrode-electrolyte interface, *Biosensors & Bioelectronics*, vol. 10, pp. 67-74

Miklavčič, D. & Puc M. (2006) Electroporation, in: *Wiley Encyclopedia of Biomedical Engineering*, Akay M. (Ed), pp. 1-11, John Wiley & Sons Inc., New York

Neu, J.C. & Krassowska, W. (1999) Asymptotic model of electroporation, *Phys. Review E.*, vol. 59, pp. 3471-3482

Pavlin, M.; Kanduser, M.; Rebersek, M.; Pucihar, G.; Hart, F. X.; Magjarevic, R. & Miklavcic, D. (2005) Effect of cell electroporation on the conductivity of a cell suspension, *Biophysical Journal*, vol. 88, pp. 4378-4390

Pavlin, M. & Miklavcic, D. (2003) Effective conductivity of a suspension of permeabilized cells: A theoretical analysis, *Biophys. J.*, vol. 85, pp. 719-729

Pliquett, U., Elez, R., Piiper, A. & Neumann, E. (2004) Electroporation of subcutaneous mouse tumors by rectangular and trapezium high voltage pulses, *Bioelectrochemistry*, vol. 62, pp. 83–93

Ramos, A. (2010) Improved numerical approach for electrical modeling of biological cell clusters, *Med. Biol. Eng. Comput.*, vol. 48, pp. 311-319

Ramos, A., Suzuki D.O.H. & Marques J.L.B. (2004) Numerical simulation of electroporation in spherical cells, *Artificial Organs*, v. 28, pp. 357-361

Reed, S. D. & Li, S. (2009) Electroporation Advances in Large Animals, *Curr Gene Ther.*, vol. 9, pp. 316–326

Rice, J., Ottensmeier, C. H. & Stevenson, F. K. (2008) DNA vaccines: precision tools for activating effective immunity against cancer, *Nature Reviews Cancer*, vol. 8, pp. 108-120

Saulis, G., Satkauskas, S. & Praneviciute, R. (2007) Determination of cell electroporation from the release of intracellular potassium ions, *Analytical Biochemistry*, vol. 360, pp. 273-281

Sersa, G., Miklavcic, D., Cemazar, M., Rudolf, Z., Pucihar, G. & Snoj, M. (2008) Electrochemotherapy in treatment of tumours, *EJSO*, vol. 34, pp. 232-240

Suzuki, D. O. H.; Ramos, A.; Ribeiro, M. C. M.; Cazarolli, L. H.; Silva, F. R. M. B.; Leite, L. D. & Marques, J. L. B. (2011) Theoretical and Experimental Analysis of Electroporated Membrane Conductance in Cell Suspension, *IEEE Transactions on Biomedical Engineering, In press*

Swagerty, D.L.; Walling, A.D. & Klein, R.M. (2002) Lactose Intolerance, *Am Fam Physician*, vol. 65 (9), (Mai 1), pp. 1845-1851

Teissié, J.; Golzio, M.; & Rols M. P. (2005). Mechanisms of cell membrane electropermeabilization: A minireview of our present (lack of ?) knowledge, *Biochim. Biophys. Acta.*, vol. 1724, (Aug. 2005), pp. 270-280

Teissie, J. & Rols, M.P. (1993) An experimental evaluation of the critical potential inducing cell membrane electropermeabilization, *Biophysical Journal*, vol. 65, pp. 409-413

Weaver, J.C. & Chizmadzhev, Y. A. (1996) Theory of electroporation: a review, *Bioelectrochemistry*, vol. 41: pp. 135-160

4

Protocol of a Seamless Recombination with Specific Selection Cassette in PCR-Based Site-Directed Mutagenesis

Qiyi Tang[1,*,#], Benjamin Silver[2,*] and Hua Zhu[2,#]

[1]Department of Microbiology/AIDS Research Program, Ponce School of Medicine, Ponce, PR,
[2]Department of Microbiology and Molecular Genetics, New Jersey Medical School, Newark, NJ, USA

1. Introduction

Genetic mutation has made great contributions to determining protein structure and function, viral pathogenesis, biological engineering, and vaccine development [1-9]. In order to alter genetic information, gene mutation (also called mutagenesis) can be achieved by many different methods. The process of mutagenesis can occur naturally, as the root of evolution; by mutagens, such as chemicals or radiation; and experimentally, by laboratory techniques [10-13]. Current experimental mutagenesis methods can be generally classified into random and directed mutation.

Experimentally directed mutagenesis methods include: insertional mutagenesis [14], PCR mutagenesis [15, 16], signature tagged mutagenesis [17], site-directed mutagenesis [18, 19] and transposon mutagenesis [20-22]. Different procedures are involved in traditional mutagenesis, including molecular cloning that depends on preparations of vector and inserted DNA fragments and ligations. This traditional mutagenesis method is not only time-consuming, but also labor intensive, and modern advancements in mutagenesis have overcome these obstacles. The development of the Bacterial Artificial Chromosome (BAC) clone of large DNA viruses was a breakthrough in the viral mutagenesis field. The BAC clone of a virus can be maintained stably and propagated inside a bacterial cell, which allows for easy manipulation of the viral genome. Any required mutation can be easily and rapidly achieved inside the E. coli cell and this mutation can be verified before making recombinant virus [23-27]. Since the development of this novel method for construction of recombinant viruses, detailed functional study of virus genome has been done using this specific BAC recombinatory mutagenesis approach [23-27].

Recombination is an important procedure in experimental mutagenesis. There are two recombination systems that are often used: DNA sequence-specific recombinase-driven recombination and homologous recombination. The DNA sequence-specific recombinase-driven recombination includes Cre-Lox and FLP-FRT recombination systems [28-36]. Cre-

---

* These authors contributed equally
# Corresponding Authors

Lox system depends on Cre recombinase that specifically recognizes lox P (locus of X-over bacteriophage P1) sites, 34-basepair (bp) specific sequences and results in recombination by the excision of intervening DNA sequence and joining of site-specific ends. FLP-FRT involves the recombination of sequences between short flippase recognition target (FRT) sites (also a 34-bp DNA sequence) by the flippase recombination enzyme (FLP or Flp) derived from *Saccharomyces cerevisiae* yeast. After recombination, one copy of the specific DNA sequence will remain in the vector, which will cause potential problems for further mutagenesis, DNA stability and, therefore, is not suitable for some applications, such as vaccine development. In addition, BAC sequence excision methods should be carefully considered, depending upon the design of the BAC. For example, it will become problematic if both the BAC vector and gene-specific antibiotic marker contain the same flanking excision sequences (such as loxP) because this is likely to cause the removal of a much larger sequence of genomic DNA than originally intended.

Homologous recombination is a type of genetic recombination in which nucleotide sequences are exchanged between two identical sequences of DNA [37, 38]. Wild-type *E. coli* is ineffective at inducing homologous recombination in foreign DNA because linear DNA is commonly degraded by RecBCD exonuclease. In order to circumvent this problem, SW102 strains were developed that contain a temperature-sensitive λ prophage encoding one gene to temporarily repress RecBCD (*gam*), as well as two genes (*exo* and *beta*) utilized for homologous recombination via double-strand break repair [38]. More specifically, exonuclease degrades DNA from the 5′ end of double-strand break sites, while Beta binds to and protects the 3′ from further degradation [39, 40]. These overhangs from double-strand breaks allow recombination between viral and plasmid DNA. Because the λ phage is temperature sensitive (due to the expression of a temperature-sensitive λ cI-repressor), linear DNA uptake and recombination can occur within a few minutes when the cell-culture temperature is increased from 32 to 42°C [38]. This allows the bacterial cells to function normally when grown at 32°C. *E. coli* also require thousands of homologous base pairs in order for recombination to occur. Addition of the modified temperature-sensitive λ phage is important because this allows homologous recombination to occur within a relatively small region of the homologous sequence, which is important because BAC mutants are usually created using PCR-amplified gene sequences with about 40 base pairs of flanking sequences that are homologous to the viral BAC [37].

PCR-based preparation for insertion of DNA provided convenience in performing mutagenesis studies. Differential PCR techniques enable researchers to make any mutation, as needed: large deletion, point mutation, substitution, insertion, non-sense mutation and shift-mutation. Allowing for the use of a positive and negative selection system is another advancement in the area of mutagenesis, especially when applied to site-directed mutagenesis when accuracy and precision are of high priority. There are multiple ways to carry out site-directed mutagenesis of a viral BAC. For example, selectable markers are necessary to isolate the successful mutation of a viral BAC, and selectable markers can be an antibiotic resistant gene (such as kanamycin or zeocin resistant genes) or a foreign metabolic gene (such as *galK*). In the case of a gene knockout, the foreign DNA usually contains a selectable marker flanked by a DNA sequence homologous to the flanking regions for the gene of interest. Around 40 bp of a flanking sequence are typically used for homologous recombination. The viral BAC and the marker with homologous sequences are then inserted into *E. coli* via electroporation. Colonies with the BAC recombinant virus can be selected based upon selection markers and verified by PCR [27].

This protocol describes the use of the *galK* positive- and counter- selection schemes to make gene mutations (*e.g.* point mutations, deletions, and insertions) within viral BACs. This system includes only two steps of recombination within a modified bacterial strain, SW102 [41], using selection and counter-selection media, and easily designed PCR. The SW102 strain, derived from DY380 *E. coli* [38, 42, 43], differs from DY380 only in that the galactokinase (*galK*) gene of the galactose operon is defective in SW102 strain. When SW102 bacteria are incubated on minimal media with galactose as the sole source of carbon, the bacteria cannot grow without *galK* supplied *in trans* [41]. When the *galK* gene is provided *in trans*, in this case it replaces a gene of interest, it can complement the defective *galK* gene in the bacteria and the bacteria can grow in minimal media with galactose, therefore, the first step is a positive selection. Later in the protocol, when *galK* is removed from the viral BAC and replaced with a mutated version of the original gene, a counterselection takes place in that clones containing *galK* will be negatively selected for by growing the bacteria on medium with a substance that produces a toxic intermediate only when a functional *galK* is present.

This protocol explains in detail how to use the *galK* positive and counterselection strategies to make any desired mutations (*e.g.* point mutations, deletions, and insertions) based on plasmid or BAC. The modified plasmid or BAC will not contain any extra DNA sequence; therefore, it is referred to as a "seamless" mutation.

2. Materials and methods

Bacterium: *E. coli* SW102 strain (free reagents from Biological Resources Branch of NCI-Frederick) [41]
Plasmid: p*galK* (free reagents from Biological Resources Branch of NCI-Frederick) [41]
Luria-Broth (LB) medium

Tryptone	10 g
Yeast Extract	5 g
NaCl	10 g

Dissolve components in distilled and deionized water (ddH$_2$O) and adjust the total volume up to 1 liter.
For LB agar: add agar to a final concentration of 1.5%.

Heat the mixture to boiling to dissolve agar and sterilize by autoclaving at 15 psi, from 121-124°C for 15 minutes.

1X M9 medium (1 liter)
6 g Na$_2$HPO$_4$
3 g KH$_2$PO$_4$
1 g NH$_4$Cl
0.5 g NaCl

Dissolve components in ddH$_2$O and make the total volume up to 1 liter. Autoclave

M63 minimal plates

1L 5X M63
10 g (NH$_4$)$_2$SO$_4$
68 g KH$_2$PO$_4$
2.5 mg FeSO$_4$ ·7H$_2$O
Dissolve components in ddH$_2$O and make the total volume up to 1 liter. Adjust to pH 7 with KOH.
Autoclave

Other Reagents

0.2 mg/ml D-biotin (sterile filtered) (1:5000)
20% galactose (autoclaved) (1:100)
20% 2-deoxy-galactose (autoclaved) (1:100)
20% glycerol (autoclaved) (1:100)
10 mg/ml L-leucine (1%, heated, then cooled down and sterile filtered)
25 mg/ml chloramphenicol in Ethanol (1:2000)
1 M MgSO$_4$ ·7H$_2$O (1:1000)

Procedure of making the M63 minimal plates:

1. Autoclave 15 g agar in 800 ml H$_2$O in a 2-liter flask.
2. Add 200 ml autoclaved 5X M63 medium and 1 ml 1 M MgSO$_4$ ·7 H$_2$O.
3. Adjust volume to 1 liter with H$_2$O if necessary.
4. Let cool down to 50°C ("hand touchable hot"), add 10 ml carbon source (final concentration 0.2%), 5 ml biotin (1 mg), 4.5 ml leucine (45 mg), and 500 μl chloramphenicol (final concentration 12.5 mg/ml) or other appropriate antibiotics. Pour the plates, 33-40 plates per liter.

MacConkey indicator plates:

Prepare MacConkey agar plus galactose according to manufacturer's instructions. After autoclaving and cooling to 50°C, to one liter, add 500 μl chloramphenicol (final concentration 12.5 mg/ml) or other appropriate antibiotics, and pour the plates, 33-40 plates per liter.

3. Protocol

3.1 Preparing SW102 that harbors plasmid or BAC

Select the plasmid or BAC that contains the target gene and will be used as a vector for making mutations. Information about the vector, including antibiotic-resistance and the targeted gene, needs to be known. For example, to make any mutation of a gene in murine cytomegalovirus (MCMV), the BAC of SM3fr [46] is usually used. SM3fr contains whole genome of MCMV with chloramphenicol resistance. The DNA sequence and gene structure of SM3fr have been published and can be accessed in public gene bank.

3.1.1 Preparation of electrocompetent cells (Fig. 1. 1a)

1. A 5-ml overnight culture of *E. coli* SW102 in LB medium will be prepared either from the frozen stock or from a single colony at 32°C or lower. The SW102 strain is resistant

to tetracycline (12.5 µg/ml), it is not necessary, but safe to include tetracycline in this step in order to exclude any possible contamination.

2. The overnight LB culture of SW102 will be diluted by 1:50 by adding 0.5 ml of the overnight culture to an autoclaved 50 ml Erlenmeyer baffled flask with 25 ml LB, but first save 1 ml LB that will be used as a reference for measuring the OD_{600nm}. The diluted SW102 in the flask will be incubated for 3-5 hrs with appropriate antibiotic selection in a 32°C shaking incubator until the density reaches an OD_{600nm} of 0.6. At this point, a bottle of ice-cold 10% glycerol or autoclaved ddH_2O needs to be prepared. (If the competent cells are to be used right away, use autoclaved ddH_2O). The competent cells can be used right away or stored at -70°C for later use.

3. When the OD_{600nm} is 0.6, the flasks containing the bacteria will be cooled down in the ice water bath slurry for a minute or two and subsequently transferred into pre-cooled 15 ml Falcon tubes.

4. The bacteria will be centrifuged in a cold centrifuge (4°C) for 5 min at 5000 RPM (standard Beckman or Eppendorf centrifuge).

5. All supernatant will be poured off and the centrifuge tubes will be briefly inverted on a paper towel, and 1 ml ice-cold ddH_2O or 10% glycerol will be added to resuspend the pellet. And the tube will be kept in the ice. The pellet will be resuspended in the ddH_2O or 10% glycerol by gently shaking the tube in the ice-water bath (gently move the tubes around in circles while keeping them in the ice water slurry, this can take a while for the first time). When the cells are completely resuspended, another 9 ml ice-cold ddH_2O or 10% glycerol will be added, the tube will be inverted for a couple of times, and centrifuged again for 5 minutes.

6. The supernatant will be poured off and the pellet will be resuspended with 10 ml ice-cold ddH_2O or 10% glycerol, as in Step 5, resuspension should be much faster this time).

7. The resuspended cells will be centrifuged once more, as in Step 5.

8. The supernatant will be completely removed by inverting the tube on a paper towel (be careful that you do not lose the pellet). The pellet will be resuspended in 250 µl ice-cold 10% glycerol. The competent cells can be used immediately for transformation or aliquoted and stored at -80°C (volume should be around 50 µl).

3.1.2 Transformation by electroporation (Fig. 1. 1b)

1. Plasmid or BAC DNA will be transformed into the SW102 competent cells made in Step 1.1. The freshly made or stored electrocompetent cells will be mixed with 1-5 µg DNA. The mixture of cells and DNA will be transferred to a pre-cooled 0.1 cm cuvette. It is important that the DNA solution contains low salt because high salt can cause electric shock during electroporation.

2. The cuvette will be placed into the electroporator power source and cuvette holder (Bio-Rad). The conditions for transformation are set according to the strain. For SW102 cells, use 25 mFD, 200 W, and 1.8 kV. The time constant (tau value) should be 3-4 msec. After the electroporation, the bacteria will be transferred immediately to a tube with 1 ml LB medium and incubated at 32°C in a shaking water bath for 1 hr.

3. The incubated bacteria will be smeared onto LB agar plates that contain appropriate antibiotics. The transformed bacteria on selective LB agar will grow as colonies after incubation at 32°C for 18-24 hrs.

4. The generated SW102 harboring plasmid or BAC will be verified by picking up a colony, isolating the DNA by BAC Minipreparation (see Step 1.3.) and detecting the DNA with PCR analysis of the BAC (Fig. 3) and/or restriction enzyme digestion (Fig. 4). This plasmid- or BAC-harboring SW102 will be given a name and used to make desired mutation of the target gene. For example, if the SW102 harbors MCMV BAC SM3fr, it will be called SW102_SM3fr.

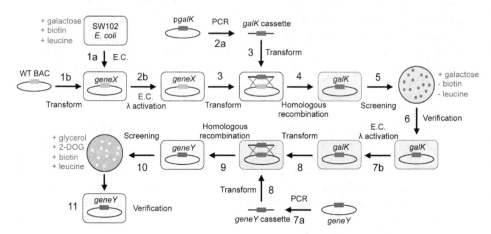

Fig. 1. **Summary of the *galK*-based mutagenesis in *E. coli* SW102. 1a.** Prepare electrocompetent (E.C.) SW102 *E. coli*. **1b.** Electroporate WT virus BAC into electrocompetent SW102. **2a.** Prepare *galK* cassette by PCR with a set of primers conferring sequence homology to the viral BAC sequences flanking *geneX*. **2b.** Prepare electrocompetent SW102 *E. coli* harboring WT viral BAC and activate defective λ phage recombination system by shaking in a 42°C water bath for 15 minutes. **3.** Electroporate the *galK*-expressing cassette into recombination-activated electrocompetent SW102 strain harboring WT BAC. **4.** Upon homologous recombination, *geneX* is replaced by *galK*. **5.** Confirm presence of *galK* by growing bacteria on M63 plates with galactose as the sole source of carbon and antibiotic, selecting colonies to screen on MacConkey agar with galactose as the sole source of carbon. The *galK*-containing recombinant clones will produce red colonies on MacConkey agar with galactose. **6.** Select red colony from screening process to verify by PCR and continue. **7a.** Prepare *geneY* cassette (PCR cassette containing desired mutation in *geneX*, referred to as *geneY*). **7b.** From red colony selected in Step 5, prepare electrocompetent SW102 harboring *galK* mutant BAC and activate defective λ phage recombination system, as in 2b. **8.** Electroporate *geneY* cassette into electrocompetent and recombination-activated SW102 containing *galK* mutant clone. **9.** Upon homologous recombination, *galK* is replaced by *geneY*. **10.** Grow bacteria on M63 with glycerol, DOG and antibiotic agar plates. **11.** Select colony and verify for production of virus.

3.1.3 BAC minipreparation

The following protocol is usually used and works very well for generating BAC DNA for initial analysis:

1. 5 ml overnight LB culture with chloramphenicol (almost all known BACs are chloramphenicol resistant) in a 15-ml Falcon tube is pelleted, and the supernatant is removed.
2. Then the pellet is dissolved in 250 µl buffer P1 (Miniprep kit, Qiagen, CA) and transferred to an Eppendorf tube.
3. The bacteria are lysed in 250 µl P2 buffer with gently mixing and incubating for 5 min at room temperature.
4. The lysate is neutralized with 250 µl buffer P3 (also called N3 buffer), followed by mixing and incubating on ice for 5 min.
5. The supernatant is cleared by two rounds of centrifugation at 13,200 RPM for 5 min in a small Eppendorf centrifuge (or other model tabletop centrifuge). Each time the supernatant is transferred to a new tube.
6. The DNA is precipitated by adding 750 µl isopropanol, mixing and incubating on ice for 10 min, and centrifugation for 10 min at 13,200 RPM in a small Eppendorf centrifuge (or other model tabletop centrifuge).
7. The pellet is washed once in 70% ethanol and the air-dried pellet is dissolved in 50-100 µl TE buffer. The isolated BAC DNA can be used for 1) restriction analysis, 2) PCR analysis and 3) DNA sequencing analysis.

3.2 Positive selection to replace the targeted DNA with *galK* gene

3.2.1 PCR to generate the *galK* cassette (Fig. 1. 2a)

1. Design primers with 50 bp homology flanking the desired site to be modified. The 3' end of these primers will bind to the *galK* cassette. For example, if a single base pair (bp) mutation will be generated, the homology arms should extend 50 bp on either side of the target bp. If the target bp is not included in the arms, it will result in a deletion of that base pair in the first step. If the target bp was changed into another bp in the arm, it will result in single bp mutation. If a small or a large deletion will be made, design the *galK* primers so that the deletion is made already in the first step. The primers should be as follows:

Forward: 5' 50bp homology: CCTGTTGACAATTAATCATCGGCA-3'
Reverse: 5' 50bp homology complementary strand: TCAGCACTGTCCTGCTCCTT-3'

2. Plasmid p*galK* will be used for PCR to amplify the *galK* gene using the primers designed as above and a proofreading DNA polymerase. It is important to use this type of DNA polymerase because the two-step substitution in the procedure will use PCR products and accuracy is a high priority. DNA template for PCR is the p*galK* plasmid. PCR can be started at 94°C for 4 min to denature the template, followed with 30 cycles of three temperatures: 94°C, 15 sec; 60°C, 30 sec; 72°C, 1 min; and the PCR will be extended for 7 min at 72°C.
3. When the PCR is finished, 1-2 µl DpnI (New England Biolabs, MA) should be added into each 25 µl reaction that is mixed and incubated at 37°C for 1 hour. This step serves to remove any plasmid template; plasmid is methylated so that DpnI can degrade it, PCR products are not methylated so DpnI cannot degrade it. Finally, the DpnI-digested PCR product will be separated in agarsoe gel and purified from gel. The PCR product will be eluted with 50 µl ddH$_2$O, 10-30 ng will be used for transformation.

3.2.2 Preparation of electrocompetent cells harboring BAC

SW102 *E. coli* strain has a defective λ prophage that, when activated, can produce recombinase. These activated and electrocompetent SW102 harboring plasmid or BAC DNA (*e.g.* SW102_SM3fr) (Fig. 1. 2b) will be prepared from the electrocompetent SW102 harboring plasmid or BAC DNA prepared in Step 1.2 of the protocol. The procedure is as follows:

1. Inoculate an overnight culture of SW102 cells containing the BAC in 5 ml LB with appropriate antibiotics (*e.g.* SW102_SM3fr in LB with chloramphenicol) at 32°C.
2. Next day, add 0.5 ml of the overnight SW102 harboring plasmid or BAC DNA in 25 ml LB with antibiotics in a 50-ml baffled conical flask and incubate at 32°C in a shaking water bath to an OD_{600nm} of approximately 0.6 (0.55-0.6). This usually takes 3-4 hrs. During this time, turn on two shaking water baths: one at 32°C, the other at 42°C. Make ice/water slurry and pre-chill 50 ml of ddH_2O.
3. When the OD_{600nm} of SW102 harboring plasmid or BAC DNA culture reaches 0.6, transfer 10 ml of the culture to another baffled 50-ml conical flask and heat-shock at 42°C for exactly 15 min in a shaking water bath. This step is to induce SW102 to produce recombinase. The remaining culture is left at 32°C as the un-induced control.
4. After 15 min induction, the SW102 in two flasks (10 ml induced and 10 ml un-induced) are briefly cooled in ice-water bath and then transferred to two 15-ml Falcon tubes and centrifuged at 4°C. It is important to keep the bacteria as close to 0°C as possible in order to get high efficiency competent cells.
5. Pour off all of the supernatant and resuspend the pellet in 1 ml ice-cold autoclaved ddH_2O by gently swirling the tubes in the ice-water bath slurry (no pipetting). This step may take a while. When the cells are resuspended, add another 9 ml ice-cold autoclaved ddH_2O. Pellet the samples again as in Step 1.4.
6. Resuspend the pellet again with I ml of ice-cold autoclaved ddH_2O by gently swirling and then add 9 ml ddH2O. The bacteria are centrifuged again, as above.
7. After the second washing and centrifugation step, all supernatant must be removed by inverting the tubes on a paper towel, and the pellet is resuspended in approximately 50 µl of ice-cold ddH_2O and is kept on ice until electroporated with PCR product. If the electrocompetent cells are prepared with 10% glycerol, the aliquots can be made and saved at -80°C for later use.

3.2.3 Electroporation of *galK* cassette

1. 25 µl of electrocompetent SW102 cells harboring plasmid or BAC DNA and 10-30 ng of PCR amplified *galK* cassette will be added to a 0.1 cm cuvette (BioRad). Electroporation will be carried out at 25 mF, 1.75 kV, and 200 ohms (Fig. 1. 3). After reaction, 1 ml LB will be immediately added to the cells, and the cells will be cultured at 32°C for 1 hour in a shaking water bath.
2. After the recovery incubation, the bacteria are washed twice with M9 buffer as follows: 1 ml of the culture is centrifuged in an Eppendorf tube at 13,200 RPM for 15 sec. Then the supernatant is removed with a pipette. The pellet is resuspended with 1 ml M9 buffer, and pelleted again. This washing step will be performed once more. After the second wash, the supernatant will be removed and the pellet will be resuspended in 1 ml M9 buffer. A serial of dilutions in M9 buffer will be made (100 µl, 100 µl of a 1:10 dilution, and 100 µl 1:100) and plated onto M63 minimal media plates with galactose,

leucine, biotin, and appropriate antibiotics (Fig. 1. 5). It is important to remove any rich media from the culture prior to selection on minimal media by washing the bacteria in M9 buffer. The uninduced SW102 will be plated as a control.

3. The plates will be incubated for 3 days at 32°C in a cabinet-type incubator. Colonies will be visible at the beginning of third day and be able to be picked up at the end of that day.

4. Pick up a few colonies and streak the colonies onto MacConkey agar plates with galactose, indicator and appropriate antibiotics to obtain single colonies (Fig. 1. 5). The colonies appearing after the 3 days of incubation should be *galK* positive, but in order to get rid of any *galK* negative contaminants (usually called hitch-hikers), it is important to obtain single, bright red colonies before proceeding to next step (Fig. 1. 6). *galK* negative colonies will be white or colorless and the *galK* positive bacteria will be bright red or pink due to a pH change resulting from fermented galactose after an overnight incubation at 32°C (Fig. 2).

5. Pick up a few bright red (*galK* positive) colonies and inoculate in 5 ml LB + antibiotics and incubate overnight at 32°C. There is normally no need to further characterize the clones. But the *galK* positive clones can be Miniprepared (Step 1.3) and verified by PCR using primers that flank the site of the *galK* gene (Fig. 3):

Forward: 5' CTGTTGACAATTAATCATCGGCA-3'
Reverse: 5' TCAGCACTGTCCTGCTCCTT-3'

The recombinant *galK* clones should be named for storage. Take MCMV BAC as an example: SW102_SM3fr_X*galK* is the name of the bacteria and SM3fr_X*galK* is the name of the BAC. "X" stands for the site of the *galK* in the BAC.

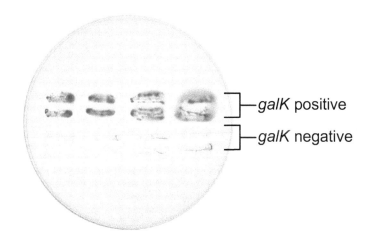

Fig. 2. **Positive selection for *galK* mutant clones and negative selection for WT/mutant/rescue clones.** SW102 bacteria with WT, *galK* mutant (*geneX* replaced by *galK* gene), *geneY* mutant (*galK* replaced by mutated *geneX*), or rescue (*geneX* restored) BACs were plated on MacConkey agar with galactose as the sole source of carbon and antibiotic. Recombinant BAC clones containing *galK* appear as red colonies; clones with WT BAC, *geneY* mutant BAC, or rescue BAC appear as white/colorless colonies.

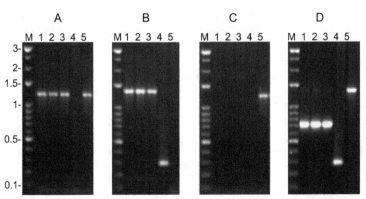

Fig. 3. **PCR verification. A.** PCR using *galK* primers with *galK* mutant clones (*geneX* replaced by *galK* – clones produced from Fig. 1. 4) as template. 1.3kb band indicates presence of *galK* gene: Lanes 1-3 are *galK* mutant clones #1-3, respectively; Lane 4 is original BAC (negative control); Lane 5 is p*galK* (positive control). **B.** PCR using primers that override the homologous sequences with *galK* mutant clones as template. 1.4kb band indicates presence of *galK*+BAC sequence, 300 bp band indicates BAC sequence only: Lanes 1-3 are *galK* mutant clones #1-3, respectively; Lane 4 is original BAC (positive control); Lane 5 is p*galK* (negative control). **C.** PCR using *galK* primers with mutant clones (*galK* replaced by *geneY*) as template (mutant clones #1-3 were created, respectively, from *galK* mutant clones #1-3 in Panels 1+2). 1.4kb band indicates presence of *galK*+BAC sequence: Lanes 1-3 are mutant clones #1-3, respectively; Lane 4 is original BAC (negative control); Lane 5 is a *galK* mutant clone used to derive a mutant clone (i.e. *galK* mutant clone #1 from Panels 1+2) (positive control). **D.** PCR using primers that override the homologous sequences with mutant clones as template. 700 bp band indicates presence of mutant gene (*geneY*), 300 bp band indicates WT BAC sequence, 1.4kb band indicates presence of *galK*+BAC sequence: Lanes 1-3 are mutant clones #1-3, respectively; Lane 4 is original BAC (positive control); Lane 5 is a *galK* mutant clone used to derive a mutant clone (positive control). All gel electrophoreses were on 1% agarose gels. M = 1 kb-Opti DNA Marker (ABM, Canada). Marker units are kilobases.

3.3 Counterselection to replace the *galK* gene for desired mutant generation

3.3.1 PCR to generate the DNA fragment to substitute *galK* gene (Fig. 1. 7a)

Design primers with 50 bp homology flanking the desired site to be modified. This 50 bp homology is usually the same as that in making *galK* gene. Usually the mutations are contained in the templates. The DNA template is a WT BAC or plasmid with the gene of interest mutated in the required fashion. The PCR product should therefore contain the desired mutations.

Forward: 5' 50bp homology target gene (18-20 bp)
Reverse: 5' 50bp homology complementary strand target gene (18-20 bp)

3.3.2 Electroporation of DNA fragment (same conditions as in Step 2.2)

1.   Both the preparation of electrocompetent cells from the SW102 harboring plasmid or BAC with *galK* gene and the activation of defective λ prophage recombinase are

necessary (Fig. 1. 7b), and the procedure is a repetition of the prepration and activation of electrocompetent SW102 harboring WT BAC or plasmid in Step 2.2.

2. Transform the PCR product into electrocompetent SW102 harboring *galK* mutant plasmid or BAC by electroporation with 0.2-1 µg of PCR product that contains the desired mutation and with homology to the area flanking the *galK* gene (Fig. 1. 8). After electroporation, the bacteria will be recovered in 10 ml LB in a 50 ml baffled conical flask by incubating at 32°C in a shaking water bath for 4.5 hrs. This long recovery period serves to obtain bacteria that only contain the desired recombined BAC or plasmid, and thus have lost any BAC still containing the *galK* cassette.

3. 1 ml of the recovery culture will be centrifuged and the pellet will be resuspended with M9 buffer and washed twice with M9 buffer. After second washing, the pellet will be resuspended in 1 ml of M9 buffer and the bacteria will be plated in a serial dilution (100 µl, 100 µl of a 1:10 dilution and 100 µl 1:100) onto M63 minimal media plates that contain glycerol, leucine, biotin, 2-deoxygalactose (DOG), and appropriate antibiotics (Fig. 1. 10).

4. Incubate at 32°C for three days. Further verification can be accomplished by PCR and/or DNA sequencing (Fig. 1. 11). A name is required for any mutation. For MCMV BAC, we name the bacterial strain SW102_SM3fr_XY and the BAC SM3fr_XY. "X" stands for the site of the mutation in the BAC genome, and "Y" stands for the mutations (deletion, insertion, point mutation or other). For BAC mutagenesis, especially for making any mutation on viruses, a rescue BAC for each mutation is very important.

3.4 Generation of rescue clone of each mutation

The mutagenesis protocol, for some purposes, does not need to progress to this step. However, for some studies, especially for viral mutagenesis, it is necessary to make rescue viruses so that the observed phenotype can be compared and confirmed. The same *galK* method will be used to make a rescue clone that was used to generate the original mutation, except for starting with the mutant BAC and working backwards.

1. The SW102_XY will be used to make electrocompetent cells, using the same procedure as in Step 2.2.

2. The competent cells will be electroporated with the PCR product of *galK* cassette (exactly the same cassette as the one produced in Step 2.1), which will result in SW102_X*galK*.

3. After growth and verification, SW102_X*galK* will undergo electrocompetent cell preparation, as before, and electroporated with a DNA fragment that will be made using the same primers, but with a template that contains no mutation.

4. Therefore the rescue clone is made using the backwards steps. We can also give a name such as SW102_XYRes for the bacteria.

3.5 Maxipreparation of BAC DNA (all recombinant clones)

Now, all of the plasmids or BAC DNA can be isolated for any purpose, *e.g.* in the case of herpesviruses, the BAC DNA needs to be isolated and purified for making viruses. Since BAC is usually a large DNA vector, its isolation and identification are different from that of regular plasmid protocol. Fortunately, several high quality kits have been commercially available for preparation of BAC DNA on a large-scale. In this case, the BAC DNA needs to be extracted from the SW102_XY (or SW102_X as a control) for transfection and production of virus.

1. Select a colony to inoculate 5 ml LB with chloramphenicol (final concentration 12.5 mg/ml) and shake overnight in an incubator at 32°C.

2. The following day, save 200-500 µl of culture to make a stock (see below), and add the remainder of the 5 ml culture to 500 ml LB with chloramphenicol (final concentration 12.5 mg/ml) and shake overnight in an incubator at 32°C. To make a bacterial stock, add 100% glycerol to the saved culture to a final concentration of 15% glycerol and store at -80°C.

3. Use the Nucleobond Maxiprep BAC DNA isolation kit (Clontech Laboratories Inc., CA) to extract BAC DNA from the culture. Because of their large size, BAC DNAs need to be handled in a way that avoids any harsh physical shearing force, including vortexing or passing quickly through fine pipette tips. Freeze-and-thaw should also be avoided.

4. The final DNA products are resuspended in 250 µl sterile ddH₂O and quantified by spectroscopy.

5. BAC DNA solutions should always be stored at 4°C.

3.6 Verification of BAC DNA integrity [Varicella zoster virus (VZV) is used as an example in the following sections]

1. Digest 3 µg of mutant BAC DNA, with WT digestion in parallel as control, with 20 U of restriction enzyme. Example: 3 µl of 1 µg/µl VZV WT BAC DNA, 20 U (1 µl) HindIII restriction enzyme (New England Biolabs, MA), 1 µl 10X NEBuffer 2 (New England Biolabs, MA), 5 µl ddH₂O.

2. Incubate overnight at 37°C and run gel electrophoresis on 0.5% agarose gel. Fig. 4 highlights the pattern observed when digesting WT VZV BAC, ORF7 Deletion VZV BAC, and ORF7 Rescue VZV BAC, respectively.

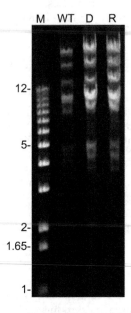

Fig. 4. **Verification of BAC DNA integrity.** For all BACs, 3 µg BAC DNA was digested with 20 U HindIII restriction enzyme overnight at 37°C and ran electrophoretically on 0.5% agarose gel. M: 1 kb Plus DNA Ladder (Invitrogen, CA), units are kb; WT: VZV WT BAC DNA; D: VZV ORF7 Deletion BAC DNA; R: VZV ORF7 Rescued BAC DNA.

3.7 Transfection of BAC DNA for virus production

BAC DNA from Maxi-preparations is transfected into human cells (*e.g.* MeWo, ARPE-19) using the FuGene 6 transfection kit (Roche, Indianapolis, IN), according to manufacturer's standard protocol. 1.5 µg of BAC DNA and 6 µl of transfection reagent are used for a single reaction in one well of 6-well tissue culture plates. Highly concentrated (>250 µg/µl) BAC DNA solutions are viscous, and BAC DNA molecules easily precipitate out of the solution when added to transfection reagent solutions. When such precipitation becomes visible, it is irreversible; predictably, the results of the transfection assays are often poor. Therefore, pre-dilute each BAC DNA before gently mixing it with the transfection reagent.

1. For each reaction, 1.5 µg of BAC DNA is diluted in serum-free medium, and the volume of DNA solution is adjusted to 50 µl with the serum-free medium.
2. For each reaction, 6 µl of transfection reagent is combined with 94 µl of medium.
3. Using pipettor tips, gently stir the DNA solution into the transfection reagent.
4. Incubate mixture for 25 minutes at room temperature.
5. Add DNA-transfection reagent solution to culture plate.

3.8 Infected cell culture

1. Transfected cells are grown in a 6-well tissue culture plate in 2 ml DMEM supplemented with 10% fetal calf serum, 100U of penicillin-streptomycin/ml, and 2.5 ug of amphotericin B/ml [47, 48].
2. Upon visualization of infection, Usually viral plaques will be developed and visualized under microscopy as shown in Fig. 5.

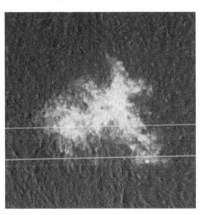

Fig. 5. **Generation of VZV by transfection of viral BAC DNA.** Human ARPE-19 cells were chemically transfected with WT VZV BAC DNA. Infected cells expressing EGFP (inserted into BAC vector) and form green plaques. One plaque is visualized by fluorescent microscope.

3.9 Summary of the protocol

The protocol of the seamless recombination with specific selection cassette in PCR-based site-directed mutagenesis is summarized in Fig. 1, using BAC as an example. Firstly, SW102 *E. coli* is made to be electrocompetent (E.C.) (Fig. 1. 1a). Then, the WT BAC is electroporated into the electrocompetent SW102 (Fig. 1. 1b). The *galK* cassette is prepared by PCR with a set of primers conferring sequence homology to the viral BAC sequences flanking *geneX* (Fig. 1. 2a). Electrocompetent SW102 *E. coli* harboring WT viral BAC is prepared, and the defective λ phage recombination system is activated by shaking in a 42°C water bath for 15 minutes (Fig. 1. 2b). The *galK*-expressing cassette is electroporated into recombination-activated SW102 strain harboring WT BAC (Fig. 1. 3). Upon homologous recombination, *geneX* is replaced by *galK* (Fig. 1. 4). The presence of *galK* in the recombinant clones is selected by growing bacteria on M63 plates with galactose as the sole source of carbon and the proper antibiotic. Colonies are then selected to screen on MacConkey agar with galactose as the sole source of carbon. The *galK*-containing recombinant clones will produce red colonies on MacConkey agar with galactose (Fig. 1. 5). Fig. 2 highlights the easily perceived selection model of the *galK* mutagenesis approach, as red colonies are indicative of *galK* presence, whereas the colorless colonies do not express *galK*. A red colony from the screening process is chosen to verify by PCR and continue on with the rest of the protocol (Fig. 1. 6). A *geneY* cassette (PCR cassette containing desired mutation in *geneX*, referred to as *geneY*, or other gene) is prepared by PCR with primers conferring homologous sequences to the *galK* region in the mutant BAC (Fig. 1. 7a). From the red colony, SW102 harboring *galK* mutant BAC, selected in Step 5, are prepared to be electrocompetent and recombination-activated as in Fig. 1. 2b (Fig. 1. 7b). The *geneY* cassette is electroporated into the electrocompetent and recombination-activated SW102 strain harboring *galK* mutant clone (Fig. 1. 8). Upon homologous recombination, *galK* is replaced by *geneY* (Fig. 1. 9). Bacteria are grown on M63 with glycerol, DOG and antibiotic agar plates to counterselect for recombinant mutant BACs (Fig. 1. 10). Since the recombinants will now lack *galK*, selection takes place against the *galK* cassette by resistance to 2-deoxy-galactose (DOG) on minimal plates with glycerol as the carbon source (Fig. 1. 10) [27]. DOG is harmless, unless phosphorylated by functional *galK*. Phosphorylation by *galK* turns DOG into 2-deoxy-galactose-1-phosphate, a toxic intermediate [44]. From the resulting DOG-resistant colonies, some will be background colonies, where the bacteria have lost the *galK* cassette by a deletion, and the rest will be truly recombinant clones. Therefore, recombinant colonies containing the modified gene sequence can be quickly selected due to the negative or counterselection of colonies with *galK*, which makes this timesaving system also highly efficient. The resulting recombinant clones will be verified by PCF (Fig. 3) and restriction enzyme digestion (Fig. 4). Transfection of the mutated viral BAC into mammalian cells then produces an infectious virus (Fig. 5) [43].

4. Discussion

Molecular cloning vectors were once plasmids that could only carry and replicate small-sized DNA and have been developed to yeast artificial chromosome (YAC), bacterial artificial chromosome (BAC) and bacteriophage P1-derived artificial chromosome (PAC) that can carry and replicate large DNA molecules. Based on these vectors, recombineering techniques have contributed profoundly to investigating protein structure, function, and to elucidate viral pathogenesis, among others. However, the methods for BAC mutagenesis we use today were not easily attained as the large size of BACs posed a serious obstacle for their exact manipulation, as is required for viral research. To overcome these obstacles, techniques were developed that utilized the power of homologous recombination in order to create recombinant viruses, with the ability to safeguard every step in the process. We can now replace, and therefore delete, large DNA fragments with selectable marker cassettes, and within the described *galK* mutagenesis system there is also a counterselection model as well, using new systems that circumvent the problems associated with conventional genetic engineering as there is no longer a size restriction as seen when using restriction enzymes or other previous methods [37,44].

As previously mentioned, the generation of both the mutant and rescued BACs within the *galK* system take advantage of counterselection. For all such counterselection schemes, any event that leads to the loss of the counterselectable marker during negative selection will mean the survival of unwanted bacteria, leading to trace amounts of background. In the *galK* system, BAC replication appears to be the epicenter of background formation [41]. Despite the reliable stability of BACs, rare deletions do occur during counterselection, leading to background. Although there is spontaneous deletion background, it is insignificant relative to the great percentage of correct recombinants due to the high frequency of recombination from the λ prophage system. Furthermore, increasing the length of homology arms used for recombination can increase specific homologous recombination efficient and reduce the number of background deletions relative to the increased number of correct recombinants.

So far, most mutagenesis studies using the BAC system require generation of a rescue clone of the mutation to assure that no mutations occur anywhere other than the target site. The procedure of making rescue BAC DNA requires an original DNA fragment to be inserted back into the viral genome. This can be achieved by PCR amplification of the DNA fragment along with homology arms flanking the mutated region and insertion of the amplified fragment back into the mutant genome by homologous recombination. This method is analogous to procedure for generating recombinant mutants, as the only difference lies in the fact that there is no designed point mutation or deletion achieved via PCR. Antibiotic-resistance selection systems are the most extensively used method, however, removal of the antibiotic-resistance gene is not only necessary for functional analysis of a viral gene, but also required for preparing viral vaccine strains. The *galK* mutagenesis protocol outlined above is a seamless mutagenesis model because there is no requirement for further excision of the flanking homologous sequences that contain LoxP or FRT that might cause the loss of viral DNA during the process of making viruses in mammalian cells, as there is in antibiotic-resistance based mutagenesis. Thus, the process of making mutants by the above-mentioned protocol is considered to be accurate as there are no unintended mutations that could occur as a result, and even if there are, the strong counterselection strategy and troubleshooting strategies help to clear any background mutants.

5. Conclusion

These and other various advantages are responsible for the increased efficacy of the homologous recombination-based method for the construction of recombinant viruses. Outdated procedures for traditional mutagenesis, including molecular cloning that depends on preparations of vector and inserted DNA fragments and ligations, are laboring and time-intensive. The development of the Bacterial Artificial Chromosome (BAC) clone of large DNA viruses was an innovation that advanced the viral mutagenesis field to a new peak for global mutation and pathogenesis studies as any required mutation can be easily and rapidly achieved by the novel recombineering method for construction of recombinant viruses by homologous recombination. This mutagenesis method has led to a great expansion in the field of molecular research, whereas *galK* mutagenesis takes this recombineering strategy to a heightened level of accuracy and, therefore, results.

6. Future directions

The amount of biomedical research utilizing plasmids and BACs has grown rapidly during the past decade, resulting in invaluable knowledge about protein structure and function, viral pathogenesis, vaccine development and gene therapy. Since the construction of the first herpesvirus BAC 12 years ago, BACs have been generated for all major human and animal herpesviruses, and this technology has greatly facilitated genetic and functional studies of herpesviruses, because recombinant viruses, especially herpesviruses, were previously difficult to produce due to their large size. Soon, we may have BACs for not only all herpesviruses [27], but all DNA viruses, as well as novel global mutational studies for several virus BACs thanks to accurate and seamless mutagenesis procedures such as the *galK* recombineering protocol. Global and local studies of virus pathogenesis should help identify new antiviral targets and produce more effective and safe vaccines. In short, future virus BAC-based mutagenesis studies achieved by the seamless *galK* mutagenesis protocol should help provide exciting new discoveries about viral pathogenesis, protein structure and function, as well as therapeutics for both viral and non-viral diseases.

7. References

[1] Ahmed, R., et al., *Genetic analysis of in vivo-selected viral variants causing chronic infection: importance of mutation in the L RNA segment of lymphocytic choriomeningitis virus.* J Virol, 1988. 62(9): p. 3301-8.

[2] Chow, L.H., *Studies of virus-induced myocardial injury in mice: value of the scid mutation on different genetic backgrounds and combined with other mutations.* Lab Anim Sci, 1993. 43(2): p. 133-5.

[3] Cutter, J.L., *All that was needed for a pandemic to occur was a genetic mutation in the H5N1 virus that would enable efficient human-to-human transmission. Afterword.* Ann Acad Med Singapore, 2011. 39(4): p. 343-2.

[4] Figlerowicz, M., P.D. Nagy, and J.J. Bujarski, *A mutation in the putative RNA polymerase gene inhibits nonhomologous, but not homologous, genetic recombination in an RNA virus.* Proc Natl Acad Sci U S A, 1997. 94(5): p. 2073-8.

[5] Garcia-Lerma, J.G., et al., *A novel genetic pathway of human immunodeficiency virus type 1 resistance to stavudine mediated by the K65R mutation.* J Virol, 2003. 77(10): p. 5685-93.

[6] Taddie, J.A. and P. Traktman, *Genetic characterization of the vaccinia virus DNA polymerase: cytosine arabinoside resistance requires a variable lesion conferring phosphonoacetate resistance in conjunction with an invariant mutation localized to the 3'-5' exonuclease domain.* J Virol, 1993. 67(7): p. 4323-36.

[7] Treanor, J., et al., *Evaluation of the genetic stability of the temperature-sensitive PB2 gene mutation of the influenza A/Ann Arbor/6/60 cold-adapted vaccine virus.* J Virol, 1994. 68(12): p. 7684-8.

[8] Chavali, S., et al., *Protein molecular function influences mutation rates in human genetic diseases with allelic heterogeneity.* Biochem Biophys Res Commun, 2011. 412(4): p. 716-22.

[9] Liu, Z., et al., *Beyond the rotamer library: genetic algorithm combined with the disturbing mutation process for upbuilding protein side-chains.* Proteins, 2003. 50(1): p. 49-62.

[10] Mason, A.B., et al., *Evolution reversed: the ability to bind iron restored to the N-lobe of the murine inhibitor of carbonic anhydrase by strategic mutagenesis.* Biochemistry, 2008. 47(37): p. 9847-55.

[11] Mohan, U. and U.C. Banerjee, *Molecular evolution of a defined DNA sequence with accumulation of mutations in a single round by a dual approach to random chemical mutagenesis (DuARCheM).* Chembiochem, 2008. 9(14): p. 2238-43.

[12] Symonds, N., *Evolution. Anticipatory mutagenesis.* Nature, 1989. 337(6203): p. 119-20.

[13] van Duin, M., et al., *Evolution and mutagenesis of the mammalian excision repair gene ERCC-1.* Nucleic Acids Res, 1988. 16(12): p. 5305-22.

[14] Bessereau, J.L., *Insertional mutagenesis in C. elegans using the Drosophila transposon Mos1: a method for the rapid identification of mutated genes.* Methods Mol Biol, 2006. 351: p. 59-73.

[15] Vallejo, A.N., R.J. Pogulis, and L.R. Pease, *PCR Mutagenesis by Overlap Extension and Gene SOE.* CSH Protoc, 2008. 2008: p. pdb prot4861.

[16] Yang, Y.X., et al., *PCR-based site-specific mutagenesis of peptide antibiotics FALL-39 and its biologic activities.* Acta Pharmacol Sin, 2004. 25(2): p. 239-45.

[17] Lehoux, D.E. and R.C. Levesque, *PCR screening in signature-tagged mutagenesis of essential genes.* Methods Mol Biol, 2002. 192: p. 225-34.

[18] Luo, S., et al., *Site-directed mutagenesis of gentisate 1,2-dioxygenases from Klebsiella pneumoniae M5a1 and Ralstonia sp. strain U2.* Microbiol Res, 2006. 161(2): p. 138-44.

[19] Balliet, J.W., et al., *Site-directed mutagenesis of large DNA palindromes: construction and in vitro characterization of herpes simplex virus type 1 mutants containing point mutations that eliminate the oriL or oriS initiation function.* J Virol, 2005. 79(20): p. 12783-97.

[20] Largaespada, D.A., *Transposon-mediated mutagenesis of somatic cells in the mouse for cancer gene identification.* Methods, 2009. 49(3): p. 282-6.

[21] Kim, Y.C., et al., *Transposon-directed base-exchange mutagenesis (TDEM): a novel method for multiple-nucleotide substitutions within a target gene.* Biotechniques, 2009. 46(7): p. 534-42.

[22] Largaespada, D.A., *Transposon mutagenesis in mice.* Methods Mol Biol, 2009. 530: p. 379-90.

[23] Dunn, W., et al., *Functional profiling of a human cytomegalovirus genome.* Proc Natl Acad Sci U S A, 2003. 100(24): p. 14223-8.

[24] Murphy, E., et al., *Coding potential of laboratory and clinical strains of human cytomegalovirus.* Proc Natl Acad Sci U S A, 2003. 100(25): p. 14976-81.

[25] Yu, D., M.C. Silva, and T. Shenk, *Functional map of human cytomegalovirus AD169 defined by global mutational analysis.* Proc Natl Acad Sci U S A, 2003. 100(21): p. 12396-401.

[26] Dolan, A., et al., *Genetic content of wild-type human cytomegalovirus.* J Gen Virol, 2004. 85(Pt 5): p. 1301-12.

[27] Warden, C., Q. Tang, and H. Zhu, *Herpesvirus BACs: past, present, and future.* Journal of biomedicine & biotechnology, 2011. 2011: p. 124595.

[28] Sauer, B., *Functional expression of the cre-lox site-specific recombination system in the yeast Saccharomyces cerevisiae.* Mol Cell Biol, 1987. 7(6): p. 2087-96.

[29] Sauer, B. and N. Henderson, *Site-specific DNA recombination in mammalian cells by the Cre recombinase of bacteriophage P1.* Proc Natl Acad Sci U S A, 1988. 85(14): p. 5166-70.

[30] Orban, P.C., D. Chui, and J.D. Marth, *Tissue- and site-specific DNA recombination in transgenic mice.* Proc Natl Acad Sci U S A, 1992. 89(15): p. 6861-5.

[31] Zhu, X.D. and P.D. Sadowski, *Cleavage-dependent ligation by the FLP recombinase. Characterization of a mutant FLP protein with an alteration in a catalytic amino acid.* J Biol Chem, 1995. 270(39): p. 23044-54.

[32] Dixon, J.E., A.C. Shaikh, and P.D. Sadowski, *The Flp recombinase cleaves Holliday junctions in trans.* Mol Microbiol, 1995. 18(3): p. 449-58.

[33] Schlake, T. and J. Bode, *Use of mutated FLP recognition target (FRT) sites for the exchange of expression cassettes at defined chromosomal loci.* Biochemistry, 1994. 33(43): p. 12746-51.

[34] Jayaram, M., *Phosphoryl transfer in Flp recombination: a template for strand transfer mechanisms.* Trends Biochem Sci, 1994. 19(2): p. 78-82.

[35] Xu, T. and S.D. Harrison, *Mosaic analysis using FLP recombinase.* Methods Cell Biol, 1994. 44: p. 655-81.

[36] Stricklett, P.K., R.D. Nelson, and D.E. Kohan, *The Cre/loxP system and gene targeting in the kidney.* The American journal of physiology, 1999. 276(5 Pt 2): p. F651-7.

[37] Lee, E.C., et al., *A highly efficient Escherichia coli-based chromosome engineering system adapted for recombinogenic targeting and subcloning of BAC DNA.* Genomics, 2001. 73(1): p. 56-65.

[38] Yu, D.G., et al., *An efficient recombination system for chromosome engineering in Escherichia coli.* Proceedings of the National Academy of Sciences of the United States of America, 2000. 97(11): p. 5978-5983.

[39] Carter, D.M. and C.M. Radding, *The role of exonuclease and beta protein of phage lambda in genetic recombination. II. Substrate specificity and the mode of action of lambda exonuclease.* J Biol Chem., 1971. 246(8): p. 2502-12.

[40] Little, J.W., *An exonuclease induced by bacteriophage lambda. II. Nature of the enzymatic reaction.* J Biol Chem. , 1967. 242(4): p. 679-86.

[41] Warming, S., et al., *Simple and highly efficient BAC recombineering using galK selection.* Nucleic Acids Res, 2005. 33(4): p. e36.

[42] Zhang, Y., et al., *DNA cloning by homologous recombination in Escherichia coli.* Nature biotechnology, 2000. 18(12): p. 1314-7.

[43] Zhang, Z., Y. Huang, and H. Zhu, *A highly efficient protocol of generating and analyzing VZV ORF deletion mutants based on a newly developed luciferase VZV BAC system.* Journal of virological methods, 2008. 148(1-2): p. 197-204.

[44] Dulal, K., Z. Zhang, and H. Zhu, *Development of a gene capture method to rescue a large deletion mutant of human cytomegalovirus.* Journal of virological methods, 2009. 157(2): p. 180-7.

[45] Sharan, S.K., et al., *Recombineering: a homologous recombination-based method of genetic engineering.* Nature protocols, 2009. 4(2): p. 206-23.

[46] Borst, E.M., et al., *Cloning of the human cytomegalovirus (HCMV) genome as an infectious bacterial artificial chromosome in Escherichia coli: a new approach for construction of HCMV mutants.* J Virol, 1999. 73(10): p. 8320-9.

[47] Marchini, A., H. Liu, and H. Zhu, *Human cytomegalovirus with IE-2 (UL122) deleted fails to express early lytic genes.* Journal of virology, 2001. 75(4): p. 1870-8.

[48] Moffat, J.F., et al., *Tropism of varicella-zoster virus for human CD4+ and CD8+ T lymphocytes and epidermal cells in SCID-hu mice.* Journal of virology, 1995. 69(9): p. 5236-42.

5

Extraction of Drug from the Biological Matrix: A Review

S. Lakshmana Prabu[1] and T. N. K. Suriyaprakash[2]
[1]Anna University of Technology, Tiruchirappalli, Tamil Nadu,
[2]Periyar College of Pharmaceutical Sciences,
The Tamil Nadu Dr. M.G.R. Medical University, Chennai, Tamil Nadu,
India

1. Introduction

The assessment of therapeutic compliance is commonly done by either indirect method or direct method. In indirect method, the assessments are done by indirect measurement of parameters such as discussion with patients and pill counts at intervals during the treatment which don't measure the drug concentration in a matrix such as blood or urine. In direct method, the assessments which rely upon evidence provided by the patients or care giver on the presumptive compliance based upon electronic medical event monitoring system and this is based upon either the qualitative or quantitative measurement of the drug under investigation in a biological matrix provided by the system.

A more objective assessment of patient compliance has its own limitations. The most widely used matrix is plasma, but the major limitation with this approach is that it provides the clinician with the concentration of drug within the systematic circulation at the time of sample collection. This concentration is primarily related to the time interval between the administration of the last dose and the sample collection time although other pharmacokinetic parameters can also influence the concentration. Plasma samples should be collected prior to the administration of the next dose. Plasma monitoring is a useful adjunct for the assessment of therapeutic compliance (Williams, 1999).

Bioanalysis is a sub-discipline of analytical chemistry covering the quantitative measurement drugs and their metabolites in biological systems. Bioanalysis in the pharmaceutical industry is to provide a quantitative measure of the active drug and/or its metabolite(s) for the purpose of pharmacokinetics, toxicokinetics, bioequivalence and exposure–response (pharmacokinetics /pharmacodynamics studies). Bioanalysis also applies to drugs used for illicit purposes, forensic investigations, anti-doping testing in sports, and environmental concerns. Bioanalytical assays to accurately and reliably determine these drugs at lower concentrations. This has driven improvements in technology and analytical methods.

Some techniques commonly used in bioanalytical studies include:

- Hyphenated techniques
    - LC–MS (liquid chromatography–mass spectrometry)
    - GC–MS (gas chromatography–mass spectrometry)

- LC–DAD (liquid chromatography–diode array detection)
- CE–MS (capillary electrophoresis–mass spectrometry)
- Chromatographic methods
  - HPLC (high performance liquid chromatography)
  - GC (gas chromatography)
  - UPLC (ultra performance liquid chromatography)
  - Supercritical fluid chromatography

The area of bioanalysis can encompass a very broad range of assays which support the clinical and nonclinical studies. Many factors influence the development of robust bioanalytical methods for analyzing samples from clinical and nonclinical studies which includes the matrices of interest, the range over which analytes need to be measured, the number and structures of the analytes, the physicochemical properties of the analytes, and their stability in the biological matrices from the time of sample draw to analysis also needs to be measured.

Because biological samples are extremely complex matrices comprised of many components that can interfere with good separations and or good mass spectrometer signals, sample preparation is an important aspect of bioanalytical estimation. This is important whether samples originate as tissue extracts, plasma, serum or urine. The question sometimes arises as to whether serum or plasma should be collected for analysis. In general, from a bioanalytical perspective, there are few advantages to choosing serum over plasma except where chemical interferences in a given assay might require processing blood all the way to serum.

Nearly all bioanalytical assays involve the use of an internal standard processed along with the sample. The area ratio of analyte to internal standard is used in conjunction with a standard curve to back calculate the concentration of analyte in the sample. The internal standard is expected to behave similarly to the analyte with respect to extraction efficiency across the range of concentration, which helps compensate for sample to sample differences in sample preparation. Often, an analogue or similarly behaving compound is used as an internal standard.

To produce adequate and precise results, certain criteria like minimal loss or degradation of sample during blood collection, the appropriate sample cleanup and internal standard, chromatographic conditions that minimize the interference and ion suppression, and sufficient signal to noise to allow for reproducible peak integration.

However, an examination of the individual plasma samples could show one or more with an unacceptable interference that is effectively washed out when the samples are pooled. Thus, it is important to use several individual matrix samples when evaluating matrix effects. If unacceptable matrix interferences are observed, it is necessary to further clean up the sample to eliminate the interference.

A key component of the sample preparation is the emphasis on analyte stability which needs to be carefully assessed from the time of sample drawn till the analysis is complete. It is important to study the analytes stability in blood at the appropriate temperatures from the time the sample is drawn until after centrifugation when plasma or serum are separated and stored in a new vial. It is also important to ensure that the anticoagulant or other components of the blood collection do not interfere with the sample preparation (Destefano & Takigiku, 2004).

Blood is the transporter of many vital substances and nutrients for the entire body and thus contains many endogenous and exogenous compounds in different concentrations. Drug determination in human plasma is often complicated by low concentrations (0.1 – 10ng mL^{-1} level). An extra problem posed by blood sample is the complex sample matrix due to proteins, which can lead to protein binding of the analyte and by limited sample volumes normally 0.5 – 1ml will be available for the determination. The magnitude of the challenge of protein purification becomes clearer when one considers the mixture of macromolecules present in the biological matrices. Hence, sample preparation is crucial in drug analysis which includes both analyte pre-concentration and sample cleanup (Ho *et al.*, 2002).

To produce meaningful information, an analysis must be performed on a sample whose composition faithfully reflects that of the bulk of material from which it was taken. Biological samples cannot normally be injected directly into the analyzing system without sample preparation. Sample pretreatment is thus of utmost importance for the adequate analysis of drugs. However, as sample pretreatment can be a time consuming process, this can limit the sample throughput. The proper selectivity can be obtained during the sample preparation, the separation and the detection. A major differentiation between the analyte of interest and the other compounds is often made during the first step. Sensitivity is to a large extent obtained by the detector. Thus, sample pretreatment is required for achieving sufficient sensitivity and selectivity, whereas the time should be kept to a minimum in order to obtain adequate speed. Therefore, there is a clear trend towards integration of sample pretreatment with the separation and the detection. Numerous sample preparation techniques have been developed for bioanalytical purpose. Sample preparation is a difficult step, especially in clinical and environmental chemistry and generally involves filtration, solid phase extraction with disposable cartridges, protein precipitation and desalting. Sample preparation prior to chromatographic separation is performed to dissolve or dilute the analyte in a suitable solvent, removing the interfering compounds and pre-concentrating the analyte.

The principle objectives of sample preparation from biological matrix are;

a.   Isolation of the analytes of interest from the interfering compounds
b.   Dissolution of the analytes in a suitable solvent and pre-concentration.

In an analytical method sample preparation is followed by a separation and detection procedure. In spite of the fact that sample preparation, in most of the analytical procedures, takes 50-75% of the total time of the analysis, most technical innovations of the last 5 years are related to separation and detection.

Extraction is one of humankind's oldest chemical operations. Extraction is the withdrawing of an active agent or a waste substance from a solid or liquid mixture with a liquid solvent. The solvent is not or only partially miscible with the solid or the liquid. By intensive contact between analyte and the extraction medium this leads the analyte transfers from the solid or liquid mixture into the extraction medium (extract). After thorough mixing the two phases can be separated either by gravity or centrifugal forces (Gy, 1982; Arthur & Pawliszyn, 1990; Zief & Kiser, 1990).

2. Sampling

The sampling and sample preparation process begins at the point of collection and extends to the measurement step. The proper collection of sample during the sample process (called primary sampling), the transport of this representative sample from the point of collection to the analytical laboratory, the proper selection of the laboratory sample itself (called secondary sampling), and the sample preparation method used to convert the sample into a form suitable for the measurement step can have a greater effect on the overall accuracy and reliability of the results than the measurement itself.

Primary sampling is the process of selecting and collecting the sample to be analyzed. The objective of sampling is a mass or volume reduction from the parent batch, which itself can be homogeneous or heterogeneous. If the wrong sample is collected or the sample is not collected properly, then all the further stages in the analysis become meaningless and the resulting data are worthless. Unfortunately, sampling is sometimes left to people unskilled in sampling methodology and is largely ignored in the education process, especially for non-analytical chemists (Smith & James, 1981). Once the primary sample is taken, it must be transported to the analytical laboratory without a physical or chemical change in its characteristics. Even if a representative primary sample is taken, changes that can occur during transport can present difficulties in the secondary sampling process. Preservation techniques can be used to minimize changes between collection and analysis. Physical changes such as adsorption, diffusion and volatilization as well as chemical changes such as oxidation and microbiological degradation are minimized by proper preservation.

Common preservation techniques used between the point of collection and the point of sample preparation in the laboratory are

- Choice of appropriate sampling container
- Addition of chemical stabilizers such as antioxidants and antibacterial agents
- Freezing the sample to avoid thermal degradation
- Adsorption on a solid phase

In secondary sampling, if the sample has made it to the laboratory, a representative subsample must be taken. Statistical appropriate sampling procedures are applied to avoid discrimination, which can further degrade analytical data. Clearly speeding up or automating the sample preparation step will reduce analysis time and improve sample throughput (Keith, 1990).

Before sample preparation, solid or semisolid substances must be put into a finely divided state. Procedures to perform this operation are usually physical methods, not chemical methods. The reasons for putting the sample into a finely divided state are that finely divided samples are more homogeneous, so secondary sampling may be carried out with greater precision and accuracy, and they are more easily dissolved or extracted because of their large surface-to-volume ratio.

2.1 Sample preparation

Sample preparation is necessary for at least two reasons:

a. To remove as many of the endogenous interferences from the analyte as possible
b. To enrich the sample with respect to the analyte, thus maximizing the sensitivity of the system.

It also serves to ensure that the injection matrix is compatible with the selected column and mobile phase.

2.2 Goal and objectives of sample preparation

Two of the major goals of any sample pretreatment procedure are

- Quantitative recovery
- A minimum number of steps.

Successful sample preparation has a threefold objective.

- In solution
- Free from interfering matrix elements
- At a concentration appropriate for detection and measurement

The most common approach in analyte separation involves a two phase system where the analyte and interferences are distributed between the two phases. Distribution is an equilibrium process and is reversible. If the sample is distributed between two immiscible liquid phase, the techniques is called liquid-liquid extraction. If the sample is distributed between a solid and a liquid phase, the technique is called liquid-solid adsorption.

Often, when analysis involves the measurement of trace amounts of a substance, it is desirable to increase the concentration of the analyte to a level where it can be measured more easily. Concentration of an analyte can be accomplished by transferring it from a large volume of phase to a smaller volume of phase. Separation can be carried out in a single batch, in multiple batches or by continuous operation.

2.3 Types of samples

Sample matrices can be classified as organic and inorganic. Biological samples are a subset of organic samples but often require different sample preparation procedures in order to preserve biological integrity and activity. Compared to volatile compounds or solid, liquid samples are much easier to prepare for analytical measurement because a dissolution or an extraction step many not be involved. The major consideration for liquid samples is the matrix interferences, the concentration of analyte, and compatibility with the analytical techniques. When a sample is a solid, the sample pretreatment process can be more complex. There are two specific cases: the entire sample is of interest and must be solubilized or only a part of the solids is of interest and the analytes must be selectively removed. If the solid is a soluble salt or drug tablet formulation, the only sample preparation that may be required is finding a suitable solvent that will totally dissolve the sample and the components of interest. If the sample matrix is insoluble in commonly solvents but the analytes of interest can be removed or leached out, then sample preparation can also be rather straightforward. In these cases, techniques such as filtration, Soxhlet extraction, supercritical fluid extraction, ultrasonication or solid-liquid extraction may be useful. If both the sample matrix and the sample analytes are not soluble in common solvents, then more drastic measures may be needed.

3. Physicochemical properties of drug and their extraction from biological material (Chang 1977; Moore 1972; Barrow 1996; Wiltshire 2000)

3.1 Molecular phenomena for solubility and miscibility

To dissolve a drug, a solvent must break the bonds like ionic bond, hydrogen bond and Van der Waals forces which inter links the compound to its neighbors and must not break substantial intermolecular bonds of the solvent without replacing them with drug solvent interaction. Because breaking of bonds is an endothermic process, requires energy and causing an increasing in enthalpy. Similarly, if the gain in entropy from the dissolution of two solvents in each other is insufficient to counteract any reduced amount of intermolecular bonding, they will not be completely miscible.

3.2 Water miscibility and water immiscibility

Commonly alcohols can have hydrogen bonding with water and also dipole-dipole interactions will aid miscibility. On the other hand, presence of alkyl groups will reduce the solubility with water and the interaction may be by means of dispersive force. Hydrocarbons are hydrophobic in nature, which dissolves the compounds by dispersive forces. Whereas halogenated hydrocarbons are more polar and dissolve the compounds by dispersive forces and dipolar interactions.

Hydrophilic groups, which are polar in nature, will encourage the solubility in water, whereas C-C, C-H and C-X bonds are hydrophobic in nature will encourage the solubility in organic solvents. Drug with several aromatic rings will have poor solubility in water due to lack of hydrogen bonding with water and the strong intermolecular dispersive forces of the solid drug will encourage the ready solubility in organic solvents.

3.3 Distribution coefficient

Drug which are in ionised forms are hydrophilic in nature than the unionized form because of the hydration of the ions, therefore the ionized forms are difficulty to extract into organic solvents whereas the unionized forms will dissolve in the organic solvents which can be extracted into organic solvents.

3.4 Choice of solvent

Several factors are to be considered while choosing a solvent to extract a drug from the matrix in addition to its powder to dissolve the required compounds which includes selectivity, density, toxicity, volatility, reactivity, physical hazards and miscibility with aqueous media.

- Ethyl acetate is a powerful solvent for many organic compounds and will therefore extract a considerable amount of endogenous material with the required drug.
- If the drug is relatively non-polar, a more selective extraction could be obtained by using a hydrocarbon solvent.
- Halogenated hydrocarbons like chloroform and dichloromethane are excellent, volatile solvents. However they are denser than water which makes them difficult to use for analysis.

- Benzene is a useful solvent, reasonably volatile, inert and immiscible with water, but its toxicity precludes its use.
- Toluene has similar properties as a solvent to benzene is not particularly toxic, however its boiling point is 111°C and it is not really sufficient volatile for use as a solvent in bio-analysis.
- Chloroform is an excellent solvent but reactivity with bases reduces its uses with basic drugs that need to be extracted at high pH.
- Di-isopropyl ether is less miscible with water than di-ethyl ether but is much more likely to form explosive peroxides and is best avoided.
- Diethyl ether is a good, volatile solvent but it is quite soluble (~4%) in water and difficult to blow to complete dryness.

3.5 Mixed solvents

In some cases pure solvents will not be satisfactory for the extraction of the compound of interest. Alcohols are excellent solvent but those with lower boiling points are too soluble in water whereas less miscible one are having high boiling points, but the use of mixed solvents containing alcohols can solve the problem. A 1:1 mixture of tetrahydrofuran and dichloromethane is a powerful solvent for the extraction of polar compounds from aqueous solutions.

3.6 Plasma proteins and emulsions

Presence of proteins can cause difficulties in extracting the drug from plasma. Emulsions are often formed and partial precipitation can unclear the interface between the two layers. The proteins can be precipitated by addition of 10-20% trichloroacetic acid or five volumes of a water-miscible solvent like acetonitrile.

3.7 Role of pH for solvent extraction

Organic acids and bases are usually much less soluble in water than its salts. As a general rule, extraction of bases into an organic solvent should be carried out at high pH usually about 2 pH units above the pKa and extraction of acids carried out at low pH. If the drug is reasonably non-polar base, it could be back extracted from the organic solvent into acid, basified and re-extracted into the organic solvent and vice versa for the acidic drugs.

4. Sample pretreatment in different biological matrices (Horne, 1985; Christians & Sewing, 1989; McDowall et al., 1989; Ingwersen, 1993; Krishnan & Ibraham 1994; Simmonds et al., 1994; Allanson et al., 1996; Plumb et al., 1997)

4.1 General concern with biological samples

Extraction of biological samples before injection into an HPLC system serves a number of objectives.

- Concentration
- Clean-up
- Prevention of clogging of analytical columns

- Elimination of protein binding
- Elimination of enzymatic degradation of the analyte

4.2 Serum, plasma, and whole blood

Serum and plasma samples may not need to be pretreated for SPE. In many cases, however, analytes such as drugs may be protein-bound, which reduces SPE recoveries. To disrupt protein binding in these biological fluids, use of one of the following methods for reversed phase or ion exchange SPE procedures.

- Shift pH of the sample to extremes (pH<3 or pH>9) with acids or bases in the concentration range of 0.1M or greater. Use the resulting supernatant as the sample for SPE
- Precipitate the proteins using a polar solvent such as acetonitrile, methanol, or acetone (two parts solvent per one part biological fluid is typical). After mixing and centrifugation, remove the supernatant and dilute with water or an aqueous buffer for the SPE procedure
- To precipitate proteins, treat the biological fluid with acids or inorganic salts, such as formic acid, perchloric acid, trichloroacetic acid, ammonium sulfate, sodium sulfate, or zinc sulfate. The pH of the resulting supernatant may be adjusted prior to use for the SPE procedure
- Sonicate the biological fluid for 15 minutes, add water or buffer, centrifuge, and use the supernatant for the SPE procedure

4.3 Urine

Urine samples may not require pretreatment for reversed phase or ion exchange SPE, but often is diluted with water or a buffer of the appropriate pH prior to sample addition. In some cases, acid hydrolysis (for basic compounds) or base hydrolysis (for acidic compounds) is used to ensure that the compounds of interest are freely solvated in the urine sample. Usually a strong acid (e.g. concentrated HCl) or base (e.g. 10M KOH) is added to the urine. The urine is heated for 15- 20 minutes, then cooled and diluted with a buffer, and the pH adjusted appropriately for the SPE procedure. Enzymatic hydrolysis that frees bound compounds or drugs also may be used.

4.4 Solid samples

Solid samples ex. tissues, faeces are normally homogenized with a buffer or an organic solvent, then remaining solids removed by centrifugation, and the diluted sample applied to the cartridge.

5. Methods of extraction

The aim of the sample preparation process is to provide a suitable sample, usually for chromatographic analysis, which will not contaminate the instrumentation and where the concentration in the prepared sample is reflective of that found in the original. The method of sample preparation selected is generally dictated by the analytical technique available and the physical characteristics of the analytes under investigation (Watt et al., 2000). The two main sample preparation methods are matrix cleanup or direct injection. In a matrix cleanup procedure, the aim is to remove as much endogenous material as possible from the drug sample.

Sample preparation is traditionally carried out (a) by liquid-liquid extraction, (b) solid-phase extraction or (c) by precipitation of the plasma proteins, while the final analysis in most cases is accomplished by liquid chromatography interfaced with mass spectrometry or tandem mass spectrometry or capillary gas chromatography.

5.1 Liquid-Liquid Extraction (LLE) (Christian & O'Reilly, 1986; Harris, 1994; Majors & Fogelman, 1993; Wells, 2003)

One of the most useful techniques for isolating desired components from a mixture is liquid-liquid extraction (LLE). LLE is a method used for the separation of a mixture using two immiscible solvents. In most LLEs, one of the phases is aqueous and the other is an immiscible organic solvent. The concept "like dissolves like" works well in LLE. The ability to separate compounds in a mixture using the technique of LLE depends upon how differently the compounds of the sample mixture partition themselves between the two immiscible solvents. Selective partitioning of the compound of interest into one of two immiscible or partially miscible phases occurs by the proper choice of extraction of solvent. In this technique sample is distributed in two phases in which one phase is immiscible to other. LLE separates analytes from interferences by partitioning the sample between two immiscible liquids or phases. First, the component mixture is dissolved in a suitable solvent and a second solvent that is immiscible with the first solvent is added. Next, the contents are thoroughly mixed (shaking) and the two immiscible solvents allowed separating into layers. The less dense solvent will be the upper layer, while the more dense solvent will be the lower layer. The components of the initial mixture will be distributed amongst the two immiscible solvents as determined by their partition coefficient. The relative solubility that a compound has in two given solvents can provide an estimation of the extent to which a compound will be partitioned between them. A compound that is more soluble in the less dense solvent will preferentially reside in the upper layer. Conversely, a compound more soluble in the more dense solvent will preferentially reside in the lower layer. Lastly, the two immiscible layers are separated, transferred and the component in that solvent is isolated. Generally after extraction hydrophilic compounds are seen in the polar aqueous phase and hydrophobic compounds are found mainly in the organic solvents. Analyte is extracted into the organic phase are easily recovered by evaporation of the solvent, the residue reconstituted with a small volume of an appropriate solvent preferably mobile phase while analyte extracted in to the aqueous phase can be directly injected into a RP column. LLE technique is simple, rapid is relative cost effective per sample as compared to other techniques and near quantitative recoveries (90%) of most drugs can be obtained by multiple continuous extraction (Fig.1).

Several useful equations can help illustrate the extraction process. The Nernst distribution law states that any neutral species will distribute between two immiscible solvents so that the ratio of the concentration remains constant.

$$K_D = C_o / C_{aq}$$

Where K_D is the distribution constant, C_o is the concentration of the analyte in the organic phase, and C_{aq} is the concentration of the analyte in the aqueous phase.

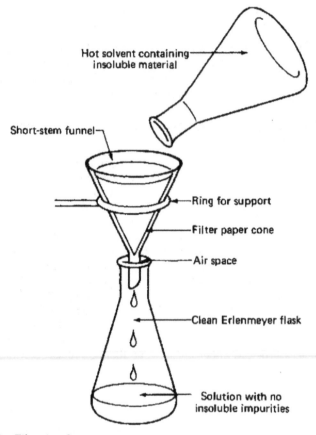

Hot solvent containing insoluble material

Short-stem funnel

Ring for support

Filter paper cone

Air space

Clean Erlenmeyer flask

Solution with no insoluble impurities

Fig. 1. Gravity Filtration Setup

To increase the value of K_D, several approaches may be used:

-   The organic solvent can be changed to increase solubility of the analyte
-   If the analyte is ionic or ionizable, its K_D may be increased by suppressing its ionization to make it more soluble in the organic phase. It can also be extracted into the organic phase by the formation of an ion pair through the addition of a hydrophobic counter-ion.
-   Metal ions can form a complex with hydrophobic complexing agents.
-   The salting out effect can be used to decrease analytes concentration in the aqueous phase.

If the K_D value is unfavorable, addition extraction may be required for better solute recovery. In this case, a fresh portion of immiscible solvent is added to extract additional solute. Normally, the two extracts are combined. Generally, for a given volume of solvent, multiple extractions are more efficient in removing a solute quantitatively than a single extraction. Sometimes, back extractions can be used to achieve a more complete sample cleanup.

If K_D is very low or the sample volume is high, it becomes nearly impossible to carryout multiple simple extractions in a reasonable volume. Also, if the extraction rate is slow, it may take a long time for equilibrium to be established. In these cases, continuous liquid-liquid extraction is used, where pure solvent is recycled through the aqueous phase.

Benefit of this technique is that, with a judicious choice of solvents and pH, very clean extracts can be obtained with good selectivity for the targeted analyte. The drug is extracted from the aqueous phase to the organic phase. One point of note is that LLE system unlike solid-phase systems, are more likely to give consistent results year after years, as there is usually less batch to batch variation with solvents.

The extraction of drug from the aqueous phase is mainly depends on the following factors:

- Solubility of analyte in the organic solvent
- Polarity of the organic solvent
- pH of the aqueous phase

In some cases there is a possibility that interferences may present in the extracted sample. In that case back liquid-liquid extraction can be performed, this gives a clear extracts. Here two times organic solvent is used for the extraction of analyte from the matrix. Often, however, it is not possible to find the optimum condition that provides both high recovery and purity of the analyte in one extraction step. Low recoveries may require further extraction to achieve acceptable value. About the purity it may require second extraction procedure with different solvent or pH of the aqueous phase. Each successive extraction increase the analytical time, also the resulting large volume of extraction solvent must be evaporated to recover the product. If extraction requires many steps, techniques such as Craig Counter Current distribution can be used to increase recovery and purity. However this technique increases the cost and time of the analysis.

5.1.1 Selection of the solvent

There are also practical concerns when choosing extraction solvents. As mentioned previously, the two solvents must be immiscible. The properties of an ideal solvent is that it should withdraw the active agent from a mixture by liquid-liquid extraction are

- Selectivity: Only the active agent has to be extracted and no further substances which mean that a high selectivity is required.
- Capacity: To reduce the amount of necessary solvent, the capacity of the solvent has to be high.
- Miscibility: To achieve simple regeneration of the solvent the miscibility of solvent and primary solvent has to be low.
- Difference in density: After extraction, the two phases have to be separated in a separator and for this a high positive difference in density is required.
- Optimal surface tension: σ low $\rightarrow$ low amount of energy for dispersing required; if surface tension < 1 mN/m stable emulsions are produced; σ > 50 mN/m $\rightarrow$ high amount of energy for dispersing and high tendency to coalesce.
- Recovery: The solvent has to be separated from the extract phase easily to produce solvent free active agents.

- Corrosion: If the solvent is corrosive prices for construction increases.
- Low price
- No or low toxicity and not highly flammable
- Flame temperature: 25°C higher than operating temperature
- Vapour pressure: To prevent loss of solvent by evaporation a low vapour pressure at operating temperature is required.
- Viscosity: A low viscosity of the solvent leads to low pressure drop and good heat and mass transfer.
- Chemical and thermal stability
- Environmentally acceptable or easily recoverable
- Convenient specific gravity
- Suitable volatility
- High chemical stability and inertness
- Not prone to form an emulsion
- Dissolves the neutral but not the ionized form of the analyte

The stoichiometric ratio of the analyte in the organic phase compared to that in the aqueous phase is known as the distribution ratio D. Ideally, this ratio should approach 100% in order to minimize the losses through the effects of small changes in sample composition, temperature and pH. Reproducibility also increases with increasing extraction efficiency, although a consistent low recovery may be acceptable if an internal standard is used to compensate for changes in efficiency.

5.1.2 Extraction under basic and acidic conditions

As mentioned above, the ability to separate compounds of a mixture using liquid-liquid extraction procedures depends upon the relative solubility that each compound has in the two immiscible solvents. A change in the pH of the solvent (the addition of acid or base) can change the solubility of an organic compound in a solvent considerably.

Liquid/liquid extraction is the most common technique used to separate a desired organic product from a biological matrix. The technique works well if your target compound is more soluble in one of two immiscible solvents. Extraction usually involves shaking a solution that contains the target with an immiscible solvent in which the desired substance is more soluble than it is in the starting solution. Upon standing, the solvents form two layers that can be separated. The extraction may have to be repeated several times to effect complete separation.

In general liquid-liquid extractions can separate four different classes of compounds:

a. **Organic bases:** Any organic amine can be extracted from an organic solvent with a strong acid such as 1M hydrochloric acid
b. **Strong acids:** Carboxylic acids can be extracted from an organic solvent with a weak base such as 1M sodium bicarbonate
c. **Weak acids:** Phenols can be extracted from an organic solvent with a strong base such as 1M sodium hydroxide
d. **Non-polar compounds** stay in the organic layer

5.1.3 Disadvantages

- Large solvent consumption is needed for extraction of drug.
- LLE is time consuming process when compare to other methods.
- LLE require an evaporation step prior to analysis to remove excess of organic solvent.
- LLE technique is not a suitable one for the estimation of several analytes.
- Emulsion formation may be possible when two immiscible phases were used in the extraction procedure.

5.2 Solid Phase Extraction (Zwir Ferenc & Biziuk, 2006; James, 2000; Krishnan & Abraham, 1994; Moors *et al.*, 1994; Plumb *et al.*, 1997; Arthur & Pawliszyn, 1990; Zief & Kiser, 1990; MacDonald & Bouvier, 1995; Wells, 2003: Scheurer & Moore, 1992)

Since the 70's SPE has become a common and effective technique for extracting analytes from complex samples. Solid phase extraction is the very popular technique currently available for rapid and selective sample preparation. Many sample preparation methods today rely on solid-phase extractions, an advantage being that SPE is amenable to automation and parallel processing. SPE evolved to be a powerful tool for isolation and concentration of trace analysis in a variety of sample matrices. The versatility of SPE allows use of this technique for many purposes, such as purification and trace enrichment (Rawa *et al.*, 2003).

The objectives of SPE are to reduce the level of interferences, minimize the final sample volume to maximize analyte sensitivity and provide the analyte fraction in a solvent that is compatible with the analytical measurement techniques. As an added benefit, SPE serves as a filter to remove sample particulates.

The degree of enrichment achievable for a particular sample is dependent upon:

a. The selectivity of the bonded phase for the analyte
b. The relative strength of that interaction

SPE prepares multiple samples in parallel (typically 12-24) and uses relatively low quantities of solvents and the procedures can be readily automated. As the name implies the principle of SPE is an extraction technique similar to that of LLE, involving a partitioning of solutes between two phases. However, instead of two immiscible liquid phases, as in LLE, SPE involves partitioning between a liquid (sample matrix or solvent with analytes) and a solid (sorbent) phase (Table 1).

SPE is a more efficient separation process than LLE, easily obtains a higher recovery of analyte by employing a small plastic disposable column or cartridge, often the barrel of a medical syringe packed with 0.1 to 0.5 g of sorbent which is commonly RP material (C_{18}-silica). The components of interest may either preferentially adsorbed to the solid, or they may remain in the second non-solid phase. Once equilibrium has been reached, the two phases are physically separated by decanting, filtration, centrifugation or a similar process. If the desired analyte is adsorbed on the solid phase, they can be selectively desorbed by washing with an appropriate solvent. If the component of interest remains in a liquid phase, they can be recovered via concentration, evaporation and or recrystallization. When SPE is performed in this single step equilibrium batch mode, it will be similar to LLE, where the solid sorbent simply replaces one of the immiscible liquids. By passing a liquid or gas through the uniform solid bed, the liquid solid phase extraction technique becomes a form of column chromatography, now commonly called phase extraction that is governed by liquid chromatographic principle.

Polarity			Solvent	Miscible in Water?
Non-polar	Strong Reversed Phase	Weak Normal Phase	Hexane	No
↑↓	↑	↑↓	Isooctane	No
			Carbon tetrachloride	No
			Chloroform	No
			Methylene Chloride	No
			Tetrahydrofuran	Yes
			Diethly ether	No
			Ethyl acetate	Poorly
			Acetone	Yes
			Acetonitrile Yes	Yes
			Isopropanol	Yes
			Methanol	Yes
			Water	Yes
Polar	Weak Reversed Phase	Strong Normal Phase	Acetic Acid	Yes

Table 1. Characteristics of Solvents Commonly Used in SPE

Liquid samples are added to the cartridge and wash solvent is selected to either strongly retain or un-retain the analyte. Interferences are eluted or washed from the cartridge, even as the analyte is strongly retained, so as to minimize the presence of interferences in the final analyte fraction, the analyte is then eluted with a strong elution solvent, collected and either injected directly or evaporated to dryness followed by dilution with the HPLC mobile phase. Conversely, when interferences are strongly held in the cartridge, the analyte can be collected for the further treatment. Advantages of SPE over LLE include a more complete extraction of the analyte, more efficient separation of interferences from analyte, reduced organic solvent consumption, easier collection of the total analyte fraction, more convenient manual procedures, better removal of particulates and more easy automation.

5.2.1 Mechanism of Solid Phase Extraction process (Font *et al.*, 1993; Sabik *et al.*, 2000)

The ideal process is to create a digital chromatography system on the SPE cartridge such that the analyte is at first fully bound, then the interferences are completely eluted, and then the analyte is entirely eluted; it is either all on or all off the cartridge. The separation mechanism is the function due to the intermolecular interactions between analyte and the functional groups of the sorbent.

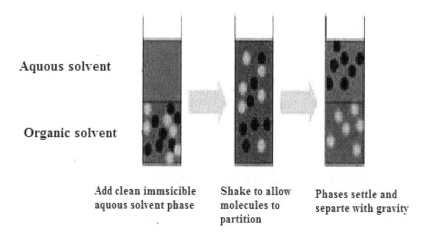

Aquous solvent

Organic solvent

| Add clean immsicible aquous solvent phase | Shake to allow molecules to partition | Phases settle and separte with gravity |

Fig. 2. Partitioning of drugs towards aqueous and organic solvent

The selection of an appropriate SPE extraction sorbent depends on understanding the mechanism(s) of interaction between the sorbent and analyte of interest. That understanding in turn depends on knowledge of the hydrophobic, polar and inorganic properties of both the solute and the sorbent. The most common retention mechanisms in SPE are based on van der Waals forces ("non-polar interactions"), hydrogen bonding, dipole-dipole forces ("polar" interactions) and cation-anion interactions ("ionic" interactions).

Each sorbent offers a unique mix of these properties which can be applied to a wide variety of extraction problems (Fig 2). Four general theory interactions exist (Yu *et al.*, 2004):

a.   **Reversed phase** involves a polar or moderately polar sample matrix (mobile phase) and a non-polar stationary phase. The analyte of interest is typically mid- to non- polar.

Retention occurs via non-polar interaction between carbon-hydrogen bonds of the analyte and carbon-hydrogen bonds of the sorbent function groups due to Van der Waals or dispersion forces.

The materials that are used as reversed phases are carbon-based media, polymer-based media, polymer-coated and bonded silica media.

Carbon-based media consist of graphitic, non-porous carbon with a high attraction for organic polar and non-polar compounds from both polar and non-polar matrices. Retention of analytes is based primarily on the analyte's structure, rather than on interactions of functional groups on the analyte with the sorbent surface.

Polymer-based sorbents are styrene/divivinylbenzene materials. It is used for retaining hydrophobic compounds which contain some hydrophilic functionality, especially aromatics.

Polymer-coated and bonded silica media is hydrophobic-bonded silica that is coated with a hydrophilic polymer. The pores in the polymer allow small, hydrophobic organic compounds of interest (e.g. drugs) to reach the bonded silica surface, while large interfering compounds (e.g. proteins) are shielded from the bonded silica by the polymer and are flushed through the SPE tube.

Several SPE materials, such as the alkyl- or aryl-bonded silicas (LC-18, ENVI-18, LC-8, ENVI-8, LC-4, and LC-Ph) are in the reversed phase category.

b. **Normal phase** involve a polar analyte, a mid- to non-polar matrix (e.g. acetone, chlorinated solvents and hexane) and a polar stationary phase. Retention of an analyte under normal phase conditions is primarily due to interactions between polar functional groups of the analyte and polar groups on the sorbent surface. Ex. Hydrogen bonding, dipole-dipole, induced dipole-dipole and pi-pi.

These phases can offer a highly selective extraction procedure capable of separating molecules with very similar structures. The main drawback is that the analyte must be loaded onto the sorbent in a relatively non-polar organic solvent such as hexane.

Polar-functionalized bonded silicas (e.g. LC-CN, LC-NH2, and LC-Diol), and polar adsorption media (LC-Si, LC-Florisil, ENVI-Florisil, and LC-Alumina) typically are used under normal phase conditions.

c. The **bonded silicas** have short alkyl chains with polar functional groups bonded to the surface. These silicas, because of their polar functional groups, are much more hydrophilic relatively to the bonded reversed phase silicas.

The bonded silicas LC-CN, LC-NH2, and LC-Diol - have short alkyl chains with polar functional groups bonded to the surface.

d. **Ion exchange SPE** can be used for compounds that are in a solution. Anionic (negatively charged) compounds can be isolated on an aliphatic quaternary amine group that is bonded to the silica surface. Cationic (positively charged) compounds are isolated by using the silica with aliphatic sulfonic acid groups that are bonded to the surface.

Biofluids can usually be applied directly to ion-exchange sorbents following dilution of the sample with water or a buffer, and possibly adjustment of pH. However, elution from strong ion-exchange sorbents can be a problem as high ionic strength or extremes of pH, may be required which may affect analyte stability or further processing of sample.

Anionic (negatively charged) compounds can be isolated on **LC-SAX** or **LC-NH2** bonded silica cartridges. Cationic (positively charged) compounds are isolated by using **LC-SCX** or **LC-WCX** bonded silica cartridges.

5.2.2 General properties of bonded silica sorbents

Although other materials are available ex. polymeric resins and alumina, the vast majority of SPE extractions are carried out by using bonded silica materials similar to those used in HPLC columns except the particle size and diameter. Bonded silica materials are rigid and do not shrink or swell like polystyrene-based resins in different solvents. Use of too high flow rate when processing cartridges may affect retention of analytes, particularly during the sample loading and elution steps. Potentially the capacity of the cartridge will be affected by all the components from the sample not only the analytes of interest.

5.2.3 Steps of Solid Phase Extraction

Generally following steps are followed for developing the method for extracting the analyte from plasma (Fig 3).

a. Pretreatment of sample - which includes dilution of sample or pH adjustment, filtration to avoid the blocking of the SPE cartridge and for better adsorption.
b. Conditioning of the cartridge - which is the main step in case of reverse phase SPE cartridges. Preconditioning is mainly done by solvent such as methanol, acetonitrile, isopropyl alcohol or tetrahydrofuran which is necessary to obtain reproducible result. Without this step, a highly aqueous solvent cannot penetrate the pores and wet the surface. Thus, only a small fraction of the surface area is available for interaction with the analyte. For the same reason, it is important not to let the cartridge dry out between the salvation step and addition of the sample.
c. Loading the sample - Sample size must be scaled to suit the size of the cartridge bed. A typical reverse phase cartridge may have capacity for up to 100 mg of very strongly retained substances.
d. Wash - very important step in case of the sample treatment by SPE. In this step a suitable solvent or water mixture is passed through SPE bed to remove the contaminants.
e. Elution of fraction - in this a suitable solvent or buffer is used to elute the analyte from the SPE bed for analysis.

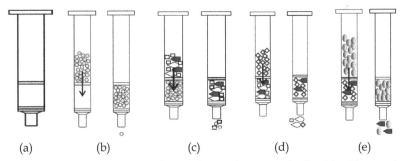

| (a) | (b) | (c) | (d) | (e) |

Fig. 3. Five steps of SPE: (a) selection of tube, (b) conditioning of tube, (c) addition of sample, (d) washing and (e) elution

5.2.4 A strategy for method development for plasma samples

-   Rationale
-   Practical consideration
-   Pretreatment of samples and cartridges
-   Screening sorbents and pH values
-   Optimizing the washing procedure
-   Optimizing the elution procedure
-   Final optimization and simplification
-   Strategies for removing persistent interferences

5.2.5 Developing SPE methods

In developing a SPE method, the properties of the isolate, sorbent and the matrix should be considered. The mechanisms of retention and the various secondary interactions may occur, and that a selective extraction from biofluid also requires the sorbent not to extract a large number of unknown compounds from the matrix. The best conditions are not easily predicted and the method needs to be developed experimentally in combination with the analytical method being used.

The following stages are recommended for method development.

a. Consider physic-chemical properties of analyte, nature of matrix and known chromatographic properties of analytes and the possible interaction with SPE sorbents.
b. Screen a range of cartridges (ex. C_{18}, C_8, C_2, Ph, CBA, SCX, SAX, PRS, NH_2, DEA, CH, Si and polymeric sorbent) under simple conditions (ex. from aqueous buffer solutions) looking for good retention of the analyte. If radiolabelled analyte is available this can conveniently be used to track analyte in such screening experiments. This experiment should not only identify likely cartridges for use in the assay, but should be used to try to confirm possible mechanism of interaction between the analyte and sorbent.
c. Select a more limited number of sorbents and examine conditions for loading/wash/elution (consider if pH control is needed, possible strength (% organic solvent) of wash solvents). Try the extraction from bio-fluid and select sorbent for final development.
d. Final development of extraction conditions and chromatographic analysis. Consider the robustness of the assay when finalizing extraction conditions. Ex. Do not select a wash solvent where a small change in composition could lead to elution of analyte from the cartridge.

The choice of different cartridges, manufactures and formats is becoming so extensive that it can appear almost overwhelming. It is generally better to build experience with a more limited set of sorbents, perhaps concentrating on cartridges from only one manufacturer. Also concentrate on investigating the use of cartridges with different functional groups (i.e. test C_{18}, C_8, C_2, Ph, CBA, SCX, SAX, PRS, NH_2, DEA, CH, Si, etc.), and those that use a contrasting mechanism to the analytical separation.

5.2.6 Characteristic features of SPE

-   Complete flexibility
-   Longer column lifetimes
-   Powerful contaminant removal
-   Greater recovery
-   Better reproducibility
-   More sensitivity

5.2.7 Advantages of SPE over LLE

-   In SPE by choosing selective adsorbent analyte can be driven completely by adsorption and desorption
-   In single stage LLE each extraction step equivalent to one chromatographic plate on the other hand by SPE in single step one can generate 10-50 plates

- Higher plate numbers in SPE leads to higher recoveries and purer of the analyte as compared to LLE
- For extracting more than one component from a mixture of component of different desorption solvent are required in case of SPE. To achieve similar results with LLE, one must perform several liquid extractions
- SPE is less time consuming and not tedious as compare to LLE

5.2.8 Limitation

- Depending on the nature of the analyte, SPE may not always be the method of choice, and liquid-liquid extraction may be a more viable solution.

6. Protein precipitation method (Backes, 2000; Wells, 2003)

This method is least one in bioanalytical. This is a very simple technique for extraction of the analyte from the matrix. If protein binding is suspected, then protein precipitation prior to sample extraction may be considered. Reagents to evaluate include perchloric, trichloroacetic and tungstic acids, and organic solvents such as acetonitrile or methanol. With all of these it is necessary to bear in mind the ability of the analytes and the matrix requirements of the extraction procedures. If protein binding is believed to be through a covalent linkage, then there is very little change of breaking it since this is the strongest of the intermolecular forces.

The main requirement for this technique is that the analyte should be freely soluble into reconstituting solvent. Preparation of sample through protein precipitation achieves separation by conversion of soluble proteins to an insoluble state by salting out or by addition of water miscible precipitation solvent or organic solvents such as acetone, ethanol, acetonitrile or methanol. Ideally, precipitation results in both concentration and purification, and is often used early in the sequence of downstream purification, reducing the volume and increasing the purity of the protein prior to any chromatography steps. In addition, precipitating agents can be chosen that provide a more stable product than found in the soluble form.

Proteins might stick to each other through one of three forces: electrostatic, hydrophobic, and van der Waals. The last one is difficult to distinguish from hydrophobic and operates over only in a very short range. Electrostatic forces operate at long range, but between like molecules are repulsive rather than attractive, so molecules have the same charge and repel each other.

Proteins are made insoluble by altering their surface properties, charge characteristics or changing the solvent characteristics; but changing the solvent characteristics being preferred. Greater the initial concentration of the desired protein, greater is the efficiency of precipitation; proteins are least soluble as its isoelectric point (pI) ranges from 4 - 10. The selection of a buffer at or near the pI of the protein is recommended. However some proteins may denature at their pI and above the pI. The solubility of a protein increases with the addition of salt and reaches a maximum after which there is a rapid linear decrease in solubility. There are several methods to reduce the solubility of proteins, which are ionic precipitation ex. ammonium sulphate, sodium chloride; metal ions ex. Cu^{+2}, Zn^{+2} and Fe^{+2};

non-ionic polymers ex. polyethylene glycol; organic solvents ex. ethanol, acetone; tannic acids, heparin, dextran sulphates, cationic polyelectrolytes ex. protamines; short chain fatty acids ex. caprylic acid; trichloroacetic, lecithins ex. concanavalin A and group specific dyes ex. procion blue. The use of temperature, pH or organic solvents can lead to denaturation and should be performed with care to minimize any decrease in yield or activity.

6.1 Type of protein precipitation

Salting out: Ammonium sulphate is the salt usually used for salting out, because of its high solubility and high ionic strength (which is proportional to the square of the charge on the ion, so that the ionic strength of 1M $(NH_4)_2SO_4$ is 3 times that of 1M NaCl). Neither ion associates much with proteins, which is good since such association usually destabilizes proteins. Its solubility changes little with temperature, it is cheap, and the density of even a concentrated solution is less than that of protein, so that protein can be centrifuged down from concentrated solutions.

Solvent Precipitation: When large amounts of a water-miscible solvent such as ethanol or acetone are added to a protein solution, proteins precipitate out. The conventional wisdom is that this is due to decrease of the dielectric constant, which would make interactions between charged groups on the surface of proteins stronger. Water miscible solvents associates with water much more strongly than do proteins, so that its real effect is to dehydrate protein surfaces, which then associate by van der Waals forces, at least if they are isoelectric or reasonably close to it. Removal of water molecules from around charged groups would also deshield them and allow charge interactions to occur more strongly, if there are areas of opposite charge on the surfaces of two proteins.

In practice, solvent precipitation is usually performed at low temperature. The condition for the protein is at 0°C and the solvent colder, -20°C in an ice-salt bath, because proteins tend to denature at higher temperatures though if sufficient control can be achieved and your protein is more stable than others, this can be selective and achieve greater purification.

Solvent precipitation can be done with polyethylene glycol at concentrations between 5 and 15%. It probably works the same way, by competing with the protein for water, but is less likely to inactivate the protein and does not require such low temperatures, but it tends to give an oily precipitate.

Commonly the sample is centrifuged at high speed for sufficient time, all the precipitated components of plasma will be settled at the bottom and clear supernatant liquid will be separated out. The obtained supernatant liquid can be injected directly into the HPLC or it can be evaporated and reconstituted with the mobile phase and further clean up of the sample can be carried out by using micro centrifuge at very high speed.

6.2 Advantage

- Now protein precipitation plates are available, able to remove the unwanted plasma proteins from plasma fluid samples prior to analysis
- Protein precipitation plates can be used in a wide range of aqueous and organic sample preparation including total drug analysis and sample preparation prior to HPLC or LC-MS/MS

- Protein precipitation plates are compatible with small volume of solvent
- Protein precipitation plate contains hydrophobic PTFE membrane as a prefilter removes the unwanted precipitated proteins prior to analysis
- Traditionally in this method plasma is mixed with protein precipitating agent and diluting solvent then the whole mixture is vertex, mixed, centrifuge and filter
- By using the new protein precipitate filter plate, precipitating solvent is added first followed by the plasma sample. This method does not require any mixing. Generally these plates are fitted to 96 well extraction plates. This new process showed 90% removal of plasma proteins when compare to the old method 60-65%

6.3 Disadvantage

- May increase the back pressure of the HPLC system
- Some components of plasma which are soluble in diluting solvent that bound to stationary phase permanently that will affect the column performance.

7. Solid Phase Microextraction (SPME) (Pawliszyn *et al.*, 1997; Pawliszyn, 1995, 1998, 1999, 2003; Wercinski 1999)

Miniaturization of sorbent technology and the concomitant decrease in solvent purchase, exposure and disposal has also taken a further giant step with the development of SPME. Solid-Phase Microextraction (SPME) is a very simple and efficient, solventless sample preparation method. SPME integrates sampling, extraction, concentration and sample introduction into a single solvent-free step. In this technique a fused silica fiber coated with polyacrylate, polydimethylsiloxane, carbowax or other modified bonded phase is placed in contact with a sample, exposed to the vapour above a sample (solid, liquid) or placed in the stream of a gaseous sample to isolate the analyte and concentrate analytes into a range of coating materials. After extraction, the fibers are transferred with the help of syringe like handling device to analytical instrument like gas chromatography (GC) and GC/mass spectrometry (GC/MS) for separation and quantification of the target analyte. In such a manner SPME has been used frequently to analyze volatile and semi-volatile compounds and may be used to purge and trap procedures. The method saves preparation time and disposal costs and can improve detection limits. The most interesting aspect of this technology involves the ability to use the exposed fiber as extraction and sample delivery device. Phases for SPME are available in a range of polarities and properties for analyses of volatile and semi-volatile compounds as well as application to extraction of analytes from liquid samples. One of the limitations of SPME is the low capacity of the fiber and the perturbation of equilibria that can occur in the presence of sample components or analytes at very high concentration versus those of lesser concentration. Dilution of the sample can overcome some of these problems but not all, as limits of detetion for trace analytes are compromised. Formaldehyde, Trition X-100, Phenylurea, pesticides and amphetamines are some of the analytes which are successfully extracted using SPME.

7.1 SPME basics (Vas & Vekey, 2004)

The concept of SPME may have been derived from the idea of an immersed GC capillary column. The SPME apparatus looks like modified syringe containing a fiber holder and a fiber assembly, the latter containing a 1–2 cm long retractable SPME fiber. The SPME fiber itself is a thin fused-silica optical fiber, coated with a thin polymer film conventionally used as a coating material in chromatography. In SPME the amount of extraction solvent is very small compared to the sample volume. As a result, exhaustive removal of analytes to the extracting phase does not occur; rather equilibrium is reached between the sample matrix and the extracting phase. In practical, the extracting phase is a polymeric organic phase commonly poly(dimethylsiloxane) and polyacrylate which is permanently attached to rod. The rod consists of an optical fiber made of fused silica, which is chemically inert. A polymer layer is used to protect the fiber against breakage. Poly(dimethylsiloxane) behaves as a liquid, which results in rapid extraction compared to polyacrylate, which is a solid. When the coated fiber is placed into an aqueous matrix the analyte is transferred from the matrix into the coating. The extraction is considered to be complete when the analyte has reached an equilibrium distribution between the matrix and fiber coating. The amount of extracted analyte is independent of the volume of the sample. Therefore, there is no need to collect a defined amount of sample prior to analysis. Thus, the fiber can be exposed directly to the ambient air, water, production stream, etc., and the amount of extracted analyte will correspond directly to its concentration in the matrix.

7.2 Principle modes of SPME

7.2.1 Direct extraction

In the direct-extraction mode, the coated fiber is inserted directly into the sample, and analytes are extracted directly from the sample matrix to the extraction phase. For gaseous samples, natural air convections and high diffusion coefficients are typically sufficient to facilitate rapid equilibration. For aqueous matrices, more efficient agitation techniques are required, such as forced flow, rapid fiber or vial movement, stirring or sonication.

7.2.2 Headspace configuration

In the headspace mode, the vapor above the bulk matrix is sampled. Thus, analytes must be relatively volatile in order to be transported from the bulk matrix to the fiber coating.

7.2.3 Advantages

Sampling of the headspace protects the fiber coating from damage by hostile matrices, such as those at very high or low pH, or those with large molecules, such as proteins which tend to foul the coating.

7.2.4 Types of extraction in SPME

-    Fiber extraction
-    In-tube extraction
-    Stir BAR sorptive extraction (SBSE)

7.2.5 Advantage of SPME

-    In SPME volatile, semi-volatile and non-volatile organic and inorganic analytes can be used for analyzed

- During desorbtion of the analyte, the polymeric phase is cleaned and therefore ready for reuse. The absence of solvent in SPME is an important feature, as it is not only environmentally friendly but makes the separation faster
- Important feature of SPME is its small size, which is convenient for designing portable devices for field work
- Solvent free environment, fast extraction, convenient automation and easy hyphenation with analytical instrument

8. Matrix Solid-Phase Dispersion (MSPD) (Barker *et al.*, 1989, 1993; Walker *et al.*, 1993)

Matrix solid phase dispersion is a sample preparation technique for use with solid sample matrices. MSPD is a microscale extraction technique, typically using less than 1g of sample and low volumes of solvents. It has been estimated to reduce solvent use by up to 98% and sample turnaround time by 90%.

Conventional extraction of organic analytes from tissue usually begins with a homogenization of a small amount of sample tissue with bulk bonded silica based sorbent in a pestle and mortar. The mechanical shearing forces produced by the grinding process disrupt the structure of the tissue, dispersing the sample over the surface of the support sorbent by hydrophilic and hydrophobic interaction which produces the mixture to become semi-dry and free-flowing, and a homogenous blend of sample. The bound solvent in the sorbent will aid complete sample disruption during the sample blending process and the sample disperses over the surface of the bonded phase-support material to provide a new mixed phase for isolating analytes from various sample matrices. The blend is then transferred into a pre-fitted SPE cartridge and elution of interference compounds and analytes of interest. This technique has recently been applied, using acid alumina, to extract the organic analyte. However, MSPD procedure needs longer analytical time generally and its limit of determination (LOD) is limited.

In method development using MSPD sorbents, the following points are to be considered

- Sample pre-treatment
- Interference elution
- Analyte elution
- Sample clean up

9. Supercritical fluid extraction (Mohamed & Mansorri, 2002; Antero, 2000)

Supercritical fluid extraction is becoming a very popular technique for the removal of non-polar to moderately polar analytes from solid matrices. Supercritical fluids (SCFs) are increasingly replacing the organic solvents that are used in industrial purification and recrystallization operations because of regulatory and environmental pressures on hydrocarbon and ozone-depleting emissions. SCF processes eliminate the use of organic solvents, so it has attracted much attention in the industrial sectors like pharmaceuticals, medical products and nutraceuticals. Pharmaceutical chemists have found SCFs useful for extraction of drug materials from tablet formulation and tissue samples.

Supercritical fluids exhibit a liquid-like density, while their viscosity and diffusivity remain between gas and liquid values. The recovery of a supercritical solvent after extraction can be carried out relatively simply by reducing the pressure and evaporating the solvent. Above the critical temperature the liquid phase will not appear even the pressure is increased. The compressibility of a supercritical fluid just above the critical temperature is large compared to the compressibility of ordinary liquids. A small change in the pressure or temperature of a supercritical fluid generally causes a large change in its density. The unique property of a supercritical fluid is that its solvating power can be tuned by changing either its temperature or pressure. The density of a supercritical fluid increases with pressure and becomes liquid-like, the viscosity and diffusivity remain between liquid-like and gas-like values. Additionally, supercritical fluids exhibit almost zero surface tension, which allows facile penetration into microporous materials leads to more efficient extraction of the analyte than the organic solvents. Carbon dioxide is a relatively good supercritical solvent will dissolve many relatively volatile polar compounds. In the presence of small amounts of polar co-solvents like water and short-chain alcohols to the bulk, the carbon dioxide gas can enhance the solubility of polar, non-volatile solutes in supercritical carbon dioxide. Supercritical fluids can be used to extract analytes from samples.

The main advantages of using supercritical fluids for extractions is that they are inexpensive, extract the analytes faster and more environmental friendly than organic solvents. For these reasons supercritical fluid CO_2 is the reagent widely used as the supercritical solvent.

9.1 Advantages

- SCFs have solvating powers similar to organic solvents, with higher diffusivity, lower viscosity and lower surface tension
- The solvating power can be changed by changing the pressure or temperature for effective extraction
- Separation of analytes from solvent is fast and easy
- Polarity can be changed by using co-solvent leads to have more selective separation of the analyte
- Products are free from residual solvents
- SCFs are generally cheap, simple and safe
- Disposal costs are less
- SCF fluids can be recycled

10. Column switching (Falco *et al.*, 1993, 1994; Henion *et al.*, 1998; Lee & Kim 2011)

Column switching techniques afford an interesting and creative form of sample preparation. This approach depends on the selectivity of appropriately chosen HPLC stationary phase to retain and separate the analyte of interest while allowing unretained components to be eliminated from the column. The benefits of this technique include total automation and quantitative transfer of targeted compounds within the column switching system. In the heart cut mode a narrow retention time region containing a desired components is cut from the chromatogram and transferred onto another HPLC column for further separation. In this instance, quantitative transfer of the components without adsorptive or degradative losses can be assured.

10.1 Advantages

It is capable of having increased selectivity by the judicious choice of two or more HPLC stationary phases. A limitation of column switching system is that sample throughout will likely not be as high as for other sample clean up methods. A second limitation of the column switching approaches includes restricted sample enrichment because of the limited amount of original, untreated, crude sample that can be loaded onto the first column of the HPLC separation. These limitations can be overcome by combining online SPE and a column switching system. In this method the recovery is more but very expensive one.

11. Future directions

Bio-equivalency is an important one for the Pharmaceutical Formulations especially in the Pharmaceutical Regulatory Market. The availability of methodology for the extraction procedure of the interested analyte coupled along with the analytical method for the quantification of the interested analyte in GC-MS/MS or LC-MS/MS will greatly facilitate future studies to rigorously establish a bio-equivalency of the Pharmaceutical formulations in the Pharmaceutical regulatory market.

12. Conclusion

The subject area of bioanalysis is a sub-discipline of analytical chemistry which covers a very broad range of assays like quantitative measurement of drugs and their metabolites in various biological systems like whole blood, plasma, serum, urine, faeces and tissues. The biological samples cannot be analyzed directly into the analytical system for the quantification of the interested analyte. Estimation of the analyte in the biological matrix can be done after the isolation of the interested analyte from the interferences in the biological sample. Isolation and estimation of the analyte is based on the sample preparation and extraction of the analyte from the biological matrix. The process adopted for the isolation of the interested analyte like sample preparation procedure and isolation steps must ensure the stability of the analyte until the completion of the analytical estimation. Bioanalysis in the pharmaceutical industry is to provide a quantitative measure of the active drug and/or its metabolite(s) for the purpose of pharmacokinetics, toxicokinetics, bioequivalence and exposure–response (pharmacokinetics/pharmacodynamics) studies. In this chapter, various extraction methods of the analyte from the biological matrix have been described with its advantages and disadvantages.

13. References

Allanson J.P. Biddlecombe RA, Jones AE, Pleasance S. (1996). The use of automated solid phase extraction in the "96 well" format for high throughput bioanalysis using liquid chromatography coupled to tandem mass spectroscopy. *Rapid Communications in Mass Spectroscopy*, 10, 811-816.

Antero L. (2000). *Supercritical fluid extraction of organic compounds from solids and aqueous solution.* Ph.D. Dissertation, Helsinki University of Technology, (Espoo, Finland).

Arthur CL, Pawliszyn J. (1990). Solid phase microextraction with thermal desorption using fused silica optical fibers. *Analytical Chemistry,* 62, 2145-2148.

Backes D. (2000) Strategy for the development of quantitative analytical procedures, In: *Principles and Practice of Bioanalysis,* Venn RF, 342-358, Taylor & Francis, NY.

Barker S A, Long AR, Short CR. (1989). Isolation of drug residues from tissues by solid phase dispersion, *Journal of Chromatography,* 475, 353–361.

Barker SA, Long AR, Hines ME. (1993). Disruption and fractionation of biological materials by matrix solid phase dispersion. *Journal of chromatography,* 629, 23-24.

Barrow GM. (1996). *Physical Chemistry,* 6th Ed. NY, McGraw-Hill (ISE).

Chang R. (1977). *Physical chemistry with applications to biological systems,* Macmillan Publishing, NY.

Christian GD., O'Reilly JE. (1986). *Instrumental Analysis,* 2nd Ed. Newton, MA: Allyn & Bacon.

Christians U, Sewing KF. (1989). Whole Blood Sample Clean-Up for Chromatographic Analysis, In: *Selective sample handling and detection in high performance liquid chromatography,* Volume 39, part 2, Frei RW, Zech K, 82-132, Elsevier, Amsterdam.

Destefano AJ, Takigiku R. (2004). Bioanlaysis in a regulated environment, In: *Pharmacokinetics in Drug development: regulatory and Development Paradigms, Volume 2,* Bonate PL, Howard DR, 105-125, AAPS, Arlington, VA.

Falco PC, Hernadez RH, Cabezav AS. (1993). Column-switching techniques for high-performance liquid chromatography of drugs in biological samples. *Journal of Chromatography: Biomedical Applications,* 619(2), 177-190.

Falco PC, Hernandez RH, Cabeza AS. (1994). Column-switching techniques for screening of diuretics and probenecid in urine samples. *Analytical Chemistry,* 66 (2), 244–248.

Font G, Manes J, Moltoj C. (1993). Solid-phase extraction in multi-residue pesticide analysis of water. *Journal of Chromatography A,* 642, 135-161.

Fried K, Wainer IW. (1997). Column-Switching techniques in the biomedical analsis if stereoisomeric drugs: why, how and when. *Journal of Chromatography B: Biomedical Sciences and Applications,* 689(1), 91-104.

Gy PM. (1982). *Sampling of particulate materials Theory and Practice,* Elsevier, Amsterdam.

Harris DC. (1994). *Quantitative chemical Analysis,* 4th Ed. NY, WH. Freeman.

Henion J, Brewer E, Rule G. (1998). Sample Preparation for LC/MS/MS: Analyzing Biological and Environmental Samples. *Analytical Chemistry News & Features,* 650A-656 A.

Ho TS, Bjergaard SP, Rasmussen KE. (2002). Liquid-phase microextraction of protein bound drugs under non-equilibrium conditions. *Analyst,* 127, 608-13.

Horne KC. (1985). Analytichem International Sorbent Extraction Technology Handbook, Analytichem International Inc. January 14-15, 24201 Frampton Avenue, Harbor City, CA 90710 USA.

Ingwersen SH. (1993). Combined reverse phase and anion exchange solid phase extraction for the assay of the neuroprotectant NBQX in human plasma and Urine. *Biomedical Chromatography,* 7(3), 166-171.

James C. (2000). Solid-Phase Extraction, In: *Principles and Practice of Bioanalysis,* Venn RF, 28-43, Taylor & Francis, NY.

Keith LH. (1990). Environmental sampling: a summary, *Environmental Science and Technology,* 24 (5), 610–617.

Krishnan TR, Ibraham I. (1994). Solid phase extraction technique for the analysis of biological samples. *Journal of Pharmaceutical and Biomedical Analysis,* 12, 287-94.

Lee HJ, Kim KB. (2011). Application of Column-Switching Methods in HPLC for Evaluating Pharmacokinetic Parameters, In: *Advances in Chromatography,* Volume 49, Grushka E, Grinberg N, 291-340. CRC Press, USA.

MacDonald PD, Bouvier ESP. (1995). *Solid phase extraction application guide and bibliography,* Milford, MA, Water.

Majors RE, Fogelman KD. (1993).The Integration of Automated Sample Preparation with Analysis in Gas Chromatography, *American Laboratory,* 25(2): 40W-40GG.

McDowall RD, Doyle E, Murkitt GS, Picot VS. (1989). Sample preparation for the HPLC analysis of drugs in biological fluids. *Journal of Pharmaceutical and Biomedical Analysis, 7*(9), 1087-1096.

Mohamed RS, Mansoori G. (2002). *The use of supercritical fluid extraction technology in food processing.* The World Market Research Centre, London, UK.

Moore WJ. (1972). *Physical Chemistry,* 5th Ed. Harlow, Longman.

Moors M, Steenssens B, Tielemans I, Massart DL. (1994). Solid phase extraction of small drugs on apolar and ion exchange silica bonded phases: towards the development of a general strategy. *Journal of Pharmaceutical and Biomedical Analysis,* 12: 463-81.

Pawliszyn J. (1995). New directions in sample preparation for analysis of organic compounds. *Trends in Analytical Chemistry,* 14, 113-122.

Pawliszyn J, Pawliszyn B, Pawliszyn M. (1997). Solid Phase microextraction. *The chemical Educator,* 2(4), DOI 10.1333/s00897970137a.

Pawliszyn J. (1998). Solid phase microextraction Theory and Practice, Wiley-VCH. Inc. NY.

Pawliszyn J. (1999). Applications of solid phase microextraction. Royal Society of Chemistry, Cambridge.

Pawliszyn J. (2003). Sample Preparation: Quo Vadis?. *Analytical Chemistry,* 75, 2543-2558.

Plumb RS, Gray RD, Jones CM. (1997). Use of reduced sorbent bed and disk membrane solid phase extraction for the analysis of pharmaceutical compounds in biological fluids, with applications in the 96-well format. *Journal of Chromatography B: Biomedical Application,* 694, 123-133.

Sabikh, Jeannot R, Rondeaub. (2000). Multiresidue methods using solid phase extraction techniques for monitoring priority pesticides, including triazines and degradation products, in ground water. *Journal of Chromatography A,* 885, 217-236.

Scheurer J, Moore CM. (1992). Solid Phase extraction of drugs form biological tissue. A Review. *Journal of Analytical Toxicology,* 16, 264-269.

Simmonds RJ, James CA, Wood SA. (1994). A rational approach to the development of solid phase extraction methods for drugs in biological matrices, In: *Sample preparation for biomedical and environmental analysis,* Stevenson D, Wilson ID, 79, Plenum Press, NY.

Smith R, James GV. (1981). *The sampling of bulk materials,* The royal society of chemistry, London.

Vas G, Vekey K. (2004). Solid-phase microextraction: a powerful sample preparation tool prior to mass spectrometric analysis. *Journal of Mass Spectrometry,* 39, 233-254.

Walker CC, Lott HM, Barker SA. (1993). Matrix solid-phase dispersion extraction and the analysis of drugs and environmental pollutants in aquatic species. *Journal of chromatography*, 642, 225-242.

Watt AP, Morrison D, Evans DC. (2000). Approaches to higher throughput pharmacokinetics in drug discovery. *Drug Discovery Today*, 5(1), 17-24.

Wells DA. (2003). Protein precipitation: High throughput techniques and strategies for method development, In: *High Throughput Bioanalytical Sample Preparation - Methods and Automation Strategies*, Wells DA, 199-254, Elsevier, Amsterdam.

Wells DA. (2003). Protein precipitation: Automation strategies, In: *High Throughput Bioanalytical Sample Preparation - Methods and Automation Strategies*, Wells DA, 255-276, Elsevier, Amsterdam.

Wells DA. (2003). Liquid Liquid Extraction: High throughput techniques, In: *High Throughput Bioanalytical Sample Preparation - Methods and Automation Strategies*, Wells DA, 277-306, Elsevier, Amsterdam.

Wells DA. (2003). Liquid Liquid Extraction: Strategies for method development and optimization, In: *High Throughput Bioanalytical Sample Preparation - Methods and Automation Strategies*, Wells DA, 307-326, Elsevier, Amsterdam.

Wells DA. (2003). Liquid Liquid Extraction: Automation strategies, In: *High Throughput Bioanalytical Sample Preparation - Methods and Automation Strategies*, Wells DA, 327-360, Elsevier, Amsterdam.

Wells DA. (2003). Solid Phase Extraction: High throughput techniques, In: *High Throughput Bioanalytical Sample Preparation - Methods and Automation Strategies*, Wells DA, 361-432, Elsevier, Amsterdam.

Wells DA. (2003). Solid Phase Extraction: Strategies for method development and optimization, In: *High Throughput Bioanalytical Sample Preparation - Methods and Automation Strategies*, Wells DA, 433-484, Elsevier, Amsterdam.

Wells DA. (2003). Solid Phase Extraction: Automation strategies, In: *High Throughput Bioanalytical Sample Preparation - Methods and Automation Strategies*, Wells DA, 485-504, Elsevier, Amsterdam.

Wercinski SCS. (1999). *Solid Phase microextraction-A Practical guide*. Marcel Dekker: NY.

Williams J. (1999). The assessment of therapeutic compliance based upon the analysis of drug concentration in hair, In: *Drug Testing Technology Assessment of field application*, Mieczkowski T, 1-32, CRC Press, Boca Raton, London, NY, Washington DC.

Wiltshire H. (2000). Strategy for the development of quantitative analytical procedures. In: *Principles and Practice of Bioanalysis*, Venn RF, 1-27, Taylor & Francis, NY.

Yu J, Wu C, Xing J. (2004). Development of new solid-phase microextraction fibers by sol-gel technology for the determination of organophosphorus pesticide multiresidues in food. *Journal of Chromatography A*, 1036, 101-111.

Zief M, Kiser R. (1990). An overview of solid phase extraction for sample preparation. *American Laboratory*, 22(1), 70-83.

Zwir Ferenc A, Biziuk M. (2006). Solid Phase extraction technique Trends, opportunities and Applications. *Polish Journal of Environmental Studies*, 15(5), 677-690.

Part 2

E-Health and Educational Aspects of Bioengineering

6

Psychomagnetobiology

José María De la Roca Chiapas[1,2]
[1]Universidad de Guanajuato, División de Salud, Departamento de Psicología
[2]Organización Filosófica Nueva Acrópolis México
México

1. Introduction

The magnetobiology is known in the scientific world as being the link between magnetic fields and the behavior of particular animal species (some migratory birds, dolphins, and ants)that navigate through the magnetic currents that are generated by the Earth, and have a special sense of orientation, which has been associated with genetic memory. This arises from an increased concentration of iron in the head, or from the presence of a chemical that is activated by the presence of magnetic fields, including those generated by human construction, allowing the assumption that such magnetic fields are observable by some birds (Hevers, 2010; Mouritsen, 2005; Wiltschko, 2007; Gegear, 2010; Oliveira, 2010).

In these species, the relationship between the animal and the strength of the magnetic field allows them to remember their previous travels, and this has a wide application, because the assumptions regarding changes in these magnetic fields suggest that magnetic fields can be linked to changes in migration flows (Lohmann, 2007), reproduction, or to the extinction of a species. This relationship between animals and the properties of the Earth is studied as a part of magnetobiology (Valentinuzzi, 2004; Temurvants & Shekhotkin, 2000).

On the other hand, psychobiology, or biological psychology, studies the physiological mechanisms and evolutionary relationship between the development of the body and brain with emotion, thought, and behavior. This has been amply demonstrated in studies of psychopathology and health psychology (Sánchez-Martín, 2011). For cases such as schizophrenia (Ritsner, 2006) or depression (Gartside, 2003), it was found that some hormones and neurotransmitters that can induce some personality traits (or variations of these neural substrates that may cause mood disorders) also influence those biological emotional states of being and pleasure (Zak & Fakhar, 2006) that are associated with an increase in particular neurotransmitters that also regulate or relieve stress (Heinrichs, 2003).

There are also known specific psychoneuroimmunological responses and reactions triggering autoimmune diseases, such as those caused by severe stress or a high-intensity emotional state. For example, some patients who are notified of a terminal illness have emotions that may trigger symptoms associated with these effects, which may activate the genes of some diseases such as lupus (Schattner, 2010), heart attacks, or cancer (Stürmer, 2006).

This chapter discusses the existence of a relationship between magnetism and the physiological responses of health and disease associated with emotional states, which is not a new concept, as many traditional medicines have assumed this relationship exists, in civilizations from different geographic locations, such as China (Rosch, 2009; Gulmen, 2004), Egypt, and many Mexican cultures (Bocanegra-García, 2009). Today it is known as magnetotherapy (Zyss, 2008).

This idea is strengthened by the somatic marker proposal, which poses that as there is a biological anticipatory response to the decisions we make, and that it is reinforced by stimuli from external perception, the consciousness and emotional states that can guide accompanying positive actions. For example, in the migratory birds discussed above, a strengthening of the biological response to the magnetic field intensity is expected, as the migration of the species depends on a successful response (Martínez-Selva, 2006).

We proposed to test the hypothesis of relationships among the body, mind, and consciousness as a way of showing what the above cultures, doctors, and healers have supposed for millennia: that a relationship between consciousness, mind, emotion, and the body exists, i.e., there is a two-way communication that can transform the emotions, perception, and biology from magnetic and electromagnetic fields that can influence biology.

An example of how changing a body's electromagnetic currents can alter the emotional state can be found in the electrical stimulator, which has proven effective for people with depression (Baeken & De Raedt, 2011), auditory hallucinations (Freitas, 2011), and chronic pain, and is a promising treatment in this area.

Another example is biological neurofeedback (Dias & van Deusen, 2011) which can change depressed or obese health patterns under transcranial magnetic stimulation or deep transcranial magnetic stimulation (Rosenberg, 2011).

This hypothesis raises the question concerning which current technology we have to measure magnetic fields in human beings and what the future options might be.

Knowing technological developments that can measure magnetic fields can help in biomedicine, as they could be used in early diagnostic techniques based on the measurement of magnetic changes that can anticipate changes in the biology, immunology and endocrinology of the body. Among the current applications available for this purpose, we note the Gas Discharge Visualization (Korotkov, 2010), Magnetic Scanner (Pacheco, 2010), and SQUIDS27 (Bryant, 2011) techniques.

Psychomagnetobiology is a new area of interdisciplinary research that includes psychology, medicine, biological engineering, and physics, and the following provides an overview of how one can use magnetism and electricity in diagnosis and treatment in both medicine and psychology(see figure 1).

2. Technologies that use electricity and magnetism for diagnosis and therapy

2.1 Microcurrent stimulation and cranial electrotherapy

Cranial Electrotherapy Stimulation (CES) is an experimental psychiatric treatment that applies a small, pulsed electric current through a patient's head (Smith, 2008). It has been shown to have beneficial effects forsome conditions, such as anxiety, depression, insomnia, and stress (Klawansky, 1995).

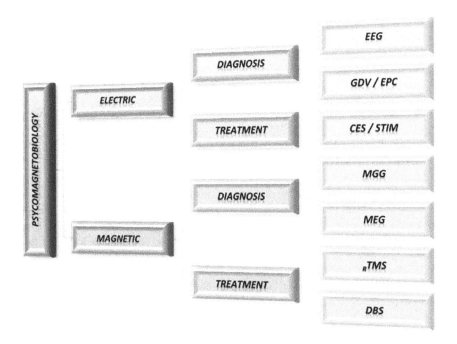

Fig. 1. Psychomagnetobiology

2.1.1 Early treatments

Electricity has been used in medicine since the ancient Greeks, who used eels for conditions such as rheumatism, gout, and pain. Unfortunately, because of a lack of legal controls, many charlatans took advantage of this situation, and captured the market. This led the Carnegie Foundation to carry out studies, and in 1910,it rejected electrotherapy as an acceptable medical method. The impact of this study has influenced physicians to this day (Fariña, 2010).

In the 1960s, Leduc and Rouxeau in France, initiated studies on the use of electricity in medicine using very low currents in the brain to induce sleep, and thus, introduced the term "electro-sleep". Note that sleep is induced after 4 to 5 hours using CES, and it is still unknown if it can induce rapid and consistent sleep (Fariña, 2010).

Electricity received another boost in its application to the brain when, in 1965,Melzack in Canada and Wall in Britain discovered a relationship between pain and the central nervous system, and tested the role of electricity in controlling pain (Melzack& Wall, 1965).

In 1967, when electrical stimulation was used to determine who really needed surgery to relieve pain, it was found that patients improved and responded therapeutically to a procedure that involved a distraction. This procedure was Transcuteneal Electrical Neuro Stimulation (TENS), a precursor of Microcurrent Electrical Therapy (MET). This method is different from MET, because TENS initially reduces pain by producing an alternative pain, and, therefore, is known as a "counter-irritant" (Debock, 2000). As the body adapts, the current is increased, even reaching levels not recommended as tolerable, or not required to stop the use of therapy. Using this method, the results are short term and recurrences are common. On the other hand, MET does not reach intolerable levels and the results are usually long lasting and positive (Fariña, 2010).

There are currently many research studies into CES in humans (Gunther & Phillips, 2010) and so far, 29 experimental animal studies have shown mostly positive results, with several using double-blind and placebo studies. No lasting adverse effects have been reported (Fariña, 2010).

CES has been approved by the US Food and Drug Administration (FDA) for use since 1979, and was reevaluated in 1986. Examples where the utility of CES/MET has been investigated are (Schmitt et al., 1986): discrete brain damage (reduction in pain and anxiety and increase in IQ); substance withdrawal syndrome (decrease in anxiety and increase in IQ), paraplegia and quadriplegia (decreased spasticity); cerebral paralysis (decrease in primitive reflexes); prisoners (reduced aggression); hypnotherapy (increases the speed and depth of the induction and reduces resistance to hypnotism); anesthesia (increases the effect of anesthesia by 37% and reduces postoperative pain); fibromyalgia (reduces pain and improves the quality of life) (Kirsch, 2006); and headaches (reduces many types of headache, such as migraine, tension, and headaches resistant to treatment where the patient has fibromyalgia and cancer) (Cork et al., 2004).

2.1.2 Mechanism of action

Studies have shown that, despite very low currents in the microampere range, 42% to 46% of the current passes through the cranial bone and enters the brain, focusing on the limbic system. It has been shown that this leads to an increase in the neurotransmitters dopamine, serotonin, and norepinephrine in the brain. The waveform used is as important as the current intensity, placement of electrodes, and exposure time (Kirsch, 2006).

2.1.3 Waveform

What really goes through the brain are unique waves that move the electrons and influence their frequency. The term used for this group of frequencies is "harmonic resonance", and using the above technology, it has been determined that the EEG changes to a more coherent resonance form, as observed in non-stressful conditions. This is comparable with results observed in deep meditation (Kirsch, 2006).

2.1.4 Microcurrent effects

Normalization of the electrical impulses reinforces the activity of the heart and blood circulation, andmicrovascular vessels are promoted; thus, cells are activated. The immune system's capacity increases, the rate of recovery doubles, healing ofwounds is promoted, and the sterilization effect suppresses bacterial growth. This technique is good for the treatment of patients with diseases associated with the feet, such as chronic pain, and reduces muscle fatigue (Kulkarni& Smith, 2000).(see figure)

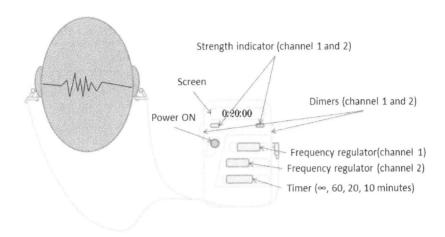

Stim diagram. The kit contains two channels, both with separate indicators and regulators. Dimmers are responsible for leveling the strength of the shock, while another regulator level the frequency. A timer is responsible for limiting the discharge time, shown on the screen. The voltage is transmitted to the patient who experiences the shock with the chosen parameters.

2.2 Gas discharge visualization and electrophotonic camera

Gas Discharge Visualization (GDV) is a technique based on the collection of signals from the body after electrical stimulation. The method is fully noninvasive, as it does not alter the physical and mental integrity of the patient.(see figure)

The functional basis of the GDV technique is the "crown effect" that occurs after passing a voltage through a conductor susceptible to discharge with the surrounding environment, creating a colorful halo around a material, based on the conducting components and the amount of volatile compounds in the environment. Using GDV, the medium is maintained at a constant voltage by a dielectric plate, and in the case of the body, the conductor is the crown of the system.

The variation of the intensity and wavelength of each signal generated can be measured according to the processes occurring within the body, and mainly the physiological state of the patient.

While it is not known exactly how this method amplifies the body's natural electromagnetic field, or how the patient's information is collected, the "entropy" is already established as a diagnostic method in the fields of medicine and psychology, with outstanding results.

GDV cameras are certified in Russia as medical instrumentation, and are freely used in hospitals and medical centers, while Europe is working on certification, and the United States is still developing the necessary research.

2.2.1 History

In 1939, Kirlian observed the formation of a halo of light after an electric shock on an X-ray machine. Researchers discovered that the phenomenon that bears his name (the Kirlian effect) was also observed in more than 30 patients, and his invention was supported in the USSR, where it was classified as "secret". In 1996, the Russian physicist Konstantin Korotkov at the University of St. Petersburg improved the Kirlian camera by adding a PC to allow processing and quantification of the data signals from the halo of light generated.

From the research of Mandel in Germany, Korotkov developed a methodology for medical diagnosis using fingers that generated a signal that could be used to diagnose the entire body's physiology.

The images obtained are known as beograms, GDVgrams or bioelectrograms.

2.2.2 Operation

The object to be scanned (e.g., a finger or plant) is placed on the surface of a glass cover that has an electroconductive layer underneath that generates luminescence by passing a discharge from the object to the conductive material. A CCD camera captures the halo of light around the object, and the signal is sent to a PC for processing and analysis (Korotkov & Popechitelev, 2002).

The computer-controlled parameters of the equipment are: pulse width (5.0ms), frequency (11.0 to 3.0kHz), amplitude voltage (1000.0 to 4000.0V), peak pulse power consumption (80watts), pulse current (1 mA), stability (0.1%), and the resolution of the CCDcamera (800 ×600pixels).

2.2.3 Applications

This technique has been used to monitor and diagnose disease. In the case of asthma, it can detect the developmentof the disease even before symptoms occur, allowing for early treatment. It can also monitor body functions during treatment and rehabilitation, providing a direct observation of drug efficacy and side effects (Alexandrova et al., 2003).

In cancer studies, statistical analysis shows differences in the beograms between healthy people and those with breast or lung cancer (Gagua et al., 2005), and during follow-up to radiation therapy, beograms show the effect on the physiological state of patients, depending on the entropic energy, which is correlated to the functional state of the organ (Gedevanishvili et al., 2004).

A study of subjects at risk of foot amputation from diabetes shows that an increase in the functional reserve of energy is a favorable sign for the recovery of the patient, a constant factor in chronic diseases (Olalde et al., 2004).

After operation, GDV can also detect and monitor the status of preoperative anxiety in patients, correlating low functional energy reserves with the Spielberger–Khaninscale (Polushin et al., 2004).

In cellular metabolism, GDV studies are correlated with metabolic processes and humoral regulatory processes at the level of the reflex nervous system. The photoelectric emission increases with the parameters of stress tolerance and decreases with a low energy index (Bundzenetal., 2003).

In the examination of blood using GDV, specifically for the electrophoretic mobility of erythrocytes, it has been used to characterize the membrane surface charge, an important factor in a wide variety of diseases that are mainly genetic in nature (Gertsenhtein & Panin, 2008).

Within psychology, there is a significant relationship between GDV beograms and the state of anxiety ,and to a lesser extent, a relationship to neuroticism. There was also a strong relationship observed with the degree of openness and empathy of a patient towards healthy subjects and athletes (Dobson & O'Keeffe, 2007; Polushinetal., 2003; Polushinetal., 2004).

Low levels of entropy in a beogram are correlated with acute stress (O'Keeffe, 2006). The GDV method can detect the influence of odorants in humans, which could be used to record the influence of environment on the psycho-emotional state (Priyatkinetal., 2006). GDV methodology can evaluate the overall improvement of an emotional state and eliminate nervous excitement during short-term rehabilitation (Sergeev & Pisareva, 2004).

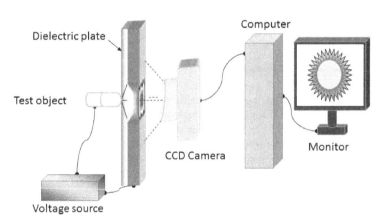

EPC Diagram. The "testing object" is placed on the "dielectric plate." Under the influence of a "voltage source", the object generates a luminous halo, which is captured by the "CCD camera" and sent to a "computer" for processing the signal.

2.3 Transcranial magnetic stimulation

Transcranial Magnetic Stimulation (TMS) is a noninvasive method that can depolarize or hyperpolarize neurons in the brain. TMS uses electromagnetic induction to induce weak electric currents in either specific or general parts of the brain.

It has been described as selective depolarization of the neurons in the neocortex or cerebral cortex, located between 1.5 and 2 cm below the cranial bone using magnetic pulses with specific intensity, either single or repetitive. This latter method is known as repetitive Transcranial Magnetic Stimulation or rTMS (Rothetal., 1994).

It is one of the latest tools in neuroscience that have been incorporated in both studies and research as therapeutic methods for the treatment of various diseases and neuropsychiatric disorders, among which are: depression, anxiety, attention deficit, hyperactivity, autism, tinnitus or unusual noise in the ear(s), post traumatic stress, phantom pain in people who have suffered limb amputation or central nervous system injuries, migraine headaches, decreased libido, some cases of schizophrenia and epilepsy, sleep disorders, obsessive–compulsive disorder, and bipolar disorder (Fitzgerald et al., 2006).

It is known to have neuro protective effects that help, at least temporarily, people affected by degenerative neurological diseases, such as multiple sclerosis, Parkinson's disease, and Alzheimer's disease, and impacts very favorably in the modulation of brain plasticity, which refers to the brain's ability to renew or reconnect neural circuits, and thus, acquire new skills and abilities and preserve memory (McDonald et al., 2011).

TMS uses a magnetic field oriented orthogonally to the plane of the coil. The magnetic field passes unimpeded through the skin and cranium, inducing a current in opposite directions in the brain to activate nearby nerve cells in the same way as currents applied directly to the cortical surface (Cacioppo et al., 2007).

The course of the field is difficult to predict, because the brain has an irregular shape and magnetic field lines are not distributed uniformly throughout the tissues or brain surface. Usually, the pulse only penetrates no more than 5 cm into the brain (Riehl, 2008). Depolarization from electromagnetic induction was originally discovered by Faraday (NIMH, 2009).

2.3.1 Clinical applications

From a therapeutic perspective, there are already a large number of studies that have shown that there are two sides to transcranial magnetic stimulation. Both TMS and rTMS have the great virtues of being harmless, but not innocuous, i.e., they are effective and can be classified as being safe. However, various measures need to be taken to ensure this (Rossini & Rossi, 2007). These techniques have been tested as a tool for treating depression and auditory hallucinations (Pascual-Leone et al., 2002) and for increasing attention (Fregni & Pascual-Leone, 2007).

2.3.2 Risks

The main contraindications for treatment are pregnant women, children under 6 years, patients with pacemakers, electrodes, or drug infusion pumps (Rossi, 2009), or patients with metallic implants in the head (Roth et al., 1992).

On the other hand, some patients subjected to cortical stimulation experience some side effects after application, which can be regarded as being minor and transient, including headaches, which can be mitigated by common analgesics (Wasserman, 1998). There have also been reports concerning patients with epilepsy or who take epileptogenic antidepressants who are unable to reach convulsive crises during treatment with transcranial magnetic stimulation (Duckworth, 2008). We have observed that at magnetic field intensities above 1Tesla and frequencies above 60 Hz can cause convulsions (Zelaya et al. 2010), and, therefore, we have suggested setting international standards for medical use.

2.4 Deep brain stimulation

Deep Brain Stimulation (DBS) was first developed as a treatment for Parkinson's disease, to reduce tremors, stiffness, difficulty in walking, and uncontrollable movements. In DBS, a pair of electrodes are implanted in the brain and controlled by a generator that is implanted in the chest. The stimulation is continuous at a frequency and level to suit the patient (Perlmutter & Mink, 2006). It has only recently been studied as a treatment for depression or OCD (obsessive–compulsive disorder).

It is currently available on an experimental basis. So far, very little research has been conducted on testing DBS for the treatment of depression (Mohr, 2011), but a few studies have been conducted showing that this treatment may be promising. In a trial using patients with severe depression that were treatment resistant, we observed that four of the six patients showed a marked improvement in their symptoms immediately after the procedure (Maybergetal., 2005). In another study on 10 patients with this disorder, we found that the improvement was consistent among the majority of patients 3 years after surgery (Greenbergetal., 2006).

2.4.1 How does it work?

DBS requires brain surgery. The head is shaved and then a screw is fixed to a frame that prevents the head from moving during surgery. Scans are taken of the head and brain using magnetic resonance. The surgeon uses these images as a guide during surgery. The patient is awake during the procedure to provide feedback to the surgeon. During this procedure,the patient feels no pain because the head is numbed using a local anesthetic.

Once ready for surgery, two well sare drilled in the head. From there, the surgeon inserts a thin tube into the brain to place the electrodes on either side of a specific part of the brain. In the case of depression, brain target area 25 is considered. This area has been found to be related to depression, hyperactivity, and mood disorders (Maybergetal., 2005). In the case of OCD, the electrodes are placed in a different part of the brain that is believed to be associated with the disorder.

After the electrodes have been implanted and the patient has provided feedback on the placement of the electrodes, the patient is placed under general anesthesia. The electrodes are connected to wires that run inside the body from the head to the chest, which are connected to two battery-powered generators. From here, electrical impulses are continuously delivered through the wires to the electrodes in the brain. Although it is unclear exactly how the device reduces depression or OCD, scientists believe that the pulses help to "restore" the area of the brain that is not working properly to function normally again (Perlmutter & Mink, 2006).

2.4.2 What are the side effects?

DBS has risks that are associated with any brain surgery. For example, the procedure can lead tobleeding in the brain or stroke, infection, disorientation or confusion, unwanted changes in mood, movement disorders, dizziness, and difficulty sleeping.

Because the procedure is still experimental, there may be other side effects that have not been identified. The long-term benefits are also unknown.

3. Diagnostic techniques for magnetism

3.1 Magnetogastrography

The use of sensors to measure magnetic activity generated by the body has been known since 1950 (Wenger et al., 1961; Wengeretal., 1957), although it has been "known" since ancient times for healing applications and diagnosis. It was in the 1970s that biomagnetic techniques began to expand their application and to become more common, with publications on ferromagnetic contamination in the lungs and other organs, as well as the detection and analysis of magnetic fields produced by bioelectric currents in human beings (Cohen, 1973; Cohen, 1969; Cohen et al., 1970; Benmair, et al., 1977; Frei et al., 1970).

Mechano-Magnetogastrography (M-MGG) is a biomagnetic technique used to determine the gastrointestinal activity of the stomach by determining the frequency of peristaltic contractions and gastric emptying half-time (Córdova et al., 2004; Córdova-Fraga et al., 2005).

Biomagnetic applications in the gastrointestinal system have resulted in the study of the transit time in different phases of the menstrual cycle in women or spatiotemporal assessment of the colon motility (Córdova-Fraga et al., 2005).

MGG is a technique that has advantages over current diagnostic techniques because it is noninvasive, lacks ionizing radiation, does not interfere with a patient's privacy, and provides reproducible results (De la Roca et al., 2007).

The reproducibility of measurements was analyzed in a study that measured the same patient over a period of several weeks evaluating the gastric emptying half-time, which showed a reproducibility coefficient above 85%. The remaining variation can be attributed to changes in motility of the same patients, depending on social and environmental conditions such as diet and health (De la Roca et al., 2007).

Several studies (De la Roca et al., 2007; Benmair & Dreyfus, 1977) have evaluated gastric emptying using magnetic tracers and the results have been presented in terms of half-time of emptying and peristaltic contractions in healthy male volunteers employing magnetic tracers in a yogurt vehicle test meal (Carneiro, 1977; Forsman, 2000), with similar results.

Gastric emptying has been evaluated in patients with functional dyspepsia, and a solid food test has been developed to compare results against benchmarks that primarily use the above type of food.

The esophageal transit time has also been evaluated. Techniques and instrumentation for biomagnetic studies permit the noninvasive functional evaluation of the gastrointestinal tract (Daghastanlietal., 1998; Córdova-Fraga et al., 2008).

We believe that the use of fluxgate sensors using magnetic markers can be used as a complement to manometric studies and are equivalent to centigraphy for clinical use.

Various studies have been performed using different test assemblies, and designs that we consider most relevant showing common patterns are presented below.

3.1.1 Magnetic stimulator

A magnetic stimulator has been used in experiments involving susceptometry (Carneiro et al., 2000) and magnetogastrography (Córdova et al., 2004). The magnetogastrograph (De la

Roca et al., 2007) was built by the Medical Physics Laboratory at the University of Guanajuato in Mexico.

The magnetic stimulator is composed of two identical coils, assembled as a Helmholtz coil array. This setting means that the coils are placed parallel to each other and share the same axis of symmetry, with the separation distance being equal to the radius of the coils.

The coils contain 60 turns composed of magnet gauge 4 copper wire (average radius =6 mm). The wire was wrapped around two aluminum supports that contain five layers of 12 turns of wire. The magnitude of the pulse field produced by this arrangement is of the order of a few millitesla.

This arrangement of coils connected in parallel has a resistance of 260 milliohms and an inductance of 0.928mH. The magnetic pulse is generated by a bank of six capacitors connected in parallel with an equivalent capacitance of 46mF. The capacitor bank is connected to the mains voltage (220V), and this voltage is rectified before reaching the capacitor bank. Once the capacitors are charged, the discharge produces a magnetic pulse of 32mT for a period of 17μs.

For field measurements, we used the above system with a Model 53 triaxial fluxgate (Physics Applet Systems),which is a solid state device used to measure the direction and magnitude of direct current magnetic fields, or variations thereof, with a frequency less than 100Hz and a magnetic field of 18mT. This device has an output in three axes that can transmit data from the measurements performed through a serial port at a frequency of 250 samples per second with a noise level of 3nT. This system is automated and operated from a PC via the LabVIEW software package both for stimulation and data acquisition.

(see figures)

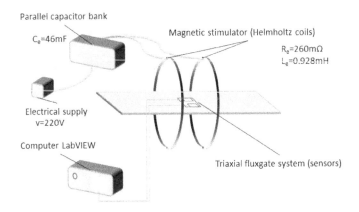

Magnet-gastrographer diagram. Helmholtz coils are responsible for generating the magnetic field, and receive the voltage from a bank of six capacitors in parallel. A triaxial fluxgate system detects the alignment of the magnetic markers, and information is recorded on a computer.

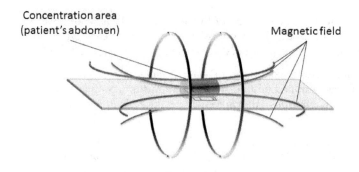

Concentration area
(patient's abdomen)   Magnetic field

The stimulator generates a magnetic field is concentrated in the abdomen

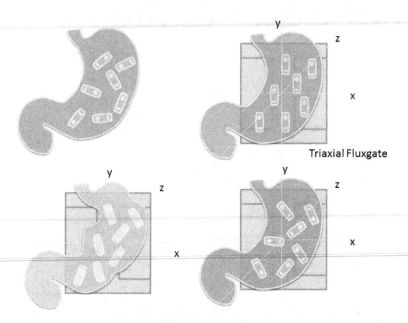

1. Stomach after eating markers, 2. Markers aligned by the magnetic field, located in triaxial plane, 3. Markers for gastric movement, 4. Misaligned markers after gastric movement.

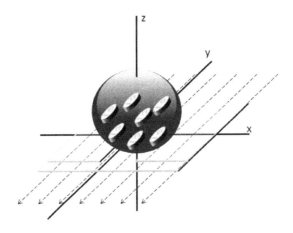

Diagram of the orientation of the markers on the triaxial sensor.

3.1.2 Magnetic contrast

Volunteers were asked to fast the previous evening and not to take any medication that would affect the gastric emptying time. The patients were asked to ingest a harmless dye and were placed in a prone position. Ten minutes after ingestion, the measurements began and were conducted every 10 minutes for a period of 1–2 hours.

In the case of semisolid food, the patients were given 250g of yogurt mixed with 4g of magnetite (Fe_3O_4). In the case of solid food in healthy volunteers (controls), this comprised a scrambled egg, a piece of bread and 250ml of peach juice. The bread contained 4g of magnetite for the MMGG and was made from milk, flour, eggs, and butter where the bread was in the form of a "hot cake" with a mass of approx. 30g (Reynaga, 2008).

3.2 Magnetic scanner

We have developed a device that is capable of creating a map of the magnetic fields obtained using magnetically marked phantoms, produced in the laboratory. This is a mobile automatic two-way device, developed to detect changes in magnetic flux, and is composed of on an array of magneto resistive sensors designed to detect magnetic fields of 100mT to 10nT (Pacheco, 2010).

4. Conclusions

Psychomagnetobiology is an interdisciplinary field that will help scientists propose strategies for the diagnosis and treatment of biological and psychological diseases, knowing that there is technology for electric or magnetic stimulation that has proven experimentally to be effective for the treatment of depression, anxiety, and in some cases, even hallucinations and addictions. It shows promise in this challenging landscape.

Both biological engineering and developments in technology are needed to implement new products in the service of human beings, but more research is required in joint efforts in psychology, medicine, and medical physics. Commitment is needed to continue working on this to help create the links needed to reach people in need.

We are currently developing new equipment and applications for research into missing areas, and the development of portable, magnetic and low-frequency equipment and software for the treatment of depression in Latin America.

Technology advances more each time to provide tools for the diagnosis and treatment of physical and mental illnesses, we have developed with several collaborators, technology to measure gastric emptying time, peristaltic contractions andtransit time in the gastrointestinal system (De la Roca-Chiapas JM et al, 2011), the influence of emotions on the symptoms of Dyspepsia (De la Roca-Chiapas et al, 2010), we have developed technology to measure magnetic fields on surfaces (Pacheco et al, 2008) , and we tried to measure tumor sizes (De la Roca-Chiapas, 2009) and now we are developing technology in electrical and magnetic microstimulation as tools that serve to treatment in people with anxiety, stress and depression.

There needs to be increasingly greater openness in science, allowing us to learn more about the man, and serve it better, this means leaving the arrogance to believe that science is "truth" and need to learn interrelated studies to do with people who think differently and in turn need projects to enter these sciences with integrators projects.

5. Future directions

Areas in which I think will developing the technology and where future research can focus are displayed in spaces that need to be covered in Figure 1. It's needed a portable and powerful magnetic technology for the treatment of depression. Lack the technological development of a scanner capable of measuring magnetic fields in the human body, and perhaps they are needed armored cameras, but I think the SQUID application for this is part of the answer, but today only deals to equip magnetoencephalography with the EEG, providing information on what brain areas are running the electric field lines. Similarly we believe that knowing how it behaves electric and magnetic fields in the human body can be used for the treatment and diagnosis.

I think in the future we will learn more about how magnetic fields of the human body and the planet affect us in the biological and psychological. We have technology that measures the order of femtoteslas, human magnetic fields, allowing us to understand how there is this mind-body-health interaction.

As for the GDV/EPC, in the future we will find a basic application that achieves characterize the behavior of fluids and tissues in humans, and their interaction with the environment.

In this sense, it is working at the University of Guanajuato to create models that seek to both basic research and the development of patents and technology with human sense.

I conclude by recalling that understanding the human being as an integral whole is an essential part of this work, since linking the mind, body and health consciousness, it is a model that seeks to achieve interdisciplinary in the service of man, speaking of the psycomagnetobiology, search of integrating different areas of expertise: medicine, psychology, physics and engineering. And this effort is retrieved from the traditional medicine of many cultures, including Mexico, China and Egypt, without being new, and that by proposing a new term is intended to entrench in a modern language, an expression that is natural multidimensional human.

As time passes, more and more will be heard: engineering biological and engineering psychological; science and technology for physical and mental health.

6. Acknowledgments

For their invaluable assistance in gathering information and format, to: BS. Jose Eduardo Lepez, and BS. Jorge Fernando Ortiz Paz for the collection of information and image design, this chapter has been possible thanks to support and UGTO/CA-37 UGTO/PTC-183 Promep, and PIFI funds from the University of Guanajuato.

7. References

Alexandrova, RA., Fedoseev, BG. & Korotkov KG. (2003). Analysis of the bioelectrograms of bronchial asthma patients, In: Proceedings of conference "Measuring the human energy field: State of the science". Baltimore: National Institute of Health, pp.70–81.

Alexandrova, RA., Nemtsov, VI., Koshechkin, DV. & Ermolev SU. (2003). Analysis of holeodoron treatment effect on cholestasis syndrome patients. In: Proceedings of VII International Scientific Congress on Bioelectrography. St. Petersburg, Russia, pp. 4–6.

Baeken, C. & De Raedt, R. (2011). Neurobiological mechanisms of repetitive transcranial magnetic stimulation on the underlying neurocircuitry in unipolar depression. Dialogues ClinNeurosci. 13(1):139–45.

Benmair, Y., Dreyfuss, F., Fischel, B., Frei, EH. & Gilat T. (1977). Study of Gastric Emptying Using a Ferromagnetic Tracer, Gastroenterology 73:1041–1045.

Benmair, Y., Fischel, B., Frei, EH. & Gilat T. (1977). Evaluation of a Magnetic Method for the Measurement of Small Intestine Transit Time, American Journal of Gastroenterology 68:470–475.

Bocanegra-García, V., Del Rayo Camacho-Corona, M., Ramírez-Cabrera, M., Rivera, G. & Garza-González, E. (2009). The bioactivity of plant extracts against representative bacterial pathogens of the lower respiratory tract. BMC Res Notes. 1(2):95.

Bryant, HC., Adolphi, NL., Huber, DL., Fegan, DL., Monson, TC., Tessier, TE. & Flynn, ER. (2011). Magnetic Properties of Nanoparticles Useful for SQUID Relaxometry in Biomedical Applications, J Magn Magn Mater, 323(6):767–774.

Bundzen, PV., Korotkov, KG. & Belobaba, O. (2003). Correlation between the parameters of induced opto-electron emission (Kirlian effect) and the processes of corticovisceral regulation. In: Proceedings of VII International Scientific Congress onBioelectrography, St. Petersburg, Russia, pp.89–91.

Cacioppo, JT., Tassinary, LG. & Berntson, GG. (2007). Handbook of psychophysiology (3rd ed.). New York, NY: Cambridge Univ. Press. pp. 121.

Carneiro, A. (1999). Study of stomach Motility using the relaxation of magnetic tracers. Phys Med Biol, 44:1691–1697.

Carneiro, AA., Ferreira, ER., Moraes, DB., Sosa, M. & Baffa, O. Biomagnetismo: Aspectos Intrumentais e Aplicacoes, Revista Brasileira de Física, 22(3):324–338.

Cohen, D. (1969). Detection And Analysis of Magnetic Fields Produced By Bioelectric Currents In Humans, *Journal Of Applied Physics* 40(3):1046–1048.

Cohen, D. (1973). Ferromagnetic Contamination in the Lungs and Other Organs of the Human Body, *Science*, 180:743–748.

Cohen, D., Edelsack, EA. & Zimmerman, JE. (1970). Magnetocardiogram taken inside a shielded room with a superconducting point-contact magnetometer, *Appl Phys Lett*, 16:278–280.

Córdova-Fraga, T., Carneiro, AA., de Araujo, DB., Oliveira, RB., Sosa, M. & Baffa, O. (2005). Spatiotemporal evaluation of human colon motility using three-axis fluxgates and magnetic markers.Med, *Biol Eng Comput*, 43(6):712–715.

Córdova-Fraga, T., Gutierrez, JG., Sosa, AM., Vargas, LF. & Bernal, AJ. (2004). Gastric activity studies using a magnetic tracer, *Institute of Physics Publishing*, 25:1261–1270.

Córdova-Fraga, T., Sosa, M., Wiechers, C., De la Roca-Chiapas, JM., Maldonado Moreles, A., Bernal-Alvarado, J. & Huerta-Ranco, R. (2008). Effects of anatomical position on esophageal transit time: A biomagnetic diagnostic technique. *World J Gastroenterol*, 14(37):5707–5711.

Cork, RC., Wood, P., Ming, N., Shepher, C., Eddy, J. & Price, L. (2004) The effect of CES on pain associated with fibromyalgia. *The Internet Journal of Anesthesiology*; 8:2.

Daghastanli, NA., Braga, FJ., Oliveira, RB. & Baffa, O. (1998). Esophageal transit timeevaluated by a biomagnetic method, *PhysiolMeas* 19(3):413–20.

De la Roca Chiapas JM, Cordova Fraga T, Barbosa Sabanero G, Macias de la Cruz JH, Cano ME, Pacheco AH, Rivera Cisneros AR, Solis S, Sosa M. Aplicaciones interdisciplinarias entre Física, Medicina y Psicología.(2009) *Acta Universitaria*; 19(2) 5:9.

De la Roca-Chiapas JM, Cordova-Fraga T.(2011). Biomagnetic techniques for evaluating gastric emptying, peristaltic contraction and transit time. *World J Gastrointest Pathophysiol* 2(5): 65-71

De la Roca-Chiapas José Ma., Solís-Ortiz Silvia, Fajardo-Araujo Martha, Sosa Modesto, Córdova-Fraga Teodoro, Zarate Alma Rosa.(2010). Stress profile, coping style, anxiety, depression, and gastric emptying as predictors of functional dyspepsia: A case-control study. *Journal of Psychosomatic Research* 68: 73–81.

De la Roca-Chipas, JM., Hernández, SE., Solis, MS., Sosa, AM. & Córdova, FT. (2007). Magnetogastrography (MGG) reproducibility assessments of gastric emptying on healthy subjects, *Physiol Meas*, 28(2):175–183.

De Oliveira, JF., Wajnberg, E., Esquivel, DM., Weinkauf, S., Winklhofer, M. & Hanzlik, M. (2010). Ant antennae: are they sites for magnetoreception, *R Soc Interface*. 7(42):143–152.

Debock, P. (2000). European perspective: a comparison between TENS and MET, *Physical Therapy Pro*, pp:28–33.

Días, AM. & van Deusen, A. (2011). A new neurofeedback protocol for depression, *Span J Psychol*, 14(1):374–384.

Dobson, P. & O'Keeffe, E. (2007). Investigation into the GDV technique and personality. *In: Proceedings of the International Scientific Conference: Measuring energy fields*, Kamnik-Tunjice, Slovenia, pp:111–113.

Duckworth, K. (2008). "Transcranial Magnetic Stimulation (TMS)" (Review). *National Alliance on Mental Illness*, pp:12-15.

Fitzgerald, PB., Fountain, S. & Daskalakis, J. (2006). A comprehensive review of the effects of rTMS on motor cortical excitability and inhibition, *Clinical Neurophysiology*, 117(12):2584–2596.

Forsman, M. (2000). Intragastric movement assessment by measurement magnetic field decay of magnetised tracer particles in a solid meal, *Med Biol Eng Comput*, 38 169–174.

Fregni, F., & Pascual-Leone, A. (2007). Technology insight: noninvasive brain stimulation in neurology-perspectives on the therapeutic potential of rTMS and tDCS, *Nature Clinical Practice Neurology*, 3(7):383-393.

Frei, EH., Benmayr, Y., Yerashlmi, S. & Dreyfuss, F. (1970). Measurements of the emptying of the stomach with a magnetic tracer, *IEEE Trans Mag*, 6:348–9.

Freitas, C., Pearlman, C. & Pascual-Leone, A. (2011). Treatment of auditory verbal hallucinations with transcranial magnetic stimulation in a patient with psychotic major depression: One-year follow-up, *Neurocase*, 25:1–9.

Gagua, PO., Gedevanishvili, EG. & Korotkov, KG. Experimental study of the GDV technique application in oncology [in Russian], *J IzvestiaVuzovPriborostroenie*, 49(2):47–50.

Gartside, SE., Leitch, MM. & Young, AH. (2003). Altered glucocorticoid rhythm attenuates the ability of a chronic SSRI to elevate forebrain 5-HT: implications for the treatment of depression, *Neuropsychopharmacology*, 28(9):1572–1578.

Gedevanishvili, EG., Giorgobiani, LG. & Kapanidze, A. (2004). Estimation of radiotherapy effectiveness with gas discharge visualization (GDV). *In: Proceedings of VIII International Scientific Congress on Bioelectrography*. St. Petersburg, Russia, pp:98–99.

Gegear, RJ., Foley, LE., Casselman, A. & Reppert, SM. (2010). Animal cryptochromes mediate magnetoreception by an unconventional photochemical mechanism, *Nature*, 463(7282):804–807.

Gertsenhtein, SY., Panin, DN. (2008). Study of Blood Electric Properties by the Method of Gas-Discharge Visualization, *Dokl Phys*. 53:107–110.

Gilula, MF. & Barach, P. (2004) CES: a safe neuromedical treatment for anxiety, depression or insomnia. *South MJ* 97(12):1269–1270.

Greenberg, BD., Malone, DA., Friehs, GM., Rezai, AR., Kubu, CS., Malloy, PF., Salloway, SP., Okun, MS., Goodman, WK. & Rasmussen, SA. (2006). Three-year outcomes in deep brain stimulation for highly resistant obsessive compulsive disorder, *Neuropsychopharmacology*, 31(11):2384–2393.

Gulmen, FM. (2004). Energy medicine, *Am J Chin Med*, 32(5):651–658.

Gunther, M. & Phillips, KD. (2010). Cranial Electrotherapy Stimulation for the Treatment of Depression. *Journal of Psychosocial Nursing*, 48(11):37–42.

Heinrichs, M., Baumgartner, T., Kirschbaum, C. & Ehlert, U. (2003). Social support and oxytocin interact to suppress cortisol and subjective responses to psychosocial stress, *Biol Psychiatry*, 54(12):1389–1398.

Heyers, D., Zapka, M., Hoffmeister, M., Wild, JM. & Mouritsen, H. (2010). Magnetic field changes activate the trigeminal brainstem complex in a migratory bird. *Proc Natl Acad Sci U S A*. 107(20):9394–9399.

Kirsch, DL. (2006). CES in the treatment of fibromyalgia. PPM, 6(6):60–64.

Kirsch, DL. (2006). Microcurrent electrical therapy (MET): A tutorial, *PPM*, 6(7):59–64.

Korotkov, KG. & Popechitelev, EP. A Method for Gas-Discharge Visualization and an Automated System for Its Implementation, *Biomed Eng*, 36(1), 23–27.

Korotkov, KG., Matravers, P., Orlov, DV. & Williams, BO. (2010). Application of electrophoton capture (EPC) analysis based on gas discharge visualization (GDV) technique in medicine: a systematic review, *J Altern Complement Med*, 16(1):13–25.

Korotkova, AK. (2006). Gas discharge visualization bioelectrography method in studies of master-sportsmen's psychophysiology, *Abstract of a Ph.D. thesis in psychology [in Russian]*, St Petersburg, Russia.

Kostyuk, N., Rajnarayanan, V., Isokpehi, RD. & Cohly, HH. (2010). Autism from a Biometric Perspective, *Int J Environ Res Public Health*, 7:1984–1995.

Kulkarni, AD. & Smith, RB. (2000).The use of MET and CES in pain control, *Cl Pr Alt Med*, 2(2):99–102.

Lohmann, KJ., Lohmann, CM. & Putman, NF. (2007). Magnetic maps in animals: nature's GPS, *J Exp Biol*, 210(21):3697–705.

Martínez-Selva, JM., Sánchez-Navarro, JP., Bechara, A. & Román, F. (2006). Mecanismos cerebrales de la toma de decisiones. *Rev Neurol*, 42(7):411–418.

Mayberg, HS., Lozano, AM., Voon, V., McNeely, HE., Seminowicz, D., Hamani, C., Schwalb, JM. & Kennedy, SH. (2005). Deep brain stimulation for treatment-resistant depression. *Neuron*. 45(5):651–660.

McDonald, WM., Durkalski, V., Ball, ER., Holtzheimer, PE., Pavlicova, M., Lisanby, SH., Avery, D., Anderson, BS., Nahas, Z., Zarkowski, P., Sackeim, HA. & George, MS. (2011). Improving the antidepressant efficacy of transcranial magnetic stimulation: maximizing the number of stimulations and treatment location in treatment-resistant depression. *US National Library of Medicine National Institutes of Health Search*.

Melzack, R. & Wall, P. (1965). Pain mechanisms: A new theory, *Science, New Series*, 150(3699):971–979.

Mercola, JM. & Kirsch, DL. (1995). The basis for MET in conventional medical practice. *J of Adv in Med*, 8(2):107–120.

Mohr, P., Rodriguez, M., Slavíčková, A. & Hanka, J. (2011). The application of vagus nerve stimulation and deep brain stimulation in depression, *Neuropsychobiology*, 64(3):170–81.

Mouritsen, H. & Ritz, T. (2005). Magnetoreception and its use in bird navigation, *Curr Opin Neurobiol*, 15(4):406–414.

Narayanamurthy, G. & Veezhinathan, M. (2011). Design of Cranial Electrotherapy Stimulator and Analyzing It with EEG. *In: Power Electronics and Instrumentation Engineering Communications in Computer and Information Science*, 102(2):42–45.

NIMH (2009) Brain Stimulation Therapies.National Institute of Mental Health.2009–11–17.Retrieved 2010–07–14.

O'Keeffe, E. (2006). The GDV technique as an aid to stress assessment and its potential application in recruitment and selection of individuals suited to positions associated with high level of stress. *In: Proceedings of X International Scientific Congress on Bioelectrography*, St. Petersburg, Russia, pp:202–204.

Olalde, JA., Magarici, M., Amendola, F. & del-Castillo, O. (2004). Correlation between electrophoton values and diabetic foot amputation risk. *In: Proceedings of Conference: Neurobiotelekom*. St. Petersburg, Russia, pp:54–58.

Pacheco, AH., Cano, ME., Palomar-Lever, E., Córdova-Fraga, T., De la Roca, JM., Hernández-Sámano, A. & Felix-Medina, R. (2010). An XY Magnetic Scanning

Device for Magnetic Tracers: Preliminary Results, *International Journal of Bioelectromagnetism,* 12(2):81–84.

Pascual-Leone, A., Davey, N., Rothwell, J., Wassermann, EM. & Puri, BK. (2002). Handbook of Transcranial Magnetic Stimulation. Hodder Arnold.

Perlmutter, JS. & Mink, JW. (2006). Deep brain stimulation. *Annual Review of Neuroscience.* 29:229–257

Polushin, US., Korotkov, KG. & Korotkina, SA. (2004). Perspectives of the application of gas discharge visualization for the estimation of organism condition at critical states. *In: Proceedings of IX International Scientific Congress on Bioelectrography.* St. Petersburg, Russia, pp:115–116.

Polushin, US., Korotkov, KG., Strukov, EU. & Shirokov, DM. (2003). First experience of using GDV method in anesthetization and resuscitation. *In: Proceedings of VII International Scientific Congress on Bioelectrography.* St. Petersburg, Russia, pp:13–14.

Priyatkin, NS., Korotkov, KG. & Kuzemkin, VA. (2006). GDV bioelectrography in research of influence of odorous substances on psycho-physiological state of man [in Russian]. *J IzvestiaVuzovPriborostroenie,* 49(2):37–43.

Reynaga-Ornelas, MG., De la Roca-Chiapas, JM., Córdova-Fraga, T., Bernal, JJ. & Sosa, M. Solid Test Meal to Measure the Gastric Emptying with Magnetogastrography, *AIP Conference Proceedings - American Institute of Physics,* 1032(1):246.

Riehl, M. (2008). TMS Stimulator Design, In Wassermann, EM., Epstein, CM., Ziemann, U., Walsh, V., Paus, T., Lisanby, SH, *Oxford Handbook of Transcranial Stimulation,* Oxford University Press, pp.13–23,25–32.

Ritsner, M., Gibel, A., Ram, E., Maayan, R. & Weizman, A. (2006). Alterations in DHEA metabolism in schizophrenia: two-month case-control study, *Eur Neuropsychopharmacol,* 16(2):137–146.

Rosch, PJ. (2009). Bioelectromagnetic and subtle energy medicine: the interface between mind and matter, *Ann N Y Acad Sci,* 1172:297–311.

Rosenberg, O., Isserles, M., Levkovitz, Y., Kotler, M., Zangen, A. & Dannon, PN. (2011). Effectiveness of a second deep TMS in depression: A brief report, *Prog Neuropsychopharmacol Biol Psychiatry,* 35(4):1041–1044.

Rossi, S. (2009). The Safety of TMS Consensus Group, Safety, ethical considerations, and application guidelines for the use of transcranial magnetic stimulation in clinical practice and research, *Clinical Neurophysiology,* 120(12):2008–2039.

Rossini, P. & Rossi, S. (2007). Transcranial magnetic stimulation: diagnostic, therapeutic, and research potential, *Neurology,* 68(7):484–488.

Roth, BJ., Maccabee, PJ., Eberle, L., Amassian, VE., Hallett, M., Cadwell, J., Anselmi, GD. & Tatarian, GT. (1994). In-vitro evaluation of a four-leaf coil design for magnetic stimulation of peripheral nerve, *Electroenceph Clin Neurophysiol,* 93(1):68–74.

Roth, BJ., Pascual-Leone, A., Cohen, LG. & Hallett, M. (1992). The heating of metal electrodes during rapid-rate magnetic stimulation: A possible safety hazard, *Electroenceph.Clin.Neurophysiol,* 85(2):116–123.

Sánchez-Martín, JR., Azurmendi, A., Pascual-Sagastizabal, E., Cardas, J., Braza, F., Braza, P., Carreras, MR. & Muñoz, JM. Androgen levels and anger and impulsivity measures as predictors of physical, verbal and indirect aggression in boys and girls, *Psychoneuroendocrinology,* 36(5):750–760.

Schattner, E., Shahar, G., Lerman, S. & Shakra, MA. Depression in systemic lupus erythematosus: the key role of illness intrusiveness and concealment of symptoms, *Psychiatry*, 73(4):329–340.

Schmitt, R., Capo, T. & Boyd, E. (1986). Cranial Electrotherapy Stimulation as a Treatment for anxiety in chemically dependent persons, *Alcoholism: Clinical and Experimental Research*, 10(2)158–160.

Sergeev, SS. & Pisareva, SA. The use of gas-discharging visualization method (GDV) for monitoring a condition of the personnel at short-term rehabilitation. *In: Proceedings of VIII International Scientific Congress on Bioelectrography*, St. Petersburg, Russia, pp:128–129.

Sidney, K. (1995). Meta-Analysis of Randomized Controlled Trials of Cranial Electrostimulation: Efficacy in Treating Selected Psychological and Physiological Conditions, *Journal of Nervous & Mental Disease* 183(7):478–484.

Smith, RB. (2002). Microcurrent therapies: emerging theories of physiological information processing, *Neuro Rehab*, 17(1):3–7.

Smith, RB. (2008). Cranial Electrotherapy Stimulation: Its First Fifty Years

Strukov, EU. Facilities of gas discharge visualization method for assessment of functional state of organism in preoperational period [in Russian]. *Abstract of a medical doctoral candidate's thesis.* St Petersburg, Russia: Military Medical Academy.

Stürmer, T., Hasselbach, P. & Amelang, M. (2006). Personality, lifestyle, and risk of cardiovascular disease and cancer: follow-up of population based cohort, *BMJ*, 332(7554):1359.

Temuryants, NA., Shekhotkin, AV. (2000). The role of the pineal gland in magnetobiological effects, *Crit Rev Biomed Eng*, 28(1–2):307–321.

Valentinuzzi, ME. (2004). Magnetobiology: a historical view, *IEEEEng Med Biol Mag*, 23(3):85–94.

Wassermann, EM. (1998). Risk and safety of repetitive transcranial magnetic stimulation: report and suggested guidelines from the International Workshop on the Safety of Repetitive Transcranial Magnetic Stimulation, *Electroencephalography and clinical Neurophysiology*, 108(1):1–16.

Wenger, MA., Engel, BT., Clemens, TL. & Cullen, TD. (1961). Stomach Motility in Man as Recorder by the Magnetometer Method, *Gastroenterology*, 41:479–85.

Wenger, MA., Henderson, EB. & Dinnin, JS. (1957). Magnetometer method for recording gastric motility, *Science*, 125:192–195.

Wiltschko, W., Ford, H., Munro, U., Winklhofer, M. & Wiltschko, R. (2007). Magnetite-based magnetoreception: the effect of repeated pulsing on the orientation of migratory birds, *J Comp Physiol A Neuroethol Sens Neural Behav Physiol*, 193(5):515–22.

Woodbury, FM. (2010). Efecto de la microcorriente sobre síntomas de ansiedad, depresión, insomnio y dolor. *Revista Galenus*, 5.

Zak, PJ. & Fakhar, A. (2006). Neuroactive hormones and interpersonal trust: international evidence. *Econ Hum Biol.* 4(3):412–429.

Zelaya, RM., Saracco-Álvarez, RA. & González J. (2010). The transcraneal magnetic stimulation for the negative symptoms in schizophrenia: a review, *Salud Ment*, 33(2):169–178.

Zyss, T. (2008). Magnetotherapy, *Neuro Endocrinol Lett*, 29(1):161–201.

7

Study on the Mechanism of Traumatic Brain Injury

Yuelin Zhang[1], Shigeru Aomura[2],
Hiromichi Nakadate[2] and Satoshi Fujiwara[3]
[1]Tokyo University of Agriculture and Technology,
[2]Tokyo Metropolitan University,
[3]Yokohama City University,
Japan

1. Introduction

Skull fracture, intracranial hemorrhage, or cerebral injury can be caused in humans due to a strong impact to the head. The following 2 types of cerebral injuries are often observed: one type is cerebral contusion which is a local brain damage to the brain, and the other is diffuse axonal injury (DAI) which is a diffuse brain damage to the brain. In various head injuries caused by external impact, cerebral contusion and DAI mainly result in direct failure of the cerebral parenchyma.

Cerebral contusions can be either coup or contrecoup contusions that occur on either the same or the opposite side of impact, respectively (Yanagida et al., 1989). Cerebral contusions are caused by rapid pressure fluctuations transmitted to the brain surface via the cerebrospinal fluid (CSF) (Fujiwara et al., 1989, Zhang et al., 2001). The hypothesis that the brain surface is destroyed when the cerebral parenchyma collides with the skull, although intuitive, has never been observed (Gurdjian et al., 1966). The cavitation theory states that the pressure gradient generated in the CSF by impact causes contrecoup negative pressure on the opposite side of impact and forms cavitation bubbles; the subsequent collapse of the bubbles causes brain tissue damage. Although this theory can be trusted, no collapse of bubbles in the head has yet been observed (Gross, 1958). Although various theories report the generation mechanism of cerebral contusion, none can sufficiently explain the entire mechanism. For predicting the dynamic response of the human head, numerous cadaver and animal experiments have been performed (Nahum & Smith, 1977; Gennarelli, 1983); however, these experiments are difficult to conduct because of cost and/or ethical concerns. Furthermore, the finite element method is widely used to predict the dynamic responses of the human head (Aomura et al., 2002). Some researchers report the reconstruction of real-world brain injury cases using multibody dynamics and finite element method (Raul et al., 2006; Riordain et al., 2003; Doorly & Gilchrist, 2006). These studies demonstrate the effectiveness of these methods but do not explain the mechanism of the brain injuries themselves.

DAI is considered to be caused by strain to the brainstem due to the rotational movement of the head. When a human head receives a external impact, the cranium moves first and then the brain follows its movement; this delay becomes more remarkable inside the brain, and a large strain is generated in the deep brain. This large strain causes damage to the axons of nerve cells and results in DAI (Fujiwara et al., 1998; Ommaya & Gennarelli, 1974).

As described above, in the past few decades, in order to clarify the mechanism of cerebral contusion and DAI, the impact experiments using physical models and animals, numerical analysis by finite element method are performed as engineering approach. In recent years, in order to clarify the tolerance of impact at the cellular level, the impact experiments are performed using cultured neuronal cells and tissues as biological approach. However, these researches often focus on the mechanism of only one type of injury; can not to evaluate all types of traumatic brain injury comprehensively. Thus, future research should be performed concurrently engineering and biological approaches. Therefore, in this study, the reconstruction analyses of real-world brain injury accident cases are performed to understand the mechanism of cerebral contusion and the impact experiments of cultured cells are performed to understand the mechanism of DAI.

2. Finite-element human head model

A computer model was constructed using cross-sectional T1 weighted MRI data of a woman's head because it was recently decided that CT should not be used for research to avoid radiation exposure. The slice thickness of the data is 3.3mm, and the slice interval is 0.0mm. Both the internal and external boundary curves of the scalp, skull, CSF, brain, and brain stem were extracted by binary image processing of the MRI data to make the internal and external surfaces of each part of the model. Three-dimensional human head models with hexahedral elements were made between the internal and external surfaces of each part (Fig. 1). The three-layered structure of the skull, which consists of an outer table, diploe, and inner table, was also reproduced. Finally, the finite element model consisted of 147,723 nodes and 114,012 elements. The material properties of each part of the model are shown in Table 1 (Willinger & Baumgartner, 2003; Nishimoto et al., 1998; Viano et al., 1998). Elastic properties were assigned to the scalp (Willinger & Baumgartner, 2003) and skull (Nishimoto et al., 1998), and viscoelastic properties were assigned to the CSF, brain, and brain stem (Viano et al., 1998).

For verifying the finite element model, the numerical results were compared with those results of the cadaver experiment by Nahum (Nahum & Smith, 1977). The impact direction was along the specimen's mid-sagittal plane, and the head was rotated forward such that the Frankfort anatomical plane was inclined 45° from the horizontal plane. The outline of the experiment is shown in Fig. 2(a). In the experiment, a 5 kg iron impactor was impacted to the head at 6 m/s (used by Nahum). However, Nahum does not definitively show what types of padding materials were interposed between the skull and impactor; therefore, in numerical calculations of this study only the time-force history(Fig.2(b)) was used as described in the literature (Nahum & Smith, 1977). The restraint condition of the head was free because the cadaver subject was seated and not restrained around the neck. The slide-type contact condition was used between the skull and CSF, CSF and brain, and brain and brain stem.

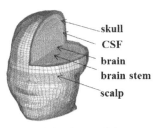

Fig. 1. Finite element human head model. The model consists of the scalp, skull (outer table, diploe, and inner table), CSF, brain, and brain stem.

	Scalp	Outer/inner table	Diploe	CSF	Brain	Brain stem
Density ρ (kg/m³)	1000	1456	850	1040	1040	1040
Young's Modulus E (MPa)	16.7	8750	4660	-	-	-
Bulk Modulus K (MPa)	-	7120	3470	2190	2190	2190
Short Time Shear Modulus G0 (MPa)	-	-	-	-	0.0125	0.0225
Long Time Modulus G∞ (MPa)	-	-	-	0.0005	0.0025	0.0045
Tangent Modulus (MPa)	-	4620	2170	-	-	-
Yield(MPa)	-	41.8	13.6	-	-	-
Poisson's Ratio v (-)	0.42	0.25	0.25	-	-	-
Time Constant (1/s)	-	-	-	500000	80	80

Table 1. Material properties of the finite element human head model. Elastic properties were assigned to the scalp (Willinger & Baumgartner, 2003) and skull (Nishimoto et al., 1998), and viscoelastic properties were assigned to the CSF, brain, and brain stem (Viano et al., 1998).

The experimental pressure response given by Nahum and pressure response calculated using finite element model in this study are shown in Fig. 3. Although slight difference was observed between the experimental and numerical results, this model was corroborated by the experimental cadaver test and sufficiently predicted intracranial pressure.

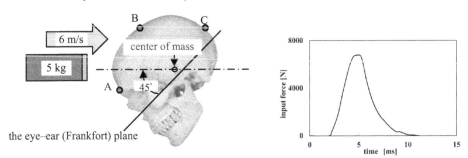

(a) The outline of the experiment of Nahum.   (b) The input force curve.

Fig. 2. Experiment by Nahum (Nahum & Smith, 1977). (a)The 5 kg iron impactor impacted the frontal region of the head at 6 m/s, and intracranial pressures were measured in the frontal (point A), parietal (point B) and occipital (point C) region of the head. (b) The input force curve obtained from the experiment.

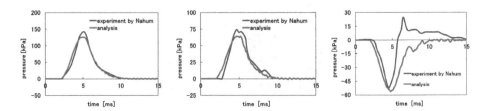

(a) Frontal region(point A)   (b) Parietal region(point B)   (c) Occipital region(point C)

Fig. 3. Comparison to the cadaver experiment by Nahum (Nahum & Smith, 1977) and numerical calculation. The validation demonstrates that this model is corroborated by an experimental cadaver test and sufficiently predicts intracranial pressure.

3. The mechanism of cerebral contusion

Fujiwara et al. analyzed 105 real-world fatal brain injury cases (Fujiwara et al., 1986). Coup contusions are caused more easily by direct blows to the head than contrecoup contusions, and contrecoup contusions are caused more easily by falls and fall-downs than coup contusions (Fujiwara et al., 1986). In this study, input force duration which is strongly related to the impact region on the human head, impact velocity, stiffness, mass or shape of the impact object, was focused. Previous studies (Aomura et al., 2008; Zhang et al., 2010) using impact experiments and finite element analysis of a water-filled acrylic container have showed that negative pressure inside the container is caused by deformation of the acrylic wall; this negative pressure tends to occur on the impact side when the force duration is short and on the opposite side of impact when the force duration is long.

In this chapter, for understanding the relationship between input force duration and dynamic response inside the skull, 9 real-world brain injury accident cases, including 3 coup contusion cases and 6 contrecoup contusion cases, were simulated using a finite element human head model. Numerical calculations were performed using LS-DYNA version 971.

3.1 Real-world brain injury accident cases

The autopsy results (performed by S. Fujiwara), in which the cause of death was cerebral contusion, are shown in Table 2. In these cases, the type of impact was classified as a blow, fall, or fall-down. In each type of impact, the contusion was classified as a coup or contrecoup contusion. Coup contusions are predominant in blows, and contrecoup contusions are predominant in falls and fall-downs. According to the postmortem data in Table 2, coup contusions tend to occur due to impacts by sharp-cornered objects (cases 1–3), and contrecoup contusions, due to impacts by objects with flat surfaces (cases 4 and 5). Contrecoup contusions are predominant in all the cases of falls and fall-downs (cases 6–9). In the simulations, the pressure threshold for causing cerebral contusion was -100 kPa. Skull fracturing was also simulated, because it was observed in all cases. The tensile stress thresholds for causing skull fracture were 70.5 MPa for the outer and inner tables and 21.4 MPa for the diploe (Nishimoto et al., 1998).

Case	Impact object	Impact region	Fracture region	Contusion region	
				Coup	Contrecoup
1	Beer bottle	Frontal region	Frontal region	Lower side of frontal lobe	
2	Wooden box (1.5 t)	Right temporal region	Right temporal region ~ right cranial fossa (semi) crushing	Right parietal lobe	
3	Sake bottle (empty: 700g)	Right parietal region	Right parietal region	Right parietal lobe	
4	Tank lorry	Parieto-occipital region	Occipital region (crushing)		Lower side of left and right frontal lobes Lower side of left temporal lobe
5	Steel Box (600 kg)	Occipital region	Occipital region		Lower side of frontal lobe Right and left temporal lobe poles

(a) Blows. The data include 3 coup and 2 contrecoup contusions.

Case	Impact object (height/weight)	Impact region	Fracture region	Contusion region	
				Coup	Contrecoup
6	Concrete (5.5 m)	Upper side of occipital region	Occipital region		Lower side of frontal lobe
7	Wooden deck (10 m)	Occipital region	Occipital region		Lower side of frontal lobe Frontal pole

(b) Falls. The data include 2 contrecoup contusions.

Case	Impact object (height/weight)	Impact region	Fracture region	Contusion region	
				Coup	Contrecoup
8	Asphalt (160 cm/58 kg)	Occipital region	Occipital region		Frontal pole
9	Asphalt (156 cm/59 kg)	Right occipital region	Right temporal region		Lower side of right frontal lobe Lower side of right temporal lobe Lateral surface of left temporal lobe

(c) Fall-downs. The data include 2 contrecoup contusions.

Table 2. Postmortem data (cause of death was cerebral contusion, 1968~1984). In these cases, the type of impact was classified as a blow, fall, or fall-down. In each type of the impact, the contusion was classified as a coup or contrecoup contusion.

3.2 Simulations of real-world brain injury accident cases

In order to begin the simulation, the relative velocity between the head and impact object, and the impact position of each case must be estimated. In this study, the impact positions are described in the postmortem data (Table 2, Impact Region). The relative velocities between the head and impact objects were estimated in order to generate negative pressure in the lesion area of the brain and fracture the skull in the fracture region described in the postmortem data. The simulations of the 9 cases are shown below.

[Case 1]

The frontal region of the head was impacted by a beer bottle (467 g), causing a skull fracture at the frontal region and a coup contusion at the lower side of the frontal lobe. In the simulation, the impact velocity (1-15 m/s) was applied to the node that constituted the beer bottle model (Fig. 4(a)). In order to determine the impact velocity of this case, the following 2 conditions had to be satisfied:

1. The negative pressure must be generated on the impact side only, because a coup contusion was observed.
2. The frontal skull must be fractured.

The impact velocity that satisfies these conditions was around 10 m/s. The intracranial pressure fluctuations of the impact side and its opposite side are shown in Fig. 4(b). The input force duration (i.e. the length of time that the impact object contacted the head) of this case was counted from the animation of the analysis result, was 1.4 ms.

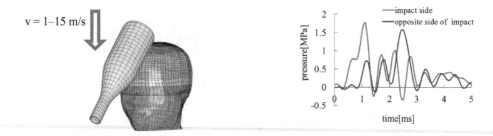

     (a) Simulation outline   (b) Intracranial pressure fluctuations (v = 10 m/s)

Fig. 4. Simulation outline and results of case 1. (a) Simulation outline. The impact velocity (1–15 m/s) was applied to the node that constituted the beer bottle model. (b) The intracranial pressure fluctuations of the impact side and its opposite side. The results show that negative pressure occurred only on the impact side.

[Case 2]

The right temporal region of the head was impacted by a wooden box (1.5 t), causing a skull fracture from the right temporal region to the right cranial fossa and a coup contusion at the right parietal lobe. In the simulation, the impact velocity (1-15 m/s) was applied to the node that constituted the wooden box model (Fig. 5(a)). The impact velocity that satisfied this case was around 15 m/s, which caused negative pressure on the impact side (Fig. 5(b)) and skull fracture in the right temporal region of the head. The input force duration (i.e. the length of time that the impact object contacted the head) of this case was counted from the animation of the analysis result, was 2.7 ms.

[Case 3]

The right parietal region of the head was impacted by a sake bottle (700 g), causing a skull fracture at the right parietal region and a coup contusion at the right parietal lobe. In the simulation, the impact velocity (1-15 m/s) was applied to the node that constituted the sake bottle model (Fig. 6(a)). The impact velocity that satisfied this case was around 15m/s, which caused the negative pressure at the impact side (Fig.6(b)) and the skull fracture at the right parietal of the head. The input force duration (i.e. the length of time that the impact object contacted the head) of this case was counted from the animation of the analysis result, was 2.2 ms.

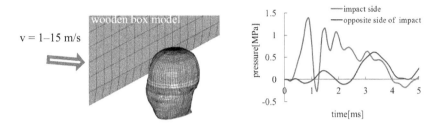

(a) Simulation outline   (b) Intracranial pressure fluctuations (v = 15 m/s)

Fig. 5. Simulation outline and results of case 2. (a) Simulation outline. The impact velocity (1–15 m/s) was applied to the node that constituted the wooden box model. For better view, only part of the wooden box is displayed. (b) The intracranial pressure fluctuations of the impact side and its opposite side. The results show that negative pressure occurred only on the impact side.

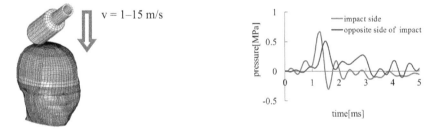

(a) Simulation outline   (b) Intracranial pressure fluctuations (v = 15 m/s)

Fig. 6. Simulation outline and results of case 3. (a) Simulation outline. The impact velocity (1–15 m/s) was applied to the node that constituted the sake bottle model. (b) The intracranial pressure fluctuations of the impact side and its opposite side. The results show that negative pressure occurred only on the impact side.

[Case 4]

A man who was walking along the street was impacted by a tank lorry (6 t) at the parieto-occipital region of the head, causing a skull fracture at the occiput region and contrecoup contusions at the lower side of the left and right frontal lobes and in the lower side of left temporal lobe. In the simulation, the impact velocity (1–15 m/s) was applied to the node which constituted the tank lorry model (Fig.7(a)). The impact velocity which satisfied this case was around 10m/s, which caused the negative pressure at the opposite side of impact (Fig.7(b)) and the skull fracture at the occiput region of the head. The input force duration (i.e. the length of time that the impact object contacted the head) of this case was counted from the animation of the analysis result, was 3.7 ms.

[Case 5]

A man who was working at a construction yard was impacted by a steel trash box (600 kg) at the occipital region of the head, causing a skull fracture at the occipital region and contrecoup contusions in the lower side of the frontal lobe and the right and left temporal lobe poles. In the simulation, the impact velocity (1–15 m/s) was applied to the node which constituted the steel box model (Fig.8(a)). The impact velocity which satisfied this case was around 9.5m/s, which caused the negative pressure at the opposite side of impact (Fig.8(b)) and the skull fracture at the occipital region of the head. The input force duration (i.e. the length of time that the impact object contacted the head) of this case was counted from the animation of the analysis result, was 3.6 ms.

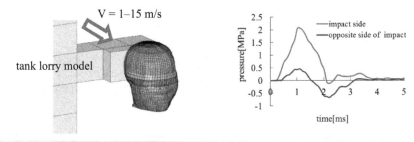

(a)Simulation outline   (b) Intracranial pressure fluctuations (v = 10 m/s)

Fig. 7. Simulation outline and results of case 4. (a) Simulation outline. The impact velocity (1–15 m/s) was applied to the node that constituted the tank lorry model. For better view, only part of the tank lorry is displayed. (b) The intracranial pressure fluctuations of the impact side and its opposite side. The results show that negative pressure occurred only on the opposite side of impact.

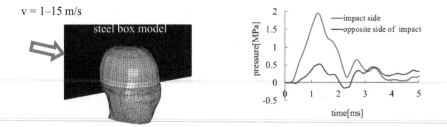

(a) Simulation outline   (b) Intracranial pressure fluctuations (v = 9.5m/s)

Fig. 8. Simulation outline and results of case 5. (a) Simulation outline. The impact velocity (1–15 m/s) was applied to the node that constituted the steel box model. For better view, only part of the steel box is displayed. (b) The intracranial pressure fluctuations of the impact side and its opposite side. The results show that negative pressure occurred only on the opposite side of impact.

[Case 6]

A man fell from 5.5 m high onto a concrete road, and the upper side of the occipital region of the head was impacted, causing a skull fracture at the occiput region and a contrecoup contusion at the lower side of the frontal lobe. In the simulation, the impact velocity (1–10.4 m/s, from the law of conservation of mechanical energy, the maximum velocity was calculated to be 10.4 m/s) was applied to the node that constituted the head model (Fig.9(a)). The impact velocity that satisfied this case was 10.4 m/s, which caused negative pressure on the opposite side of impact (Fig.9(b)) and a skull fracture in the occiput region of the head. The input force duration (i.e. the length of time that the impact object contacted the head) of this case was counted from the animation of the analysis result, was 3.7 ms.

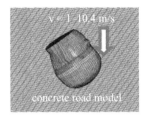   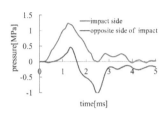

(a) Simulation outline   (b) Intracranial pressure fluctuations (v = 10.4 m/s)

Fig. 9. Simulation outline and results of case 6. (a) Simulation outline. The impact velocity (1–10.4 m/s) was applied to the node that constituted the head model. For better view, only part of the concrete road model is displayed. (b) The intracranial pressure fluctuations of the impact side and its opposite side. The results show that negative pressure occurred only on the opposite side of impact.

[Case 7]

A sailor fell from 10 m high onto a deck, and the occipital region of head was impacted, causing a skull fracture in the occiput region and contrecoup contusions in the lower side of the frontal lobe and frontal pole. In the simulation, the impact velocity (1–14m/s, from the law of conservation of mechanical energy, the maximum velocity was calculated to be 14 m/s) was applied to the node that constituted the head model (Fig.10(a)). The impact velocity that satisfied this case was around 10 m/s, which caused negative pressure on the opposite side of impact (Fig.10(b)) and skull fracture in the occiput region of the head. The input force duration (i.e. the length of time that the impact object contacted the head) of this case was counted from the animation of the analysis result, was 3.6 ms.

[Case 8]

A man fell down onto an asphalt road, and the occipital region of the head was impacted, causing a skull fracture in the occiput region and a contrecoup contusion in the frontal pole. In the simulation, the impact velocity (1–15 m/s) was applied to the node which constituted the head model (Fig.11(a)). The impact velocity which satisfied this case was around 11m/s, which caused the negative pressure at the opposite side of impact (Fig.11(b)) and the skull fracture at the occiput region of the head. The input force duration (i.e. the length of time that the impact object contacted the head) of this case was counted from the animation of the analysis result, was 2.9 ms.

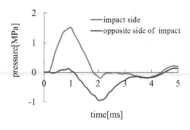

(a) Simulation outline   (b) Intracranial pressure fluctuations (v = 10 m/s)

Fig. 10. Simulation outline and results of case 7. (a) Simulation outline. The impact velocity (1–14 m/s) was applied to the node that constituted the head model. For better view, only part of the deck model is displayed. (b) The intracranial pressure fluctuations of the impact side and its opposite side. The results show that negative pressure occurred only on the opposite side of impact.

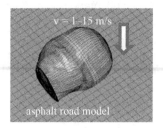
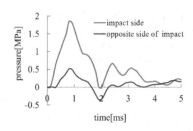

(a) Simulation outline   (b) Intracranial pressure fluctuations (v = 11 m/s)

Fig. 11. Simulation outline and results of case 8. (a) Simulation outline. The impact velocity (1–15 m/s) was applied to the node that constituted the head model. For better view, only part of the asphalt road model is displayed. (b) The intracranial pressure fluctuations of the impact side and its opposite side. The results show that negative pressure occurred only on the opposite side of impact.

[Case 9]

A man fell down onto an asphalt road, and the right occipital region of the head was impacted, causing a skull fracture in the right temporal region and contrecoup contusions in the lower sides of the right frontal and right temporal lobe and on the lateral surface of the left temporal lobe. In the simulation, the impact velocity (1–15 m/s) was applied to the node which constituted the head model (Fig.12(a)). The impact velocity which satisfied this case was around 7m/s, which caused the negative pressure at the opposite side of impact (Fig.12(b)) and the skull fracture at the right occipital region of the head. The input force duration (i.e. the length of time that the impact object contacted the head) of this case was counted from the animation of the analysis result, was 3.4 ms.

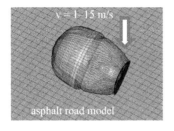 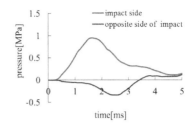

(a) Simulation outline   (b) Intracranial pressure fluctuations (v = 7 m/s)

Fig. 12. Simulation outline and results of case 9. (a) Simulation outline. The impact velocity (1–15 m/s) was applied to the node that constituted the head model. For better view, only part of the asphalt road model is displayed. (b) The intracranial pressure fluctuations of the impact side and its opposite side. The results show that negative pressure occurred only on the opposite side of impact.

3.3 Relationship between force duration and contusion type

Coup and contrecoup contusions were classified according to the force duration which was obtained from the simulations (Table 3). In coup contusion cases, the input force durations were 1.4~2.7ms, and in contrecoup contusion cases, the input force durations were 2.9~3.7ms. Short (1.4~2.7ms) and long (2.9~3.7ms) force durations were obtained from coup and contrecoup contusion cases, respectively. These results show that when the head is impacted by sharp-cornered objects, coup contusions due to the short force durations are caused more easily; meanwhile, when the head is impacted by objects with flat surfaces, contrecoup contusions due to the long force durations are caused more easily.

	Case	Impact object	Impact region	Force duration
Coup contusion	1	Beer bottle	Frontal region	1.4 ms
	2	Wooden box (1.5 t)	Right temporal region	2.7 ms
	3	Sake bottle (empty: 700 g)	Right parietal region	2.2 ms
Contrecoup contusion	4	Tank lorry	Parietal occipital region	3.7 ms
	5	Steel Box (600 kg)	Occipital region	3.6 ms
	6	Concrete (5.5 m)	Upper side of occipital region	3.7 ms
	7	Wooden deck (10 m)	Occipital region	3.6 ms
	8	Asphalt (160 cm/58 kg)	Occipital region	2.9 ms
	9	Asphalt (156 cm/59 kg)	Right occipital region	3.4 ms

Table 3. Force durations obtained from the simulations. Short (1.4~2.7ms) and long (2.9~3.7ms) force durations were obtained from the coup and contrecoup contusion cases, respectively.

3.4 Summary of chapter 3

In this chapter, the relationship between the input force duration and the dynamic response of the human head were focused by reconstructing the real-world brain injury accident cases using the finite element human head model based on the contusions were caused by the negative pressure inside the skull. The results were shown, in the coup contusion cases, the impact objects were sharp so the short force durations were obtained and the negative pressure occurred at the impact side. In contrast, in the contrecoup contusion cases, the impact objects had flat surfaces so the long force durations were obtained and the negative pressure occurred at the opposite side of impact. In the other words, coup contusion tends to occur when the force duration is shorten, and contrecoup contusion tends to occur when the force duration is lengthen, so the force duration is shown to be the parameter for separating coup or contrecoup contusions.

4. The mechanism of DAI

Gennarelli et al. characterized the strain caused by a rotational acceleration load to the head, and proposed the strain threshold of DAI by using finite element models for the crania of human and baboon (Meaney et al., 1995; Susan et al., 1990). They observed the shear deformation generated in each part of the brain with a high speed camera when the head was rotated, and assumed that the strain that caused DAI was larger than 9.4%. Pfister et al. developed a device which could generate shear deformation of cells cultured on a plane by pulling the ground substance and made it possible to produce a strain up to 70% and a strain rate up to 90/s (Pfister et al., 2003). Laplaca et al. cultured cells in a gel and generated 3D deformation of the cells by producing a shear deformation of the gel (strain < 50%, strain rate < 30/s). Neuronal cells showed a lower tolerance to this strain than the glial cells (LaPlaca et al., 2005; Cullen et al., 2007).Tamura et al. analyzed the difference in strain caused by a tensile test between porcine brain tissue and nerve fiber in the white matter, and reported that the maximum neural fiber strain was ~25% of the level in the surrounding tissue (Tamura et al., 2006). Nakayama et al. showed morphological changes of axons and the progress of this damage over time caused by one-dimensional, horizontal oscillations of nerve cells (Nakayama et al., 2001). These data suggested that an axon would receive damage with a strain of larger than 10% and a strain rate of larger than 10/s.

In this chapter, in order to clarify the influence of the axonal damage on the damage of cells, the cytotoxicity and mortality of PC12 cell (rat adrenal pheochromocytoma cell) line were evaluated by applying huge acceleration to cells. Huge acceleration was generated by an impact machine and was given to 2 kinds of cells, i.e. with and without axons.

4.1 Cell culture

In this study, PC12 cell line (obtained from Riken Cell Bank, Tsukuba, Japan) was used. Cells were cultured in DMEM (Dulbecco's Modified Eagle's Medium; Gibco, Gland Island, NY, USA) supplemented with 10% FBS (fetal bovine serum), 10% HS (horse serum), and Pennicillin-Streptmycin (10U/ml, 100ng/ml, Sigma-Aldrich, St. Louis, MO, USA) in 95%Air, 5% $CO2$ at 37℃.

In the impact experiment, the PC12 cells with and without axons were used. Axons were developed by adding 50ng/ml NGF (nerve growth factor, 2.5S; Invitrogen, Carlsbad, CA, USA). The cells were seeded in PLL (poly-l-lysine)-coated dishes (Φ35mm) at a density of 1×104/cm2, and incubated for 5 days. Phase-contrast images of cells are shown in Fig. 13.

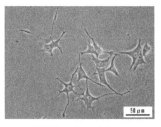 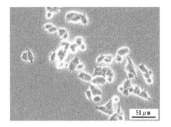

(a) PC12 cells with axons   (b) PC12 cells without axons

Fig. 13. Phase-contrast images of PC12 cells cultured for 5days. (a) PC12 cells with axons. (b)PC12 cells without axons.

4.2 Impact experiment with huge acceleration

The outline of the impact experiment is shown in Fig. 14. The impact experiments were carried out with an SM-100-3P impact machine (AVEX Electronics Inc., PA, USA). The impact machine can accelerate the specimen, with a range of acceleration of 3~20000 G and for duration of 0.1~60ms.

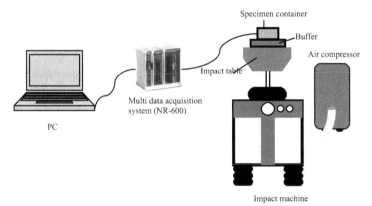

(a) Devices used for the impact experiments

(b) Impact machine (c) Cultured dish and culture solution (d) Specimen container (e) Impact table

Fig. 14. Major components of the impact experiment. A dish seeded with PC12 cells was filled with culture solution (c), and fixed in position in a stainless plate (d). The plate was then fixed on the impact table (e).The impact table was elevated and dropped by the compressor.

A dish seeded with PC12 cells was filled with culture solution (Fig. 14(c)), and fixed in position in a stainless plate (Fig. 14(d)). The plate was then fixed on the impact table (Fig. 14(e)).The impact table was elevated and dropped by the compressor. The impact table collided against the buffer generating an impact with huge velocity to the dish.

The strain applied to the cultured dish was measured with a strain gauge (KFG-2N-120-C1; Kyowa Electronic Instruments Co.), which was attached to the bottom of the dish and connected to a multi-data acquisition system (NR-600; Keyence Co.). A total of 72 impact experiments were carried out, i.e. 6 experiments per condition. The input acceleration ranged from 3000 G to 10000 G, and the duration of all accelerations was 0.1ms.

4.3 Evaluation of Injury

4.3.1 Cytotoxicity of PC12 Cells

Cytotoxicity of cells was measured by the LDH (lactate dehydrogenase) assay. LDH is a soluble cytosolic enzyme that is released into the culture medium following loss of membrane integrity resulting from either apoptosis or necrosis.

After the impact, the culture solution in the dish was centrifuged for 10 minutes at 250 G, 100 μl of the reaction solution (cytotoxicity detection kit (LDH), Roche Diagnostics) was added to 100 μl of the supernatant. The reactions were incubated at room temperature for 30 min using 96-well plates (96-well Micro Test III assay Plate, BD Falcon), and 1N-HCl was added as a stop solution. The absorbance of each specimen was measured at 490 nm by a Microplate Reader (Model 680; Bio-Rad Laboratories). Similarly, the absorbance of the low control specimens (no load cells) and high control specimens (cells dissolved by 1% TritonX-100 in PBS) were measured. The cytotoxicity of PC12 cells was calculated by the following equation.

$$\text{Cytotoxicity } (\%) = (C - LC) / HC \times 100\% \tag{1}$$

where C is the LDH quantity (IU/l) obtained from the impact experiment specimen, LC is the LDH quantity (IU/l) obtained from the low control specimen, and HC is the LDH quantity (IU/l) obtained from the high control specimen.

4.3.2 Mortality of PC12 cells

Mortality of cells was measured by the dye exclusion method with trypan blue dye. This method determines cell viability by mixing a suspension of live cells with a dilute solution of trypan blue; cells that exclude dye are considered to be alive, while stained cells are considered to be dead.

After the impact, cells were separated from the dish with 0.25% Trypsin-EDTA (Gibco), collected in a microcentrifuge tube (1.5 ml), and centrifuged for 30 seconds at 2000 G.

$$\text{Mortality } (\%) = N / M \times 100\% \tag{2}$$

where N is the number of the dead cells and M is the total number of cells.

4.3.3 Morphological change of axon

When an axon is damaged, terminal swellings coincide with the detachment of the growth cones from the substrate. The detachment of the growth cones from the substrate destroys the cytoskeletal network, which determines and maintains cell shape, resulting in a spherical deformation of the axon. Terminal swellings form in the early stages of the injury. When the cytoskeletal destruction occurs at non-terminal sites along the axon, spherical deformations develop slowly, and these appear as beads. Beadings grow in the later stages of injury (Nakayama et al., 2001; Fujiwara et al., 2004).

After the impact, the morphological changes of the axons were observed with a phase-contrast microscope.

4.3.4 Statistical analysis

Statistical analysis of the cytotoxicity and mortality of cells with axons and without axons for each experimental condition were assessed with the t-test; $p < 0.05$ was considered to be statistically significant. Data were expressed as the mean ± standard error of the mean (SEM).

4.4 Results of damage evaluation

4.4.1 Strain and strain rate obtained from the impact experiments

The average strain and strain rate at the bottom of dish are shown in Table 4. As an example, the strain fluctuation when the peak of acceleration was 7000G is shown in Fig.15.

A strain from 0.035% to 0.201% and a strain rate from 6.67/s to19.02/s were measurable on the bottom of dish with the input accelerations from 3000 G to 10000 G. The strain measured at the bottom of dish had increased linearly as the input acceleration increased.

The strain rate tended to increase linearly as the input acceleration increased. However, when the input acceleration was 5000, 6000 and 8000 G, the strain rate was 13.02/s, 13.11/s, and 13.36/s, respectively. There was no significant difference between these strain rates. The strain rate obtained at 7000 G was 14.62/s, which was larger than the strain rate obtained at 8000G. It was difficult to control the duration of the acceleration accurately in the impact experiment when the input acceleration was very powerful. Although the strain obtained at 8000 G was larger than the strain obtained at 7000G, the duration of 8000 G became longer than the duration of 7000 G, resulting in the smaller strain rate at 8000 G.

Condition number	Peak acceleration [G]	Duration [ms]	Average strain [%]	Average strain rate (1/s)
1	3000	0.1	0.035	6.67
2	5000	0.1	0.065	13.02
3	6000	0.1	0.148	13.11
4	7000	0.1	0.164	14.62
5	8000	0.1	0.187	13.36
6	10000	0.1	0.201	19.02

Table 4. Strain and strain rates obtained from the impact experiments. The strain rate tended to increase linearly as the input acceleration increased.

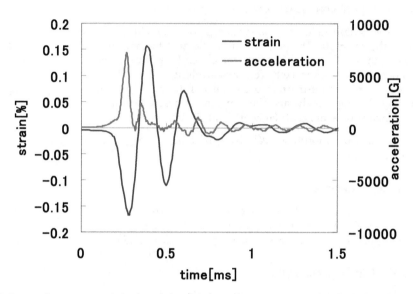

Fig. 15. Strain fluctuation when the peak of the acceleration was 7000 G. The strain of this input was 0.164%.

4.4.2 Cytotoxicity of PC12 cells

Cytotoxicity of cells with and without axons immediately after the impact experiment is shown in Fig.16. The relationship between the input acceleration and cytotoxicity of cells is shown in Fig.16 (a), the relationship between the strain and cytotoxicity of cells is shown in Fig.16 (b), and the relationship between the strain rate and cytotoxicity of cells is shown in Fig.16(c). Since the input of the impact experiment was acceleration, the strain rate could not be controlled in detail; the data obtained from 5000-8000G were concentrated on around 14/s as shown in Table 4.

Cytotoxicity of cells seemed to increase as the input acceleration and strain increased, but these relationships did not show strong correlations (Figs. 16(a), and 16(b)). Although the tendency that cytotoxicity of cells increased as the strain rate increased was shown, the correlation could not be quantitatively evaluated, because in the small strain rate range around 14/s cytotoxicity of cells did not increase monotonically.

For the two results obtained from 3000 and 5000 G (correspond to 0.035% and 0.065% in the strain, and to 6.67/s and 13.02/s in the strain rate), cytotoxicity of cells with axons was not significantly higher than in cells without axons. When the input acceleration was larger than 6000G, the strain was larger than 0.15%, and the strain rate was larger than 13.11/s, cytotoxicity of cells with axons was significantly higher than in cells without axons (Figs. 16(a), 16(b), and 16(c); *$p<0.05$).

Therefore, it appeared that cytotoxicity of cells increased as the strain rate increased, and cells with axons were more easily damaged than cells without axons when the strain rate was larger than 13.11/s.

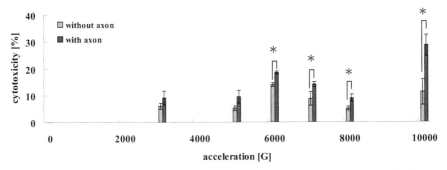

(a) Relationship between the input acceleration and cytotoxicity of PC12 cells

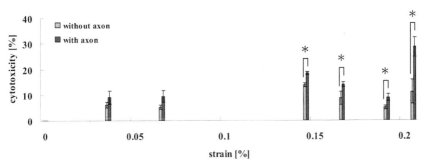

(b) Relationship between the strain and cytotoxicity of PC12 cells

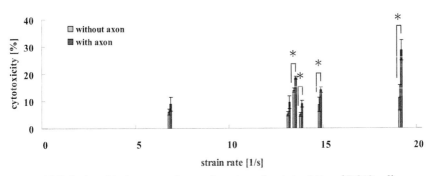

(c) Relationship between the strain rate and cytotoxicity of PC12 cells

Fig. 16. Experimental results of cytotoxicity of PC12 cells. Error bars represent SEM. (* $p < 0.05$)

4.4.3 Mortality of PC12 cells

Mortality of cells with and without axons immedieately after the impact experiment is shown in Fig. 17. The relationship between the input acceleration and mortality of cells is shown in Fig. 17(a), the relationship between the strain and mortality of cells is shown in Fig. 17(b), and the relationship between the strain rate and mortality of cells is shown in Fig. 17(c).

Mortality of cells seemed to increase as the input acceleration and strain increased, but these relationships did not show strong correlations (Fig. 17(a), and 17(b)). Although the tendency that mortality of cells increased as the strain rate increased was shown, the correlation could not be quantitatively evaluated, because in the small strain rate range around 14/s mortality of cells did not increase monotonically.

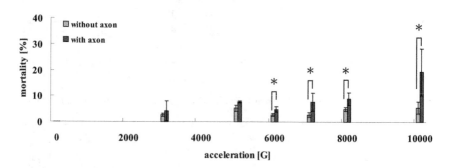

(a) Relationship between the input acceleration and mortality of PC12 cells

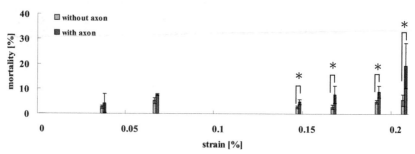

(b) Relationship between the strain and mortality of PC12 cells

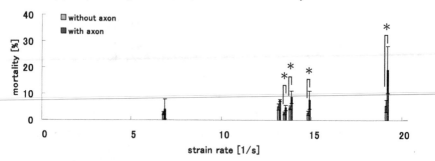

(c) Relationship between the strain rate and mortality of PC12 cells

Fig. 17. Experimental result of mortality of PC12 cells · Error bars represent SEM. (* $p < 0.05$)

For the two results obtained from 3000 and 5000 G (correspond to 0.035% and 0.065% in the strain, and to 6.67/s and 13.02/s in the strain rate), mortality of cells with axons was not significantly higher than in cells without axons. When the input acceleration was larger than 6000 G, the strain was larger than 0.15%, and the strain rate was larger than 13.11/s, mortality of cells with axons was significantly higher than in cells without axons (Figs. 17(a), 17(b), and 17(c); *$p<0.05$).

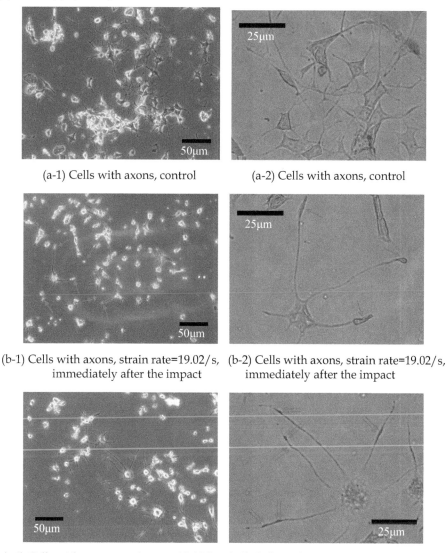

(a-1) Cells with axons, control   (a-2) Cells with axons, control

(b-1) Cells with axons, strain rate=19.02/s,   (b-2) Cells with axons, strain rate=19.02/s,
immediately after the impact   immediately after the impact

(c-1) Cells with axons, strain rate=19.02/s,   (c-2) Cells with axons, strain rate=19.02/s,
4 h after the impact   4 h after the impact

Fig. 18. Phase-contrast images of PC12 cells, control (a-1, a-2), immediately after the impact (b-1, b-2) and 4 h after the impact(c-1,c-2). For a clearer view, (a-1), (b-1), and (c-1) are enlarged in (a-2), (b-2), and (c-2), respectively. Terminal swellings are indicated with single arrows and beadings are indicated with double arrows.

Therefore, it appeared that mortality of cells increased as the strain rate increased, and cells with axons had an increased mortality than cells without axons when the strain rate was larger than 13.11/s.

4.4.4 Morphological change

In order to observe damage to axons, cells with axons were observed with a phase-contrast microscope. Phase-contrast images of cells with axons on the dish are shown in Fig.18.

Phase-contrast images of cells with axons before the experiments (control) are shown in Figs. 18(a-1) and 18(a-2). Cells extending their axons and a network creates with these axons can be observed. Cells and their axons attached to the substrate. Terminal swellings or the beadings of axons are not observed at this stage. Phase-contrast images of cells immediately after the impact with the strain rate of 19.02/s are shown in Figs. 18(b-1) and 18(b-2); terminal swellings of axons are indicated with single arrows in the figures. Terminal swellings can be observed when the terminals of axons detach from the substrate. Since beading occurs in the later stages of damage, beading can not be observed yet at this stage. Phase-contrast images of cells 4 h after the impact with the strain rate of 19.02/s are shown in Figs. 18(c-1) and 18(c-2); beadings in the damaged regions are indicated with double arrows in the figures. Although beadings can not be observed in the images immediately after the impact as shown in Figs. 18(b-1) and 18(b-2), beadings can be clearly observed in the images taken after 4 h as shown in Figs. 18(c-1) and 18(c-2).

4.5 Summary of chapter 4

In this chapter, in order to study the influence of the axonal damage on cell damage, an impact experiment with huge acceleration was performed on PC12 cell line. In order to evaluate damage to axon, the impact experiments were performed on cells with and without axons. The strain at the bottom of cultured dish was measured, and the strain rate was calculated. The cytotoxicity and mortality of PC12 cells were evaluated by the input acceleration, strain and strain rate. As a result, the strain rate seemed to be the most appropriate to evaluate the cytotoxicity and mortality of cells. The cytotoxicity and mortality of cells increased as the strain rate increased, and cells with axons were more easily damaged and had an increased mortality than cells without axons when the strain rate was larger than 13.11/s. These data suggest that the presence of axons increased the cytotoxicity and mortality of cells.

5. Conclusion

In this study, to elucidate the mechanics of head injuries, the engineering approach using finite element analysis and the biological approach using cultured cells are performed. In the engineering approach, the results of the reconstruction of the real-world brain injury accident cases are shown, in the coup contusion cases, the negative pressure occurred at the impact side and it had direct correlation with the short force durations. In contrast, in the contrecoup contusion cases, the negative pressure occurred at the opposite side of impact and it had direct correlation with the long force durations. As the result, the force duration is shown to be the parameter for separating coup or contrecoup contusions. In the biological approach, the results of the impact experiments using the cultured cells are shown, the strain rate seemed to be the most appropriate to evaluate the cytotoxicity and mortality of cells. Damage to axons was confirmed by terminal swellings and beadings of the axons. These data indicated that the presence of axons increased the cytotoxicity and mortality of cells.

6. Future directions

In the future works, in order to evaluate all types of traumatic brain injury comprehensively, the tolerance of nerve damage which is obtained from the impact experiment of cultured cells is applied to numerical analysis. Therefore, to construct a human head finite element model containing the nerve fibers, a method to accurately measure the material properties of nerve fibers have to be developed and the structure of the neural network inside the skull have to be clarified by the image processing.

7. Acknowledgement

We thank Associate Prof. Atsushi Senoo (Tokyo Metropolitan University, Faculty of Health Sciences, Division of Radiological Sciences) for his assistance in photographing MRI data to construct the finite element human head model.

8. References

Aomura, S.; Fujiwara, S & Ikoma, T. (2002). The Study on the Influence of Different Interface Conditions on the Response of Finite Element Human Head Models under Occipital Impact Loading, *JSME International Journal, Series C*, Vol.46, No.2, pp.583-593.

Aomura, S.; Zhang, Y.; Fujiwara, S. & Nishimura, A. (2008). Dynamic Analysis of Cerebral Contusion under Impact Loading, *Journal of Biomechanical Science and Engineering*, Vol. 3, No. 4, pp.499-509.

Cullen D.K.; Simon C.M. & LaPlaca M.C. (2007). Strain rate-dependent induction of reactive astrogliosis and cell death in three-dimensional neuronal-astrocytic co-cultures, *Brain Research*, Vol.1158, pp.103-115.

Doorly, M.C. & Gilchrist, M.D. (2006). The Use of Accident Reconstruction for the Analysis of Traumatic Brain Injury Due to Head Impacts Arising from Falls, *Computer Methods in Biomechanics and Biomedical Engineering*, Vol.9, No.6, pp.371-377.

Fujiwara S.; Ogawa Y.; Hirabayashi M.; Inamura K. & Aihara H. (1998). Biomechanics of Cerebral Contusion and Diffuse Axonal Injury, *Proceedings of the 6th Indo Pacific Congress on Legal Medicine and Forensic Sciences*, pp.1003-1006.

Fujiwara S.; Yanagisawa A.; Sato H.; Shindo Y.; Koide K. & Nishimura A. (2004). DAI Diagnosis Using β-APP mRNA Expression Analysis and its Application to Forensic Medicine, *Research and Practice in Forensic Medicine*, Vol.47, pp.85-90.

Fujiwara, S.; Yanagida, Y. & Mizoi, Y. (1989). Impact Induced Intra-cranial Pressure Caused by an Accelerated Motion of the Head or by Skull Deformation: An Experimental Study Using Physical Models of the Head and Neck, and Ones of the Skull, *Forensic Science International*, Vol.43, pp.159-169.

Fujiwara, S.; Yanagida, Y.; Fukunaga, T.; Mizoi, Y. & Tatsuno, Y. (1986). Studies on Cerebral Contusion in the Fatal Cases by Blow, Fall and Fall Down, *The Japanese Journal of Legal Medicine*, Vol.40, No.4, pp.377-383.

Gennarelli, T.A. (1983). Head Injury in Man and Experimental Animals: Clinical Aspects, *Acta Neurochirurgica*, Suppl.32, pp.1-32.

Gross, A.G. (1958). A New Theory on the Dynamics of Brain Concussion and Brain Injury, *Journal of Neurosurgery*, Vol.15, pp.548-561.

Gurdjian, E.S.; Lissner, H.R.; & Hodson, V.R. (1966). Mechanism of Head Injury, *Clinical Neurosurgery*, Vol.12, pp.112-128.

LaPlaca M.C.; Cullen D.K.; McLoughlin J.J. & Cargill R.S. (2005). High rate shear strain of three-dimensional neural cell cultures: a new in vitro traumatic brain injury model, *Journal of Biomechanics*, Vol.38, No.5, pp.1093-1105.

Meaney D.F.; Smith D.H.; Shreiber D.I.; Bain A.C.; Miller R.T.; Ross D.T. & Gennarelli T.A. (1995). Biomechanical Analysis of Experimental Diffuse Axonal Injury, *Journal of Neurotrauma*, Vol.12, No.4, pp.689-694.

Nahum, A.M. & Smith, R. (1977), Intracranial Pressure Dynamics during Head Impact, *Proceeding of 21st Stapp car Crash Conference*, pp.339-366.

Nakayama Y.; Aoki Y. & Niitsu H. (2001). Studies on the mechanisms responsible for the formation of focal swellings on neuronal processes using a novel in vitro model of axonal injury, *Journal of Neurotrauma*, Vol.18, No.5, pp.545-554.

Nishimoto, T.; Murakami, S.; Abe, T. & Ono, K. (1998). Mechanical Properties of Human Cranium and Effect of Cranial on Extradural Hematoma, *Transactions of the Japan Society of Mechanical Engineers, Part A*, Vol.61, No.591, pp. 2386-2392.

Ommaya A.K. & Gennarelli T.A. (1974). Cerebral concussion and traumatic unconsciousness, correlation of experimental and clinical observations on blunt head injuries, *Brain*, Vol.97, pp.633-654.

Pfister B.J.; Weihs T.P.; Betenbaugh M. & Bao G. (2003). An in vitro Uniaxial Stretch Model for Axonal Injury, *Annals of Biomedical Engin_eering*, Vol.31, No.5, pp.589-598.

Raul, J.S.; Baumgartner, D.; Willinger, R. & Ludes, B. (2006). Finite Element Modelling of Human Head Injuries Caused by a Fall, *International Journal of Legal Medicine*, Vol.120, No.4, pp.212-218.

Riordain, K.O.; Thomas, P.M.; Phillips, J.P. & Gilchrist, M.D. (2003). Reconstruction of Real Word Head Injury Accidents Resulting from Falls using Multibody Dynamics, *Clinical Biomechanics*, Vol.18, pp.590-600.

Susan S.M.; Lawrence E.T. & Thomas G. (1990). Physical Model Simulation of Brain Injury in the Primate, *Journal of Biomechanics*, Vol.23, No.8, pp.823-836.

Tamura A.; Nagayama K. & Matsumoto T. (2006). Measurement of Nerve Fiber Strain in Brain Tissue Subjected to Uniaxial Stretch (Comparison Between Local Strain of Nerve Fiber and Global Strain of Brain Tissue), *Journal of Biomechanical Science and Engineering*, Vol.1, No.2, pp.304-315.

Viano, D.C.; Casson, I.R.; Pellman, E.J.; Zhang, L.; King, A.I. & Yang, K.H. (1998). Concussion in Professional Football Brain Responses by Finite Element Analysis: Part 9, *Neurosurgery*, Vol.57, No.5, pp.891-916.

Willinger, R. & Baumgartner, D. (2003). Human Head Tolerance Limits to Specific Injury Mechanisms, *International Journal of Crashworthiness*, Vol.6, No.8, pp.605-617.

Yanagida, Y.; Fujiwara, S. & Mizoi, Y. (1989). Differences in the Intra-cranial Pressure Caused by a 'Blow' and/or a 'Fall' — An Experimental Study Using Physical Models of the Head and Neck, *Forensic Science International*, Vol.41, pp.135-145.

Zhang, L.; Hardy, W.N.; Omori, K.; Yang, K.H.; & King, A.I. (2001). Recent Advances in Cerebral Injury Research: A New Model and New Experimental Data, *Proceedings of the ASME Bioengineering Conference*, pp.831-832.

Zhang, Y.; Aomura, S.; Nakadate, H. & Fujiwara, S. (2010). Study on the Mechanism of Cerebral Contusion Based on Judicial Autopsy Report, *Proceedings of the 6th World Congress on Biomechanics*, Vol.31, pp.505-508.

8

Quality Assessment of E-Health Solutions in Primary Health Care – Approach Based on User Experience

Damir Kralj
Ministry of the Interior, PD Karlovac
Croatia

1. Introduction

The significance of primary health care (PHC) within the overall health care system of the each country is tremendous. PHC provides the first contact between patients and health care system, and keeps the most complete medical records of particular patient which could be used later for different medical secondary purposes. Well organized and computerized PHC significantly improves both the quality of care and contributes to significant savings in treatment. In order to improve the quality of such information systems, it is necessary to introduce a methodology for measuring their actual quality. Our research focuses on the creation of models for assessing the quality of IT solutions, based on users' (doctors') experiences, within the newly introduced primary health care information system in the Republic of Croatia.

The process of implementation of the national e-Health infrastructure in the Croatian public health care system started in 2006 by introduction of the Croatian primary health care information system (PHCIS or CEZIH in Croatian language) (Croatian Institute for Health Insurance [CIHI], 2010). The first areas of the system implementation includes the integration of family doctor's offices (FDO) into a comprehensive system, that includes the integration of various types of FDO specialized solutions with national infrastructure, Croatian Institute for Health Insurance and Public Health Authority. The system is generally tested "in vivo" i.e. in real production conditions and with real patients' data collected in FDO. The central part of the information system was designed by the renowned company, specialized in area of those projects, as a very stable and quality system based on well-defined business processes, legal and semantic rules, and communication and messaging standards such as EN13606 and HL7v3 (Končar & Gvozdanović, 2006). Design of the applications for managing of the electronic healthcare records (EHR) in FDO was left to a number of small IT companies competing on the Croatian market. These applications had to undergo certification process defined by the Ministry of Health and Social Welfare (MHSW). Certification included only the area of communication and basic data exchange with the central part of the system. The concept and functionality of these applications has been left to the manufacturers of these applications (Kralj & Tonković, 2009). With such situation in place, it seems very difficult to measure the quality and effectiveness of an information system which is in the early stage of development. For these reasons, our motivation was to establish a methodology to quantify and qualify overall quality criteria.

2. Methods used in the project

For the purposes of this study, we defined a methodology that consists of eight steps showed on Fig. 1. As we can see, our methodology is based on overview and analysis of domestic papers and foreign projects, studies, standards, initiatives and certification criteria. On these foundations, we constructed our assessment and built assessment tool i.e. questionnaire.

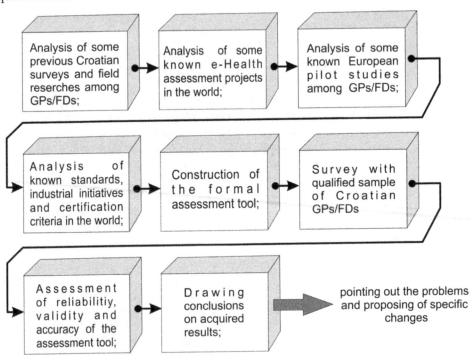

Fig. 1. Preview of used methodology

3. Overview of the relevant documents and projects

The basic idea of the formation of such a methodology arises from so-called frameworks for assessing of community readiness for the introduction of e-Health. One of the earliest references is the "Framework for rural and remote readiness in telehealth" that was conducted in 2002 by Canada's advanced research and innovation network CANARIE (CANARIE, 2002). This paper describes the basic assumptions that derive from the theory of change and stages of change. Readiness is defined as a cognitive indicator of actual conditions, with determining the factors that contribute to success and factors affecting the failure of some innovations. The next interesting example can be found in association of the

Aga Khan University in Pakistan and the University of Calgary in Canada (Khoja et al., 2007a). The subject of the project was development of tools for e-Health readiness assessment in developing countries. In this project were proposed methods for validation and reliability testing of the tool for e-Health readiness assessment. The assessment is based on a quantitative presentation of qualitative data. In order to verify the reliability of the tool and to avoid multiple control testing, there was introduced a calculation of the Cronbach's Alpha (α) coefficient of correlation for each category of readiness and for all categories combined. A third interesting example is found in the study "e-Health Readiness Framework from Electronic Health Records Perspective" (Li, 2008) conducted on the University of New South Wales in Australia. Research contribution of this study consists of three essential elements: a model of framework, methodology of assessment and evaluation of framework based on criteria and case studies. Our work is taking good reference in some basic analysis from these models. However, some early results have shown that they are not sufficient in Croatian example. More precisely, all the models referenced above are based on an analysis of isolated cases ("in vitro") by gathering the elements to assess the readiness of a small part of health system for the introduction of e-Health concept, while our framework requires experiences assessment in real production ("in vivo") and large scale deployment, which leaves us with highly challenging environment that requires careful assessment and offers less change manoeuvre space. Analysis of the Croatian papers drew our attention to specific problems before (Kern & Polašek, 2007) and shortly after (Kralj & Tonković, 2009) the beginning of implementation of e-health concept. Analysis of worldwide standards such as HL7, EN13606 and DICOM, and initiatives such as Integrating the Healthcare Enterprise (IHE) and EHR-implement, is simply unavoidable. In addition, when creating our methodology, we have also taken into account latest recommendations from European Institute for Health Records (EuroRec) EHR-Q TN project, criteria of The EuroRec EHR Quality Seal Level 1 and Level 2 (EuroRec, 2010), projects and recommendations of the American Office of the National Coordinator for Health Information Technology (ONC) and certification criteria of the Certification Commission for Health Information Technology (CCHIT) (Centers for Medicare and Medicaid Services, 2010), which has lead us to the final readiness assessment model. It should be noted that these American recommendations and criteria resulted in mid-2010 with a set of certification criteria called Meaningful Use of EHR Stage 1, which is a direct consequence of the American Recovery and Reinvestment Act (ARRA) of 2009. The contents of European studies that were conducted by research agencies Empirica (Dobrev et al., 2008) and Health Consumer Powerhouse (Björnberg et al., 2009) were also of great help in the making process of the assessment tool.

4. Construction of the assessment tool

Based on previously mentioned foundations we made a framework for the assessment tool i.e. questionnaire. The framework consists of seven main units. While the first unit contains general questions about the doctor and his/her office, the remaining six units measure major dimensions i.e. categories of experience which our work has identified as needed. As we see in Table 1, these six categories are: basic experience, technological experience, engagement, domain experience, organizational experience and societal experience. Each of these categories is a key performance indicator (KPI) of the current state of implementation of the e-Health concept in the health system as a whole. When designing this framework, we tried to include the key factors that can describe doctors' problems and attitudes, doctors' involvement in the process of adopting of new technologies, the impact of new technologies on the domain workflow, changes in communication with other health organizations and offices, and, of course, the impact on communication between doctor and patient. In addition, we have tried as much as possible to reduce any overlap between categories i.e. to ensure unambiguity of the categories.

A) BASIC	B) TECHNOLOGICAL	C) ENGAGEMENT
• Attitude about use of computers in FDO; • Organization of work on computer support; • Impact assessment of use of computers to work process; • Attitude about basic elements of e-Health.	• Problems with hardware and network support in FDO; • Quality and reliability of EHR applications in FDO; • Readiness of diagnostic equipment for use in e-Health; • Data protection and patient safety.	• Self-assessment of the IT and medical domain knowledge; • FD engagement in process of new system implementation; • Use of EHR for evaluation of doctor's work and research; • Care about the safety and security of the EHR.
D) DOMAIN	E) ORGANIZATIONAL	F) SOCIETAL
• Domain usability and functi-onality of EHR applications; • Structuring and encoding of information in EHR application; • Implementation of advanced decision support systems; • Monitoring and quality of work assessment according to working guidelines; • Overall satisfaction with the EHR applications from the domain view.	• Use of e-mail communication with other health institutions; • Possibilities of migration to a paperless business; • Elements of e-business integrated into EHR application; • Forms of electronic reporting built in existing application support; • Interoperability and compa-tibility of EHR applications with current diagnostic systems.	• The impact of use of computers and EHR applications on patients' satisfaction; • The impact of health contents available on the Internet on behaviour of patients in FDO; • Forms of electronic communication between doctors and patients.

Table 1. Structure of the framework with general description of its categories

Based on previously described framework we made a rather comprehensive questionnaire. In total there are 118 questions of which 103 issues have been defined for assessment. These 103 questions for the assessment consist of 32 multiple choice questions in a Likert scale of 1-5, 54 questions with dichotomous answers (YES-NO i.e. 1 or 0), and 17 questions with offered answers that should be marked. General questions about the doctor and his/her office are very important for assessing the quality of the population sample. In addition, based on information about the structure of the measured population we can perform comparison of the results depending on the specific groups within the population (e.g. gender, specialization, age, years of service, etc.). The time allocated for completing the questionnaire was estimated at 20 minutes. The questionnaire therefore is rather straightforward and easy to use and populate.

Given that the questionnaires as this are very cumbersome to display on the site of small format, Table 2 shows the part of the questionnaire form with general questions about the doctor and his/her office, while Table 3 shows the main part of the questionnaire form for all six categories for assessment broken on multiple pages. Each category i.e. part of the questionnaire with each other is separated by a space line. For easier orientation in the questionnaire, all items i.e. questions are numbered. Questions are marked with a combination of the ordinal number of subcategories within the main category and the ordinal number of questions within the subcategories.

Dana about FDO	County:	
	Municipal:	
	City:	
	FD office type:	a) urban b) rural c) insular
	FD office autonomy:	a) in a health center b) under lease c) private
	Practice organization:	a) standalone b) group practice
Data about FD	Age:	
	Years of working:	
	Gender:	M - F
	Specialization?	Yes - No
	Do you use in your work the computer and the application for managing of EHR?	Yes - No
	Name of the EHR application (program) that you use:	

Table 2. Part of the questionnaire form with general questions about the doctor and his office

A	Question	Answer		
1.1	How long have you been using the computer in your practice? (years, months)			
1.2	Do you use a computer to record administrative data about patients?	Yes - No		
2.1	Who performs an update of administrative patient data?	a) doctor b) nurse c) doctor and nurse d) administrator		
2.2	Do you use your computer for recording of patients' medical data?	Yes - No		
2.3	Who performs an update of medical information about patients?	a) doctor b) nurse c) doctor and nurse d) administrator		
2.4	Do you use your computer during the examination and interview with the patient?	Yes - No		
3.1	Evaluate the impact of using of the computer on the quality of your work in the office.	significantly reduces the efficiency	1 - 2 - 3 - 4 - 5	significantly increases the efficiency
3.2	Evaluate the impact of computers due to the dynamics of your practice.	significantly slows	1 - 2 - 3 - 4 - 5	significantly accelerates
4.1	Express your opinion on the impact of the degree of integration of health information systems on the efficiency of the entire health system.	significantly reduces the efficiency	1 - 2 - 3 - 4 - 5	significantly increases the efficiency
4.2	Express your opinion on electronic prescribing of drugs (e-Prescription).	I do not support at all	1 - 2 - 3 - 4 - 5	I fully support
4.3	Express your opinion on electronic referral of patients (e-Referral).	I do not support at all	1 - 2 - 3 - 4 - 5	I fully support
4.4	Express your attitude about the secondary use of medical records of patients for the purpose of development and progress of the entire health care system.	I do not support at all	1 - 2 - 3 - 4 - 5	I fully support
4.5	Express your attitude about the creation of a central registry of electronic health records for all patients ("central EHR").	I do not support at all	1 - 2 - 3 - 4 - 5	I fully support

B	Question	Answer			
1.1	Does the nurse use a separate computer that is networked with your computer?	Yes - No			
1.2	Who performed the installation of your computer and network support in the office?	a) equipment supplier b) software vendor		c) friends or acquaintances d) yourself	
1.3	Which operating system are you using on your computer?	a)MS Windows XP b)MS Windows Vista c)MS Windows 7		d)MacOS c)Linux	
1.4	On which way are you connecting to the Internet?	a) modem (PSTN) b)ISDN c)ADSL		d)GSM or similar e)through the proxy of the greater network	
1.5	Does your nurse have access to the Internet?	Yes - No			
1.6	Rate the size of the cost of the hardware and network support in your practice.	very small	1 - 2 - 3 - 4 - 5	very large	
1.7	Rate the size of your office communication costs.	very small	1 - 2 - 3 - 4 - 5	very large	
2.1	How many of the EHR applications you changed so far? (number)				
2.2	Is your current HER application certified to work on the central HIS (CEZIH)?	Yes - No			
2.3	Rate the quality and availability of contextual help system in your EHR application (e.g. press the F1 key for the current problem)?	very poor or there is no	1 - 2 - 3 - 4 - 5	high quality and accessible	
2.4	Rate telephone support system (helpdesk), if it is ensured by the vendor of your EHR application?	very poor or there is no	1 - 2 - 3 - 4 - 5	high quality and accessible	
2.5	Rate the remote administration and troubleshooting, if it is ensured by the vendor of your EHR application?	very poor or there is no	1 - 2 - 3 - 4 - 5	high quality and accessible	
2.6	Rate the remote version update, if it is ensured by the vendor of your EHR application?	very poor or there is no	1 - 2 - 3 - 4 - 5	high quality and accessible	
2.7	Rate the remote update of prescribed nomenclatures, if it is ensured by the vendor of your EHR application?	very poor or there is no	1 - 2 - 3 - 4 - 5	high quality and accessible	
2.8	Rate the size of the cost of the software support in your practice.	very small	1 - 2 - 3 - 4 - 5	very large	
3.1	Do you use any kind of equipment that provides computer-readable results (ECG, spirometry, ultrasound, holter, etc.)?	Yes - No			

3.2	Mark a letter before the name of the equipment that you use, and which can be connected to your computer. (multiple answers possible)	a) ECG b) spirometer c) blood pressure gauge	d) ultrasound e) ECG holter f) RR holter
4.1	Which of the following actions require, in your opinion, a greater degree of security and data protection?	a) transactions between bank accounts b) transmission of electronic health records c) both equally	
4.2	Which of the following categories, in your opinion, is more important for patient safety in health care system?	a) protection against of unauthorized access to patient data b) timely access to patient's data c) both equally	
4.3	Which are the safety and privacy protection elements that you use in your practice? (multiple answers possible)	a) physical access limitation (locking) b) password on the PC startup c) password on the entrance of the EHR application d) encrypting the entire contents of the hard disk c) none of the above	
4.4	Which are the elements of protection against data loss that you use in your office? (multiple answers possible)	a) regular data backup included in EHR application b) making of additional backup copies which are stored separately c) use of the devices for uninterruptible power supply (UPS)	
5.1	Do you know what the IHE certification is? (short explanation - full name)	Yes - No	
5.2	Do you know what the HL7 standard is? (short explanation - purpose)	Yes - No	
5.3	Do you know what the EN13606 standard is? (short explanation - purpose)	Yes - No	
5.4	Do you know what the DICOM standard is? (short explanation - purpose)	Yes - No	
5.5	Should the doctors be, at least roughly, familiar with the above standards and recommendations?	Yes - No	

C	Question	Answer		
1.1	Did you attend any kind of informatics schools or courses in the past five years?	Yes - No		
1.2	Rate your level of IT knowledge.	very low	1 - 2 - 3 - 4 - 5	very high
1.3	Rate the doctors' overload with necessary level of ICT knowledge.	very low	1 - 2 - 3 - 4 - 5	very high
1.4	Rate your level of domain (medical) knowledge.	very low	1 - 2 - 3 - 4 - 5	very high

2.1	Rate the frequency of your visits to bibliographic databases on the Internet.	rarely (or never)	1 - 2 - 3 - 4 - 5	often
2.2	Rate the frequency of your visits to health portals on the Internet.	rarely (or never)	1 - 2 - 3 - 4 - 5	often
2.3	Rate the frequency of your visits to e-journals on the Internet.	rarely (or never)	1 - 2 - 3 - 4 - 5	often
2.4	Rate the intensity of use of e-mail in the life and work.	rarely (or never)	1 - 2 - 3 - 4 - 5	often
2.5	Do you know what HON certificate is? (brief description - purpose)	Yes - No		
3.1	Have you been engaged in developing and testing some of the CEZIH certified EHR applications?	Yes - No		
3.2	Are you a member of an informal group of doctors who help each other in their work with purpose of better understanding and use of EHR applications?	Yes - No		
3.3	Do you have the role of "leader" in such a group?	Yes - No		
4.1	Do you use the information from your computer application for the professional evaluation of your work?	Yes - No		
4.2	Do you use the information from your computer application for administrative and financial evaluation of your work?	Yes - No		
4.3	Do you use the information from your computer application for your research work?	Yes - No		
5.1	How often do you change the password for entry into the EHR application?	a) daily b) weekly c) monthly	d) sometimes (not too often) e) still use the first one f) I do not use a password	
5.2	How the doctors, who replace you in your absence, realize access to your EHR application?	a) use my account b) I open new user account for every replacement c) I have opened a special account for replacement regardless of the physician		
5.3	How often do you make EHR data backup?	a) daily b) weekly c) monthly	d) sometimes e) after each change f) I do not perform backup	

D	Question	Answer		
1.1	Rate how your EHR application follows the domain workflow i.e. organization of the work (SOAP).	very bad	1 - 2 - 3 - 4 - 5	very well
1.2	Evaluate the usability of the user interface of your application (ease of handling and intuitiveness).	very difficult and confusing	1 - 2 - 3 - 4 - 5	very easy and intuitive
1.3	Is in your EHR application visible reason and the content of previous patient's visit(s), prior to entering data of a new visit?	Yes - No		
2.1	Mark the letter before the name of the disease classification which your application supports. (multiple answers possible)	a) ICD10 c) Read b) ICPC-2 d) SNOMED		
2.2	Is your EHR application capable for structured and atomized input of patient's vital parameters?	Yes - No		
2.3	Is your EHR application capable for menu oriented input of previously defined items (instead of typing a free text)?	Yes - No		
2.4	Does your EHR application contains regularly updated list of drugs prescribed by the Ministry of Health and Social Welfare?	Yes - No		
2.5	Does your EHR application contains regularly updated nomenclatures of diagnostic procedures prescribed by the Ministry of Health and Social Welfare?	Yes - No		
2.6	Does your EHR application contains regularly updated sets of working guidelines prescribed by the Ministry of Health and Social Welfare?	a) Yes, on the computer b) No, but, if necessary, it connects me with MHSW Internet portal	c) No, nor it connects me on external sources d) I don't know	
2.7	Does your EHR application contain regularly updated nomenclature of medical institutions as prescribed by the Ministry of Health and Social Welfare?	Yes - No		

!!!	NOTE: The set of questions 3.x refers to a systems which based on the described state of the patient helps the doctor by offering a closest solution (or several solutions) as assistance for a final decision. These systems can be part of the application and / or built in diagnostic equipment (e.g. ECG with auto-diagnostic). In the case of drug prescribing, these systems are based on possible drug interactions and side effects, etc.			
3.1	Does your EHR application have inbuilt functionalities for diagnosis decision support?	Yes - No		
3.2	If your previous answer is "Yes", rate the quality and usability of the system.	completely useless	1 - 2 - 3 - 4 - 5	fully usable
3.3	Are you using an advanced decision support system for determining a diagnosis independent of your EHR application (software or equipment with inbuilt functionalities)?	Yes - No		
3.4	Does your EHR application has inbuilt advanced helping functionalities for drug prescribing?	Yes - No		
3.5	If your answer is "Yes", rate the quality and usability of the system.	completely useless	1 - 2 - 3 - 4 - 5	fully usable
3.6	Does your EHR application has inbuilt advanced helping functionalities to assist physicians to refer patients for further treatment?	Yes - No		
3.7	If your previous answer is "Yes", rate the quality and usability of the system.	completely useless	1 - 2 - 3 - 4 - 5	fully usable
3.8	Rate how your EHR application monitors chronic diseases?	very poor or not at all	1 - 2 - 3 - 4 - 5	very good
3.9	Rate how your EHR application monitors allergies?	very poor or not at all	1 - 2 - 3 - 4 - 5	very good
4.1	Does your EHR application has some visible indicators (gauges, visual indicators) that alerts you to the current efficiency and the quality of your work (rates and coefficients)?	Yes - No		
4.2	Does your EHR application has any form of an advanced system for short-term and long-term evaluation of the quality of your work?	Yes - No		
5.1	From domain point of view, rate the overall satisfaction with your EHR application.	very unsatisfied	1 - 2 - 3 - 4 - 5	very satisfied
5.2	From domain point of view, compare your CEZIH compatible application with the application you were using before CEZIH system.	significantly worse	1 - 2 - 3 - 4 - 5	much better

E	Question	Answer
1.1	Do you use e-mail to communicate with your colleagues in primary health care?	Yes - No
1.2	Do you use e-mail for communication with specialists in clinics?	Yes - No
1.3	Do you use e-mail for receiving of laboratory results of your patients?	Yes - No
1.4	Do you use e-mail for receiving of specialist medical examination of your patients?	Yes - No
1.5	Do you use web-services or e-mail for ordering patients to specialists in clinics (e-ordering)?	Yes - No
2.1	Do you store medical documents received by e-mail in the particular folders inside of your EHR application?	Yes - No
2.2	Do you scan medical paper documentation of your patients and store them in electronic form to particular folders of your EHR application?	Yes - No
3.1	Can you perform e-ordering directly from your EHR application?	Yes - No
3.2	Does your EHR application support an electronic referral (e-referral)?	Yes - No
3.3	Does your EHR application support electronic drug prescribing (e-prescription)?	Yes - No
4.1	What is the form of the reports that you submit to the Croatian Institute for Health Insurance?	a) paper form b) paper form and floppy disk c) electronically through the web portal CEZIH d) electronically directly from the application
4.2	Does your application support electronically reporting to the Croatian Institute for Health Insurance?	Yes - No
4.3	Does your application support electronically reporting to Public Health?	Yes - No
4.4	Is your application ready for exchange (synchronization) of medical data with a central registry of EHRs?	Yes - No

4.5	Can you from your EHR application directly check the status of the insured patient in the Croatian Institute for Health Insurance?	Yes - No
5.1	What do you do when foreign and, increasingly, domestic insured, bring you the results of diagnostic procedures (X-ray, CT, MRI, etc.) made on CD or DVD?	a) I browse the pictures on the computer, and store the medium with patient's paper documents b) I browse the pictures on the computer, and store them in the particular folders of my EHR application c) I do not browse the pictures, but I ask the patient to bring me a hardcopy of findings (film or paper)
5.2	Does your application allow direct preview and storage of these digital records?	Yes - No

F	Question	Answer		
1.1	Rate the satisfaction of the patients with using of the EHR application in your practice.	very unsatisfied	1 - 2 - 3 - 4 - 5	very satisfied
1.2	Do you grant to your patients, on their request, insight into their medical history on your computer?	Yes - No		
1.3	Do you issue to your patients a complete history of his/her illness in electronic form (CD, floppy, etc.), for the purpose of the transfer to another physician and to ease work to the newly elected physician and, also, to contribute to patient's safety?	Yes - No		
2.1	Do your patients use available health contents on the Internet to be further informed about their condition?	Yes - No		
2.2	Do your patients comment with you the information about their problems that they have been collected on the Internet?	Yes - No		
3.1	Do you use e-mail to communicate with your patients?	Yes - No		
3.2	Do you use some form of electronic ordering patients for examination in your practice?	Yes - No		
3.3	Do your patients use e-mail for delivering of information about their medical condition (e.g. monitoring of chronic disease)?	Yes - No		
3.4	If your office has its own web page, which contents are available on it? (multiple answers possible)	a) advertising b) health advices c) e-ordering system d) useful links to other health contents	e) subsystem for acquisition of data for monitoring of chronic diseases f) I have no web page	

Table 3. The main part of the questionnaire form with questions for assessment purpose

5. Analysis of survey process and data collected

The survey was conducted during the period from mid-December 2009 until the end of January 2010. Questionnaire was made in electronic PDF/FDF form with the ability to automatically return to the sender via e-mail, and in the classical paper form. The questionnaires in electronic form were offered via dedicated mailing list, which has approximately 1100 formal users (assuming that the number of active users is much smaller), while about 70 questionnaires were distributed in paper form at the professional meetings and collected on spot or received by post. Random sample selection depended on FDs' free will to fill the questionnaire.

A total of 115 complete and correctly filled questionnaire forms were collected (87 or 75.7% of 115 in electronic and 28 or 24.3% of 115 in paper form). Therefore, we included approximately 4.7% of total 2450 Croatian FDs. By analysis of general data about the respondents and their offices, we got the structure of the analysed sample, which is showed in Table 4.

Category	Characteristics		
Age	Median: 49	Interquartile range: 44 – 51	
Years of working	Median: 23	Interquartile range: 18 – 26	
Gender	Male: 23,6 %	Female: 76,4 %	
Specialization	Yes: 66 %	No: 34 %	
FD office autonomy	Health center: 18.9 %	Under lease: 69,8 %	Private: 11,3 %
FD office type	Urban: 64,1 %	Rural: 32,1 %	Insular: 3,8 %

Table 4. Characteristics of tested sample of the Croatian family doctors and their offices

By comparison of data from well-known official Croatian health statistical publications (Baklaić et al., 2007), and data known from some previous analysed works (Kern & Polašek, 2007; Kralj & Tonković, 2009) with data showed in Table 4, it can be concluded that analysed sample is representative enough to draw conclusions from the study.

For the purposes of the upcoming numerical and statistical analysis, a quantification of the collected responses was performed. In addition to quantitative analysis, we performed a qualitative analysis of collected data that can assess the actual state of e-Health concept implementation, and point on the existing problems and shortcomings of the current model of e-Health concept implementation.

6. Results analysis and discussion

The results of qualitative and quantitative analysis of the categories and total experience are shown in Table 5. Due to limited space, the qualitative ratings are summarized for the most important elements, while the quantitative rates provide fairly realistic overall scores on a scale from 0 to 1. Prior to conducting of the survey, we hypothesized that actual state of the implementation of e-Health concept in the Croatian primary health care corresponds to the descriptive assessment: "somewhere halfway". The presented overall quantitative result to some extent confirms this assessment.

Category	Results
A	-40% of FDs believe that the new system and EHR applications slow down their work; -In 4.3% of FDOs nurses write medical information in the EHR, while in 10.4% of FDOs the doctor updates the administrative and demographic data of patients; -34% of FDs do not support e-prescribing, while 35% do not support e-referral; -50% of FDs are mainly against the secondary use of medical data; -66% of FDs do not believe in the security and confidentiality of data in a central EHR;
	Mean rate: 0.696 Items: 12 Cronbach α: 0.667
B	-All contracting FDOs are equipped with the necessary ICT equipment; -Automatic remote software update is provided for all EHR applications; -All EHR applications use the same formal structured and coded lists of health registers and nomenclatures that are automatically updated on a regular basis; -All EHR applications have authorized access and role specific access rights; -41% of FDs have some diagnostic devices that provides results in an electronic format suitable for inserting in EHR; -Transfer to another EHR application is rather difficult due to portability and "data lock" issues (including basic demographic data, prescriptions, referrals, and several types of reports)=> EHR is not longitudinal in its most important part;
	Mean rate: 0.555 Items: 24 Cronbach α: 0.694
C	-13% of FDs believed to be overloaded with unnecessary knowledge of IT technologies; -50% of FDs considered that they should have at least roughly knowledge of the norms and recommendations upon which is based the e-Health concept; -26% FDs attended IT schools or courses in the past 5 years; -17% of FDs assess their IT knowledge as very high; -24% of FDs are members of some informal groups for helping in better understanding of functionalities of their EHR applications; -57% of FDs use the data from their EHR applications for quality evaluation of their work; -75% FDs give to their replacement doctors to work on their user account (security risk); -Only 60% of FDs make daily backup their data (EHR);
	Mean rate: 0.427 Items: 18 Cronbach α: 0.722
D	-In only 34% cases FDs considered that EHR applications very well follow domain work flow, while in 49.6% of the cases considered that the user interface is user friendly and intuitive; -All applications support atomized entry of demographic and administrative data for uniquely identified patient, which are available from all parts of his EHR; -In 61% of EHR applications is possible atomized (structured) input of the physical status; -All applications contain ICD-10 classification of diseases, and equal central updated nomenclatures of procedures, medication and health care institutions; -In 72.6% and 75.4% cases EHR applications offer support for chronic disease and allergies monitoring, respectively -Decision support systems are in their beginnings as a simpler forms of work assistance; -In 41.5% cases EHR applications have built-in clinical and pharmacological guidelines; -In 51.9% cases EHR applications have built-in visual indicators for the financial indexes for diagnostic-therapeutic procedures, drug prescribing and the rate of sick leave; -35.7% of FDs are very satisfied with the overall domain properties of their EHR application;
	Mean rate: 0.430 Items: 23 Cronbach α: 0.822

Category	Results
E	-All EHR applications are capable for e-prescribing and e-referral (not in function in testing time); -All EHR applications have the ability to add scanned paper-based diagnostic test results into the EHR, but only 22.6% of FDs use this feature; -EHR applications support some forms of electronic reporting (not all yet fully implemented in the central system); -82.6% FDs communicate by email with their colleagues in primary health care, while only 18.3% with doctors in clinics and hospitals; -All EHR applications are capable to remotely check the patient's health insurance status;
	Mean rate: 0.324 Items: 17 Cronbach α: 0.689
F	-In 23.5% cases patients are satisfied with the implementation of the new information system; -EHR applications currently do not provide patients with reports on their health status in electronic human readable format; -Only 27% of FDs communicate with patients via e-mail and other electronic media; -Only 9.6% FDs collect information about chronic diseases of their patients via e-mail.
	Mean rate: 0.446 Items: 9 Cronbach α: 0.541
Overall	Mean rate: 0.48 Items: 103 Cronbach α: 0.886

Table 5. Major qualitative results, average quantitative results and reliability coefficients of experiences assessment presented by categories

To determine and prove the reliability of our measurement tool, we used a calculation of the Cronbach α coefficient of correlation for each of categories (Cronbach, 1951). The recommended amount of this coefficient for a high degree of reliability, i.e. internal consistency of questionnaire, is ≥0.7. Before calculating the Cronbach α coefficient, we conducted verification of the required sample size with Bonett's formula (Bonett, 2002) using null hypothesis of Cronbach α coefficient equal to 0.7, against a two-sided alternative at α=0.05 and power (1-β)=0.8. For total number of 103 items and estimated coefficient to approximately 0.8, we calculated a minimum sample size of 41, which is significantly less than our 115. Cronbach α calculation was performed with SPSS Statistics 17.0.

Our population sample was not previously prepared for the testing. For this type of testing are common slightly smaller amounts of the Cronbach α than in controlled or clinical conditions. As we see from Table 5, the lowest Cronbach α has a category of social experience (0.541), however, it is a common occurrence in the questionnaires that have fewer than ten questions. So called face validity and content validity (Khoja, 2007b) of our measurement tool were confirmed through interviews and commentaries of the doctors. Comments were positive, and confirm the relevance of all categories in over 75% of cases. To determine the detailed structural validity we should apply factor analysis. To determine accuracy, it would be necessary to carry out additional field researches and calculations of correlations. The reliability and validity do not automatically withdraw the accuracy of the collected data. Although it is theoretically possible to achieve higher reliability and internal consistency of the questionnaire with incorrect data, sufficient reliability is a prerequisite for accuracy. We see this as the subject of further research.

As we see from the results presented in Table 5, categories A, C and partly F reflect the doctors' views about essential objectives of the e-Health concepts and doctors' engagement in achieving these goals. From the results of all other categories we can see how EHR applications meet the current worldwide certification criteria. Based on identified system performance, and current Croatian certification criteria (CIHI, 2010), we can conclude that Croatian EHR applications would be able to almost entirely meet the criteria of EuroRec EHR-Q Seal 1 and in some parts even the Seal 2 criteria, which is subject to more detailed analysis. However, in domain functionality, which is better covered by American ONC Meaningful Use of HER Stage 1, is still necessary to significantly improve the functionality. Here we primarily mean the introduction of full electronic data (clinical and administrative) interchange with all health care organizations, insurers and, of course, patients. Furthermore, we see some encouraging first results in applying of the working guidelines, guideline-based decision support systems and monitoring of chronic diseases and allergies. A similar situation is with monitoring and indication of the quality of doctor's work. These are definitely significant areas of further improvement.

7. Possible directions for future research

Continued research in order to improve our measurement methodology i.e. our measurement tool, is more than essential. It is necessary to continuously align our measurement methodology with best international practices. We expect that assessment of doctors' attitudes and their engagement in acquiring of the ICT knowledge will be of minor importance in the coming period, because, as the information system evolves, awareness and ICT knowledge of medical population becomes larger, and the focus of interest becomes the functionalities of applied software solutions. Judging by the latest global trends, the greater importance will have functionalities that contribute to the e-Health privacy and security, use of decision support systems in order to increase the quality of treatment, and, of course, functionalities that will allow patients to monitor phases of their own treatment and to more easily achieve their rights. References for that have to be drawn from the European projects and thematic networks such as epSOS (Smart Open Services for European Patients) (epSOS, 2011) and CALLIOPE (Call for Interoperability) (CALLIOPE, 2011). Objective of these projects is the harmonization of functionalities of the EHR applications and legislations among the current and future EU Member States in order to achieve cross-border interoperability. As we pointed out previously in the discussion, another important area of further research is the application of appropriate statistical methods to determine the reliability and accuracy of the measurement. In addition, development of appropriate statistical methods is essential for comparison of the measurement results between different stages of development of the applied EHR systems.

8. Conclusion

In this article, we have presented some preliminary results of what is envisioned to be a comprehensive methodology and criteria to measure quality of EHR system implementations in primary health care. Lord Kelvin once said: "If you can not measure it, you can not improve it." So, the focus of this article was on a measuring tool which is the basis for data analysis that serves to identify some key areas of quality to measure. From the amount of collected survey data and results of their analysis, we can conclude that in the practical implementation of this assessment method exist certain problems. The form of the questionnaire is very complex since it is necessary to perform testing of measured population across all categories simultaneously and in one pass. This can result in a weaker survey response of the tested population. However, with a simple questionnaire we could not manage to collect enough of useful information. Furthermore, one can say that our methodology is limited because the assessment of the functionalities of EHR applications is reduced only to the functionalities that are visible to doctors and can be expressed as an experience. However, we must be aware that in the quality EHR application, all the key features must be visible, or at least well-documented in the user guide and contextual help system. Analysis of data collected by our measurement tool can be held within six basic categories, but it is possible to evaluate the categories i.e. functionalities that are derived from a combination of basic categories. For example, by combining data from several basic categories, we can analyse functionalities such as the implementation of working guidelines and decision support systems (Kralj et al., 2010), or patients' privacy and safety protection (Kralj et al., 2011). We entered in the designing process of our measurement tool with the main idea to construct and implement an open type methodology. That means that we have decided to continuously align our measurement methodology with best international practices. Croatian certification criteria are still mainly based on the local requirements and needs of current developments, and do not draw direct reference to some of the internationally recognized quality indicators and frameworks, or take into account clinical protocols, experts practice and expectations on readiness and experience by users. Since the certification of EHR applications is performed by successive stages of development, we expect to be relatively easy to fully comply with worldwide technical criteria, however it remains to be seen what additional requirements we will identify as important, or how would international certification processes apply to localized environments and large scale deployment. The preliminary results give us confidence that our assessment methodology could be used as the potential tool for monitoring of further improvements of Croatian certification criteria, also in respect to forthcoming development phases of the Croatian healthcare information system.

9. Acknowledgment

I would like to express my gratitude to prof. dr. Stanko Tonković, from the Faculty of Electrical Engineering and Computing, University of Zagreb, Croatia, and also dr. Miroslav Končar, from the ORACLE Croatia, Zagreb, Croatia, for their guidance and support during the making of this work. I also want to thank all the Croatian family doctors who responded and participated in the survey.

10. References

Baklaić, Ž.; Dečković-Vukres, V. & Kuzman M. (2008). *Croatian Health Service Yearbook 2007,* Croatian National Institute of Public Health, Zagreb, Croatia

Björnberg, A.; Garrofé, B.C. & Lindblad, S. (2009). *Euro Health Cosumer Index 2009 – Report,* Health Consumer Powerhouse, Brussels, Belgium

Bonett, D. G. (2002). Sample size requirement for testing and estimating coefficient alpha. *Journal of Educational and Behavioral statistics*, Vol.4, No.27, (2002), pp. 335-340

CALLIOPE. (2011). A European thematic network for e-Health interoperability. 10.10.2011, Available from: http://www.calliope-network.eu

CANARIE. (2002). Final report: Framework for rural and remote readiness in telehealth. Written by the alliance of building capacity, June 2002, 10.05.2010, Available from: http://www.fp.ucalgary.ca/.../Projects-Canarie-final%20Report,%20June%202002.htm

Centers for Medicare and Medicaid Services: ONC Meaningful Use of EHR. (2010). 02.11.2010, Available from: https://www.cms.gov/EHRIncentivePrograms

Croatian Institute for Health Insurance. (2010). CEZIH PZZ. 02.11.2010, Available from: http://www.cezih.hr

Cronbach, L. J. (1951). Coefficient alpha and the internal structure of tests. *Psychometrika*,Vol.3, No.16, (1951), pp. 297-334

Dobrev, A.; Haesner, M.; Hüsing. T. et al. (2008). *Benchmarking ICT use among General Practitioners in Europe – Final Report,* Empirica, Bonn, Germany

epSOS. (2011). Smart open services for european patients. 10.10.2011, Available from: http://www.epsos.eu/

EuroRec: European Institute for Health Records. (2010). 02.11.2010, Available from: http://www.eurorec.org

Kern, J. & Polašek, O. (2007). Information and Communication Technology in Familiy Practice in Croatia. *European Journal for Biomedical Informatics,* No.1, (2007), pp. 7-14

Khoja, S.; Scott, R.; Casbeer, A.; Mohsin, M.; Ishaq, A.F.M. & Gilani, S. (2007). e-Health readiness assessment tools for healthcare institutions in developing countries. *Telemedicine and e-Health,* Vol.4, No.13, (2007), pp. 425-431

Khoja, S.; Scott, R.; Ishaq, A.F.M. &; Mohsin, M. (2007). Validating eHealth Readiness Assessment Tools for Developing Countries, In: *e-Health International Journal,* 25.06.2010, Available from: http://www.ehealthinternational.net

Končar, M. & Gvozdanović, D. (2006). Primary healthcare information system – The Cornerstone for the next generation healthcare sector in Republic of Croatia. *Int J Med Inform,* No.75, (2006), pp. 306-314

Kralj, D. & Tonković, S. (2009). Implementation of e-Health Concept in Primary Health Care - Croatian Experiences, *Proceedings of 31st Int. Conf. on Information Technology Interfaces (ITI2009 Posters Abstracts),* pp. 5-6, ISBN 978-953-7138-15-8, Cavtat, Croatia, June 22-25, 2009

Kralj, D.; Tonković, S. & Končar, M. (2010). Use of Guidelines and Decision Support Systems within EHR Applications in Family Practice - Croatian Experience, *Proceedings of 12th Mediterranean Conference on Medical and Biological Engineering and Computing (MEDICON 2010),* pp. 928-931, ISBN 978-3-642-13038-0, Chalkidiki, Greece, May 27-30, 2010

Kralj, D.; Tonković, S. & Končar, M. (2011). A Survey on Patients' Privacy and Safety Protection with EMR Applications in Primary Care, *IFMBE Proceedings Volume 37: 5th European Conference of the International Federation for Medical and Biological Engineering,* pp. 1132-1135, ISBN 978-3-642-23507-8, Budapest, Hungary, September 14-18, 2011

Li, J. (2008). E-Health Readiness Framework from Electronic Health Records Perspective –
    Master Thesis. University of New South Wales, Sydney, Australia, November 2008.
    25.06.2010, Available from: http://handle.unsw.edu.au/1959.4/42930

9

Genetic Engineering
in a Computer Science Curriculum

Nevena Ackovska[1], Liljana Bozinovska[2] and Stevo Bozinovski[2]
[1]University Sts Cyril and Methodius, Institute of Informatics,
[2]South Carolina State University,
[1]Macedonia
[2]USA

1. Introduction

Traditionally genetic engineering is understood as a molecular biology discipline. The tools used in molecular biology are specific, mostly used by people who come from biological or medical background, which made the discipline distant from classical Computer Science. In this paper we would like to address computer science auditorium and point out the importance of understanding genetic engineering.

Although the term "genetic engineering" was coined 1951 in a science fiction novel by Williamson (reprinted in (Williamson, 2002)), it was not until 1970s when the first achievements in DNA modification showed that genetic engineering is actually happening. As in all of the sciences, genetic engineering has its own milestones. In this introductory part we will describe the earliest genetic engineering achievements, using computer science terminology.

Through the prism of computer science terms, the best way to look at DNA is that it represents a string of letters, written in four letter alphabet. The string might be interpreted as a text subject to processes such as transcription and translation (Watson & Crick, 1953), or as database and software for processes in a cell (Bozinovski and Bozinovska, 1987; Bozinovski et al., 2000). The first genetic engineering achievement was the *cut-and-paste* operation of a DNA segment. Using two types of enzymes, restriction enzymes for cut operation, and DNA ligase for join operation, a segment from one DNA was cut and pasted in another DNA. Actually, the achievement was greater, since the DNA of two living forms, the bacterial phage lambda and monkey virus SV40, were (re)combined into a new DNA (Jackson et al., 1972). The second achievement, in 1973, was *prepare-and-copy* operation. A DNA fragment was inserted in a plasmid (pSC101) and put into a bacterium (Escherichia coli) and was replicated inside the bacterium. This experiment has another important point: a fragment was prepared outside a cell (in vitro) and was replicated inside a cell (in vivo). The cell machinery processed (replicated) the foreign piece of software as its own. That was the first step or cell (re)programming. In 1974 the first transgenic animal was created, by inserting a foreign DNA into a mouse embryo. In 1977 and 1978 the first human proteins (somatostatin and insulin) were produced inside a bacterium. Bacteria produced human insulin become commercially available in 1982 opening market for various types of genetically engineered products, making genetic engineering a new industry. Important achievement happened in 2010 by which a complete synthesized genome was introduced into a bacterial cell which had no DNA (Gibson et al., 2010). Therefore, a new life form was created with complete software prepared outside the cell. The bacterium was named Synthia and is the first synthetic life form. All of these examples and many more, represent the milestones that the science and technology of genetic engineering have already made possible for humans to use.

As computer engineers, it is our view that genetic engineering is a type of (software) engineering, and the way of doing genetic engineering is a way of programming and reprogramming a DNA. As today many computer science curricula contain a bioinformatics course, we argue that Computer Science curriculum should contain courses in genetic engineering as well. Since Computer science in part is about programming, genetic reprogramming can be viewed as important part of the Computer Science education. We will also present our experience in teaching genetic engineering to computer science students.

In this paper we will first describe natural way of doing genetic engineering. That is the way the Nature was doing genetic engineering long before it became known to humans. Then, we will describe a metaphor that can be used in education of Computer Science students. Afterwards we will describe our approach to introducing genetic engineering into Computer Science curriculum, including also lab exercises.

This paper addresses mainly the computer science auditorium. We present some elementary knowledge in molecular biology, but the concepts are presented through (computer) engineering terms. However, the notion presented here might be of interest for other scientists coming from biological sciences background.

2. Genetic engineering before Genetic Engineering

Genetic engineering, by human definition, is a process of human produced genotypic effect in order to obtain some phenotypic effects. Phenotypic effects can be at molecular level, such as production of a new protein, or at higher level, producing visible effect either in the structure or in the behavior of an organism. Usually, it is achieved by planned DNA alternation, in order to (re)program behavior of a cell or a multicellular organism.

The nature has been doing genetic engineering for a very long time. Some of the life forms on the Earth survive using sort of "genetic engineering". The question we will start our presentation here is how we can help Computer Science students to understand the concept of Genetic Engineering? We believe that the following story of phage lambda is an inspirational approach toward understanding genetic engineering for computer science students and professionals alike, if we can level the terminology used in classical Genetic engineering to Computer Science. As we have stated before, the way the story is been transferred to Computer Science students enables them to understand the life cycle of phage lambda. The simplicity and the clarity of the terminology seem to be of high importance in order for these students to understand complex biological processes. Even more, the story explains one way of genetic engineering done by the Nature itself.

"Consider a life form, a bacterium named Escherichia coli. It is a prokaryote, it does not have a nucleus inside the cell. It is a life form that is capable of self reproduction. We call it a single cell organism. Its chromosome is a circular one. Every twenty minutes it replicates itself, provided there are favorable conditions in the environment. The new bacterium is pretty much the copy of the previous one.

Now consider another life form, namely phage λ (lambda). It is a life form which cannot reproduce itself. So, the phage lambda needs a host organism to reproduce and E. coli is such a host. A phage (or bacteriophage) is a virus to a bacterium: it can replicate inside a bacterium and eventually destroy it. It is interesting that a phage is harmless to human cells and to all other eukaryotic cells (cell that do have a nucleus).

The phage lambda is a life form with head-and-tail appearance. The head contains the DNA as a single chromosome. The phage DNA is double stranded. It contains the phage genome, i.e. the set of all phage genes. The tail is used to insert the DNA into a bacterial cell. Once inside a bacterium, the phage chromosome exhibits two possible behaviors: lysogenic and lytic. Lysogenic behavior occurs if the phage chromosome remains its linear form. It then integrates into the bacterial chromosome and becomes segment of that chromosome. When bacterium replicates its chromosome, the phage segment is also replicated. However, under some condition, such as environment stress, linear phage chromosome leaves the bacterium chromosome, and exhibits lytic behavior. It is not necessary that a phage has a lysogenic behavior; it might start its lytic behavior immediately after entering a bacterial cell. The lytic behavior starts with transforming the linear phage chromosome into a circular one. That is preprogrammed into the sticky ends of both side of the linear chromosome. The site of sticking the both strands together is named a cos site. Once into its circular form phage chromosome replicates inside bacterium and creates new phage life forms. Eventually bacterium explodes releasing phages into the environment outside the bacterium.

Since the replication inside a bacterium is lethal for the bacterium, a simple immune system was designed in bacterial evolution: restriction enzymes. Restriction enzymes (or endonucleases) are enzymes able to recognize a foreign DNA and cut it, rendering it non reproducible. Restriction enzymes are probably the first form of immune system in organisms."

The story of phage lambda seems to be very simple. However, it explains one of the Nature's life cycles in a simplified manner, understandable for engineering students. From the above example the students can learn about three mechanisms how a bacterial genome can be modified:

1. by incorporating a linear DNA segment into the bacterial chromosome,
2. by adding a separate circular chromosome into the bacterium and creating a two-chromosome system inside the bacterium,
3. by producing environment stress that would activate some genetic response of inserted DNA segments.

In the following section we elaborate on other terminology modification in order to explain processes and actors in genetic engineering to (computer) engineers.

3. Leveling the terminology: Metaphors for understanding genetics

For computer science, genetic engineering and cell (re)programming can be viewed as kind of software engineering. A software sequence is written, and then inserted into genetic machinery for compilation and execution. Now, if we "translate" the biological processes into terminology that computer science students understand, we obtain greater results and better understanding of these rather complicated processes. In the sequel we will consider some metaphors that can be used in teaching computer science students concepts of genetic engineering.

A metaphor is understood as a paradigm transformation; using knowledge from a familiar system in order to understand the phenomena in another system. The first metaphor explaining genetic processes was the biochemistry metaphor, which basically relied on the fact that DNA is an acid. Obviously the acid metaphor was not good enough to explain the life processes; the fact that the DNA is an acid cannot explain information that is stored into a DNA. Another metaphor proposed in 1953 (Watson & Crick, 1953) stated that the DNA is a sequence of letters, actually a text where information is stored. Today it is a dominant metaphor. The principal processes named transcription and translation (Crick, 1958) are linguistic, text processing terms. In 1985 the relation between genetic engineering and robotics was pointed out (Bozinovski, 1985). An observation that DNA is actually a database was first made in 1987 (Bozinovski, 1987; Bozinovski & Bozinovska, 1987; Demeester et al. 2004; Pirim, 2005). Afterwards, new metaphor for genetic engineering was proposed, the robotics and flexible manufacturing metaphor, which proposed a viewpoint that the cell is a flexible manufacturing system. According to that metaphor, some molecular structures should be viewed as cell robots, an example being the tRNA, which is a transporting robot. Related to the flexible manufacturing metaphor is the systems software metaphor (Bozinovski et al. 2000; Bozinovski et al 2001; Danchin & Noria 2004, Ackovska & Bozinovski, 2008; Ackovska et al. 2008).

The latest metaphor is very comprehendible for computer science and computer engineering students. It uses the concept of a genetic file as a logical segment on DNA. In the cell processes related to manufacturing (e.g. protein biosynthesis) the genetic files are considered existing and read-only. In this paper we are focused on files that can be altered, such as updated, created, written, inserted into another files, and otherwise manipulated. This also happens in nature, but more importantly, it is a basis of genetic engineering. In the following section we address the issue of writing in genetic files, and creating genetic file systems and genetic disks.

4. Genetic files, disks, and genetic file systems

When studying genetics engineering and genetics in general, the crucial concept is the concept of gene. Thus, a very natural question is "What is a gene?" A usual answer is that a gene is a segment of a DNA that encodes for either protein or RNA. Also, one could encounter slightly different definitions (Brown, 2002).

In computer science and engineering a usual reasoning on an information processing system considers the files of that system. So, for genetic information processing we might ask the question "what are the files of the genetic information processing system?" Is the concept of a gene corresponding to the concept of a file? Having that as a starting point, in this section we will present our understanding of DNA organization and DNA computing in terms of files and related concepts.

Looking for a concept of a file in DNA, we found that the transcription units (or scriptons (Ratner, 1975)) are analogous to cell files. A transcription unit is a segment of DNA that eventually becomes transcribed to RNA. In prokaryotes, a transcription unit often produces a transcript with several genes (so-called polycistronic RNA). In eukaryotes it produces a precursor RNA (pre-RNA), which contains the information about a single gene, but in order to obtain it, additional processing needs to be performed.

The eukaryotic files are rather complex and contain segments of a gene, interleaved with segments that do not belong to the gene. Those segments are known as introns, as opposite to exons (gene expressing segments). To the people involved with genetics, there is a standard question considering this phenomenon: how did it happen that eukaryotic genes became segmented? However, for computer engineers introduced to the concept of a file, the answer is straightforward – busy files are fragmented. Defragmentation is sometimes needed in computer file systems. Moreover, it is expected that between two fragments of a file an entire different file could be expected. This fact points to the concept of distributed file systems (Nutt, 1992; Tanenbaum, 1994; Tanenbaum & Van Steen, 2007). And indeed, this is the case in molecular genetics: After the first evidence that Tetrahymena ribozyme is actually an intron (Kruger et al., 1982), more evidence has been found that genetic files could be found within a complete different file (Been, 2006).

The cell files contain genes and other important sequences. Some files are executable ones, they will produce cell robots. Cell robots are either enzymes (protein based) or ribozymes (RNA based enzymes). Besides genes, which contain program files, the file system contains files that are not genes; they are data structures, some of which can be used as template for pattern recognition.

We believe that while the genes are the proper concept when talking about heredity, the concept of a file is very useful in describing the DNA transcription process. This makes the first step in the analogy between the computer systems and genetic systems. The cell, especially the eukaryotic cell, undergoes extensive file processing: from copying the pre-RNA file until obtaining the RNA message. This process includes operations like: cut (introns), join (exons), right append (trailer string), left append (header string), letter replacement and so on, which are standard file processing operations in every modern computer operating system (Bozinovski et al., 2001).

Under genetic disk (or cell disk) we understand a cell chromosome. For example, human genome is a distributed file system that resides on 46 (or 23 pairs) disks in each cell. There are exceptions, for example gonad cells have only 23 cell disks. In some organisms besides main genomic system there is a satellite chromosome system. An example is the bacterium E. coli which contains its chromosome, but it can also contain satellite chromosomes from life forms such as phages and plasmids. So, under the concept of genetic disk we include both main, cell-replicating chromosomes, as well as the satellite, independently replicating chromosomes, such as plasmid chromosomes or phage chromosomes. Examples of satellite chromosomes in eukaryotic cells are mitochondria DNA.

We define disk segment as a set of files that will be written on a genetic disk. Under the genetic file system we understand ordered set of genetic files. Usually, genetic disks contain files in strict order. Our approach understands genome (set of all genes in a life form) as part of the cell file system.

Therefore, this approach toward genetic engineering starts with understanding that the DNA is cell operating system which resides on cell disks (chromosomes). Set of chromosomes can be viewed as an array of genetic disks.

In the sequel we will often use the terminology of genetic disks and disk segments to refer to a chromosome or DNA sequence. Here we will first describe the phage lambda and its genetic disk. The DNA of phage lambda is double stranded and circular one. A double stranded DNA allows storing gens on both strands, so that gens can be copied and processed by reading them on both sides, in opposite directions. Figure 1 describes behaviour of a phage DNA in a cell of bacterium E. coli.

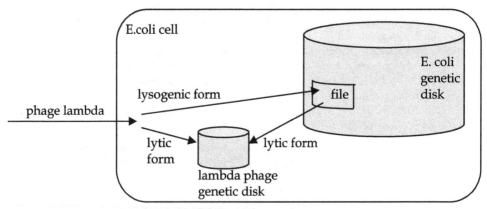

Fig. 1. Phage lambda DNA enters into an E. coli bacterium. The lambda DNA either becomes a file into the bacterium system disk or becomes own system disk.

Figure 1 shows two forms of existence of phage lambda DNA in a bacterium E. coli. It may integrate, as a disk segment (its linear form) into the E.coli DNA disk (lysogenic existence). However, it may encircle (its circular form) and form its own system disk (lytic existence). Once becoming a system disk, it replicates and forms new phages.

The phage lambda genetic disk is a file system that contains set of genes (genome) that can be divided into functional groups. According to the systems software metaphor the genes are viewed as source program files that are compiled into proteins. The phage genetic disk also contains two data files and they are not compiled.

5. Education of computer science students in genetic engineering

Contemporary education of computer science students is often related through molecular genetic through various forms of Bioinformatics courses. Bioinformatics is about genetic sequences that are stored on databases, usually available on the World Wide Web. Many institutions offer digital encyclopedias related to molecular biology. Many applications are built for using the knowledge stored in databases, such as searching various forms of similarities among genetic sequences and predicting genes in a genetic sequence (Xiong, 2006). The goal of post genome informatics (Kanehisa, 2000) is to understand the information in genetic sequences, including function of all the genes and other functional sentences. At this point, Genetic Engineering is usually not part of bioinformatics courses.

We propose that genetic engineering should be included in Computer Science curriculum. One way of doing that is through the existing bioinformatics related courses. Example is the elective undergraduate course CS495 Biocomputing and Bioinformatics which is part of the Computer Science curriculum of the South Carolina State University, or the courses Intelligent Systems (Madevska-Bogdanova & Ackovska, 2009), DNA Programming and Bioinformatics at the Institute of Informatics, University Sts Cyril and Methodius. The other approach is introducing a separate course.

In genetic engineering related course students will be able to learn programming life forms. So far, they are capable of programming robots through various type of robotics courses contemporary found in Computer Science curricula. However, Genetic engineering is a way of designing, writing, and executing programs for living beings: it is about designing new genes and genomes and consequently, their phenotypes. Therefore it is interesting for the students to learn how to program DNA in order to design life form robots.

The core of genetic engineering is creation of a file or set of files that will be written on a genetic disk. There are several ways how to obtain a genetic file. Examples are: 1) cut a file or segment from existing disk (using restriction enzymes), 2) copy a file from existing disk (copy on mRNA and then synthesize complementary DNA, cDNA), and 3) synthesize a human made (artificial) file or segment.

In this section we will be focused on genetic disk segments, and genetic tools as ways of creating genetic source programs that would be executed by the cell. We shall explain the way we represent the theoretical knowledge, including restriction enzymes, engineered genetics disks, artificial chromosomes and the process of transferring the source genes into host systems and creating genomic libraries.

5.1 Theoretical knowledge

When creating a curriculum it is always a question which topics should be primarily covered. When designing an interdisciplinary course, such as Genetics Engineering for Computer science students, it is even more difficult to make such a decision. It would depend on how much time or space the instructor has available for the course. The course can be separate, usually graduate course, for example on Physiology Engineering (Bozinovski and Bozinovska, 2011), or it can a part of an undergraduate course, for example on Bioinformatics. In any case, organization of genetic files, genetic disks, and related operating systems and robotics metaphors for understanding genetics is a good introduction to the subject. Other topics might include: sequencing, amplification, modifying enzymes, cloning, screening, applications, and state of the art. A good textbook might be Nichol's book (Nicholl, 2008).

5.1.1 A tool for genetic engineering: restriction enzymes

One of most used DNA modifying enzymes are restriction enzymes. They are tools for cutting a DNA string. Restriction enzymes, also known as endonucleases, are DNA cutting bionanorobots.

Restriction enzymes are naturally used by bacteria which use them as a natural defense mechanism to cut an invading phage DNA. Many bacterial restriction enzymes have been found and they are named according to the bacteria they are isolated from and the order (first, second, third etc.) in which they are isolated. For example EcoRI means first isolated restriction enzyme from *Esherichia coli*, HindIII means the third isolated restriction enzyme from bacteria *Haemophilus influenzae*, and PstI means the first restriction enzyme extracted from *Providencia stuartii*.

Genetic engineering relies upon ability to cleave (cut, splice) and ligate (paste) a functional piece of DNA predictably and precisely. Example is cutting a gene from a DNA. One should look for restriction sites on both sides of the gene and then use specific restriction enzymes that will cut at the observed restriction sites. As a result a DNA fragment (genetic disk segment) is obtained, which contains the gene of interest. That fragment should be inserted into another, recipient DNA. The same restriction enzymes are also used to cut the recipient DNA into which the fragment will be inserted. This cut-and-paste operation is one of the ways of engineering a genetic disk.

Each restriction enzyme recognizes a specific nucleotide sequence in the DNA, called a restriction site, and cuts the DNA molecule at only that specific sequence. Many restriction enzymes leave a short length of unpaired bases, called a "sticky" end, at the DNA site where they cut. Other restriction enzymes make a cut across both strands creating double-stranded DNA fragments with "blunt" ends. In general, restriction sites are palindromic, meaning the sequence of bases reads the same forwards as it does backwards on the opposite DNA strand. Example of a palindromic restriction site recognized by restriction enzyme HindIII is given in Figure 2. As Figure 2 shows, HindIII cuts the DNA at last letters of the palindrome and leaves two one-stranded ends of DNA, named sticky ends.

```
5'-N-N-N-A-A-G-C-T-T-N-N-N-N-
3'-N-N-N-T-T-C-G-A A-N-N-N-N-
```

Fig. 2. A two strand palindromic string in a DNA. This particular one is recognized by a HindIII restriction enzyme. Arrows show the cut sites, leaving sticky ends at the cut.

A DNA in presence of particular restriction enzymes will be cut at all the corresponding restriction sites. For example, HindIII restriction enzymes will cut the lambda phage chromosome into 8 segments (or fragments).

5.1.2 Engineered source disks

A genetic program is a file written on a genetic disk that contains a code for specific function in a cell. The program can be created and written by a human, which is essence of cell (re) programming. The written program is called source program, and the disk the program is written on is named source disk. A source disk is usually prepared outside a target organism. There are basically two types of source disks: engineered genetic compact disks and artificial chromosome disks.

A) Engineered Genetic Compact Disk. Engineered compact disk is based either on a plasmid or a phage. Usually a natural plasmid (or phage) is loaded with an engineered DNA sequence.

For computer engineers, it can be viewed as a rather small capacity disk media such as compact disk (CD). A cell can be viewed as a computer system that also contains a separate disk replicator, so that a particular CD can be replicated by the cell information processing system. A plasmid and phage can be inserted in a cell using natural way: just put a cell and plasmids into a favorable environment, and a plasmid (or phage) will enter the cell.

A typical example is the engineered plasmid pBR322, which is used for transferring a particular disk segment into bacterium E. coli. Plasmid pBR322 was engineered out of three natural E. coli plasmids: R1 plasmid, containing amp^R gene, which provides resistance to ampicillin, R6-5 plasmid, containing tet^R gene which provides resistance to tetracycline, and pMB1 plasmid, containing replication origin (ori) segment. There is an ori part in a genetic disk which makes the disk replicable. There are specific spots on the pBR322 where restriction enzymes such as EcoRI, SalI, PstI, PvuiI, and HindIII can make the disk open for inserting a file. After insertion of the disk segment, the plasmid disk has extended its used disk space. Plasmid pBR322 has capacity of accepting disk segments of about 10 Kbp (Brown, 2001).

B) *Artificial chromosome.* Artificial chromosome is an engineered genetic disk which has organization of a natural chromosome, but is much shorter. An example of artificial chromosome is the Yeast artificial chromosome (YAC). Figure 3 shows a procedure of insertion of a DNA fragment into a Yeast artificial chromosome.

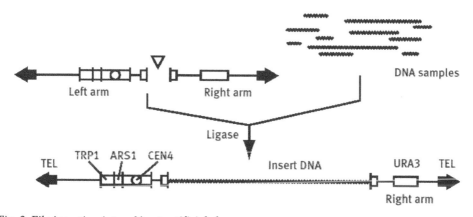

Fig. 3. File insertion into a Yeast artificial chromosome

Figure 3 shows how a DNA fragment is inserted into an artificial chromosome in order to be transferred into a cell. A chromosome of a eukaryote (yeast is a one-cell eukaryote) contains specific DNA part such as telomeric DNA (TEL1), replication origin (ori), replication sequences (ARS1), and centromeric DNA (CEN4). Centromeric DNA enables segregation of the DNA at the time of cell division. Selectable markers are genes that allow distinguishing cells that have this artificial chromosome. For example, for the pYAC2 chromosome, the genes are: amp^R, ura3, and trpI. Artificial chromosomes possess important property, they do not transfer files into the main, cell replicating disk(s) of the cell. Instead they remain as a separate disk, in addition to the cell chromosome disks. The inserted DNA is possibly an engineered one, which acts as part of the chromosome system of the cell where the artificial chromosome is inserted.

There are several types of artificial chromosomes. One is the Bacterial Artificial Chromosome (BAC), which is based on plasmid F and can accept segments between 80-300Kbp. The Yeast Artificial Chromosome (YAC) is often used for transferring files into yeast (Burke et al, 1987). A YAC is able to accept disk segments of Mbp size. Mammalian artificial chromosomes (MAC) (Grimes & Cooke, 1998) as well as Human Artificial Chromosomes (HAC) (Larin & Mejia , 2002) were also engineered. Recently a plant artificial chromosome with length of 30 Mbp was reported (Ananiev et al, 2009).

5.1.3 Transferring source disks into cell file systems

Engineered genetic compact disks are plasmids (or phages). They are transferred into a cell by simply mixing cells and plasmids under favorable conditions; the plasmids will enter the cells. In genetic engineering those source disks are named vectors, pointing out their inherent transferring ability.

There is no natural way of inserting a rather large artificial chromosome into a cell. Usual procedure is electroporation (Rebersek & Miklavcic, 2011), by which the artificial chromosome is forced into a cell as a high energy particle. Once inside a cell, an artificial chromosome behaves in principle like other cell chromosomes; it just has much less files than the natural chromosomes.

5.1.4 Arrays of genetic disks: Genomic libraries

Genetic engineering can create new life forms or modify existing life forms. Example of modified life forms are so called transgenic organisms, which in their genome contain genes from another organism. For example, it is possible to include a gene of a human in a bacterium.

There are other applications that are more oriented toward file systems rather than the primary intention of modifying an existing mechanism. One such application is creating an array of genetic disks to store a particular genome. Figure 4 (Ackovska et al., 2010) shows creation of a human genomic library. The human genome (all human genes) are kept into array of transgenic disks, each disk is a bacterial genome in which a human gene is inserted.

In humans there are 23 pairs of disks on which the cell operating system resides. They are marked F1-F22 and M1-M22 on Figure 4, pointing out that F23 disk represents the X chromosome and M23 disk represents the Y chromosome. The capacity of those disks is between 50 and 230 Mbp. The density of files on disks is low, only about 5% of the entire DNA contains genes. The other parts are control sequences, data structures, and also areas which, by today's knowledge, contain no meaningful information. Some of human cell files are longer than 100 Kbp, so BAC disks are used (Osoegawa et al., 2001). In such a case, about 30,000 disks, which means 30,000 bacteria are needed to store the human genome.

The obtained set of files contains the whole genome, but is not the true representative of the cell file system and cell operating system, as it is in the original 23 pairs of disks. However, a transgenic disk array allows access to a particular segment and set of files faster and in a way more convenient for study. So called arrayed genomic libraries, arranged as a matrix, are built for easier access of a particular segment.

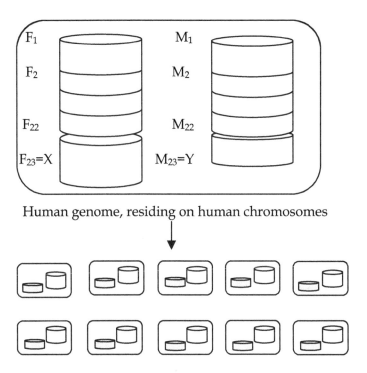

Human genome, residing on human chromosomes

Human genome, each gene residing on a plasmid disk inside a bacterium.

Fig. 4. Creating genomic library on a transgenic disks array

5.2 A lab experience in genetic engineering for computer science students

In addition to knowledge to be conveyed to students as lectures, lab experience is important part of every course. Working in labs with tools for genetic engineering is very different from the everyday practice for computer science students. If one is going to expose computer science students to work in a genetic lab, there are basically two ways of doing that: 1) lab work is carried out in a computer science lab, or similar, in which limited number of equipment and tools for genetic engineering can be installed; 2) lab work is carried out in a specialized lab, for example a molecular biology lab, which already has all the genetic engineering equipment and tools. The tools for genetic engineering include instruments such as water baths, dry baths, centrifuges, incubation ovens, spectrophotometers, electrophoresis chambers, polymerase chain reactors, and electroporators, among others. The lab activities can be carried out as regular lab activities inside a Computer Science course, such as Bioinformatics course. Another way of carrying out genetic engineering labs is an extracurricular activity, for example activity funded by a research project. In short, here we describe two lab exercises and the way to organize this specific lab practice so it can become closer to Computer science students.

5.2.1 Lab example 1: Extracting DNA from human saliva cells

First, we will describe a lab activity that does not require many specialized devices, and as such this activity can be carried out inside a computer lab. As introduction to this lab activity one should mention that, DNA is most important molecule for life. It carries hereditary information and also manages production of proteins and manages processes in a cell. Human genome is organized in linear chromosomes, and somatic cells such a saliva cells have 46 chromosomes. The total length of all chromosomes is about 2 m. However, DNA is invisible, it is a nanostructure, and its total width is 2 nm.

Understanding and working with DNA is one crucial educational task in hands-on lab experience for a Computer Science student. A DNA sample can be obtained from human blood, such as in medicine or in forensics. In lab practices for Computer Science obtaining DNA sample from human saliva is preferable approach. Therefore, in this lab task a computer science student will be able to extract her/his own DNA from her/his saliva.

The important part of "hardware" needed for this exercise, not usually used by CS students, is an incubator, for example an incubation oven, which can keep a temperature of 50°C for some time. There are devices such as water bath, that keeps water on that temperature, and students insert their lab tubs into that water for some time. The simplest approach is to take any heating source, take a thermometer, heat the water until thermometer shows around the desired temperature, and put the lab tubes in such water for needed time. If conditions allow, a water bath might be purchased. However, a lab kit that contains necessary tools, such as enzymes, tubes etc, should be purchased. There are many vendors of these types of kits, one example is BioRad.

The detailed explanation of the lab activity for extraction a human DNA from human saliva cells is given in Figure 5. It shows detailed flow of the processing of the tube in which the saliva is placed up until the DNA is visible in the same tube. It is not necessary to give the students task in detailed terms. As the reader may notice, the explanation written for CS students differs significantly than the explanation found in lab manuals for biology or medical students. Computer science students are accustomed to process thinking. Given below is a description of the task.

Step 1. **Collect cells.** You can collect thousand of cells from the inside of your mouth just by gently chewing your cheeks and rinsing your mouth with water.

Step 2. **Break open (lyse) the cells.** Use provided *lysis buffer*. It will break open the cell membrane and nucleus membrane and release DNA.

Step 3. **Remove proteins.** Use provided enzymes named *proteases*. They will remove all the proteins in your solution along with proteins that keep DNA as a thread. Proteases work best at 50°C. So you incubate (in a water bath previously warmed up to 50°C) your solution together with enzymes to that temperature.

Step 4. **Condense DNA, make it visible.** Use salt and cold alcohol. It will precipate DNA out of the solution, and you can see it as a mass of white threads.

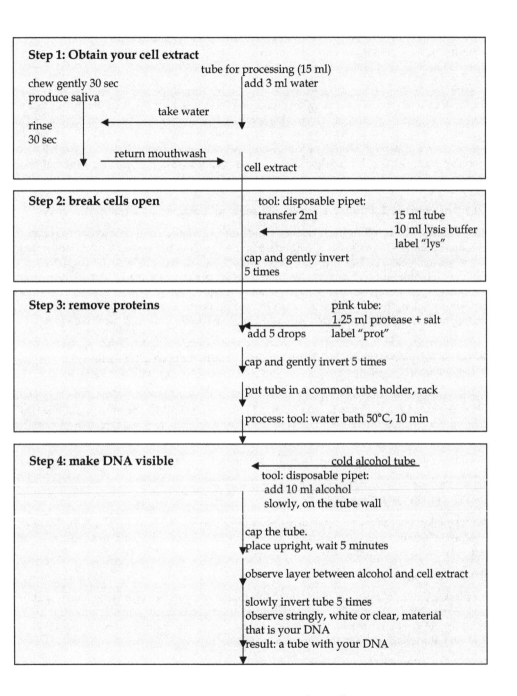

Fig. 5. Lab procedure for extracting DNA from human saliva cells

5.2.2 Organization of a lab for computer science students

Organization of the laboratory is important for any lab activity. It is especially important for computer science students working with tools such as lab tubes and pipets, which are not part of their regular computer science lab activity.

A workstation should be prepared, which can be for one student or shared by more than one student. Each student has her/his processing tube, in which she/he puts the saliva and then processes it using the provided tools such as lysis buffer and protease enzymes. The equipment, such as Water Bath, is shared by all the students. All the processing tubes are put in a rack and then placed in the Water Bath for incubation. Some tools needed for processing are kept in the refrigerator, such as alcohol. This part is very different form everything the Computer Science students are accustomed to.

5.2.3 Lab example 2: Finding shortest file segment obtained by a restriction enzyme

Here we describe a lab that introduces hands-on experience with restriction enzymes to computer science students. For Computer Science students the easiest way to understand restriction enzymes is that they are type of cell robots that are able to cut DNA at a particular point. In this lab exercise the fragments are obtained from a DNA and their length is estimated. This exercise uses a classical lab technique named agarose gel electrophoresis to estimate length of a particular DNA fragment. Since specialized equipment, electrophoresis chamber, is used, this activity might be easier if carried out in a biology lab.

The task description is as follows: DNA of a phage lambda is given. Also, given are three restriction enzymes: EcoRI, PsfI, and HindIII. Cut the lambda DNA with each enzyme. Determine which restriction enzyme will obtain DNA fragments with minimum length.

The lab procedure is performed in parallel on four processing tubes. One tube is lambda DNA, the second is lambda+EcoRI, the third is lambda+HindIII, the forth is lambda+PstI. They are centrifuged and then put into an incubator with temperature that is best for enzymes. The enzymes cut the DNA into number of fragments.

The following is more elaborated description of the lab procedure for obtaining fragments of the phage lambda genome processed by various restriction enzymes:

Step 1.   Obtain DNA fragments

Put restriction enzymes into three tubes with DNA, the fourth is just DNA. Mix the tubes, possibly in a centrifuge, heat the tubes at 37°C for 30 min, possibly in a water bath. Result: each tube contains fragmented DNA.

Step 2.   Distinguish fragments by their lengths

Put marker dye into tubes, mix, possibly with centrifuge, prepare agarose gel wells, put the tube content into the gel wells, and perform electrophoresis at 100V for 30 min. Result: The resulting gel contains information how far in the electrical field the fragments travelled. However, the result is not visible for human eyes.

Step 3.   Visualize DNA fragments

Put colouring marker on the gel, wait, then wash the gel. Result: Visible lanes of DNA fragments with different travelling paths.

If the exercise is done correctly there is a visible result of the procedure, which can be analyzed and/or photographed. On a solid rectangle piece of gel, there are 4 visible vertical lanes. Each lane has horizontal bars, each bar representing DNA fragments with various lengths. The first lane contains uncut DNA (single fragment), the second lane contains fragments cut by PstI, the third and the fourth lanes contain fragments cut by EcoRI and HindIII respectively. It can be observed that the horizontal bar at longest distance is in the lane where fragments cut by PstI are positioned. Therefore, the answer to this lab task is: If a lambda DNA is processed with three restriction enzymes, EcoRI, PstI, and HindIII, the shortest DNA fragment will be obtained by PstI restriction enzyme.

The lab exercise is carried out in groups, each group having own lab workstation with all the tubes and materials necessary. The devices like electrophoresis chamber and water bath might be common for the groups.

6. Implementation

This approach has been successfully implemented to at least two Universities: South Carolina State University in Orangeburg, SC, USA and St. Cyril and Methodius University in Skopje, Macedonia. It is applied in the courses CS495 Biocomputing and Bioinformatics (South Carolina state University), and partly in the course DNA Programming, Intelligent Systems and Bioinformatics (Sts Cyril and Methodius University). During lab activities part of which are described here, computer science students are given opportunity to obtain hands on experience with bionanorobots and other nano structures used in genetic engineering. Each lab has a lab quiz that asks students to relate the observed knowledge with their background knowledge in robotics, flexible manufacturing, and operating systems. Students show enthusiastic interest in learning steps toward genetic engineering.

7. Conclusion

The paper describes an innovative approach toward education of Computer Science students and specialists in genetic engineering. We argue that genetic engineering is good way of evolution for classical engineering and that it should be part of computer science education. Computer science is about programming and reprogramming, and cell (re)programming is essence of genetic engineering and of creating new organisms.

For a long time computer science is involved in creating artificial creatures such as robots. However, biological robots, such as bionanorobots, were not part of computer science education. We argue that the students should be familiarized with possibility of programming DNA that will compile into a bionanorobot.

Appropriate language should be used in explaining molecular genetics to computer science students. This paper proposes use of the computer science related metaphors, such as robotics and flexible manufacturing metaphor as well as operating systems and systems software metaphor.

This approach has been successfully implemented to at least two Universities: South Carolina State University (SCSU) in Orangeburg, SC, USA and Sts. Cyril and Methodius University in Skopje, Macedonia. It is applied in the course CS495 Biocomputing and bioinformatics in SCSU, and the course Bioinformatics, Intelligent Systems, and DNA

Programming at the Faculty of Natural Sciences and Mathematics at Sts. Cyril and Methodius University. The terminology used in these courses is adapted toward computer engineering way of thinking. The addressed issues are strongly connected to the terms of files, disks, operating systems, file processing, robots etc. This enables the computer science students better understanding of some of the most complicated processes known to man, the processes of life. Students are given opportunity to obtain hands on experience with bionanorobots, such as restriction enzymes. Computer Science student have shown interest toward understanding and reprogramming DNA. In addition to theoretical lectures, the students participate in extracurricular activities which give them hands on-experience with DNA manipulation. It seems that this way of reasoning makes students curious for additional knowledge in Genetic Engineering. Many of these students have expressed interest for the upcoming master's degree program related to biorobotics at SCSU. Some of the students, who graduated at the University St. Cyril and Methodius, and have taken classes that support Genetic Engineering education, are already students in some of the Europe's Genetics Engineering masters programmes.

We believe that we should continue towards further research for appropriate metaphors in relation between computer science and genetic engineering. We also believe that we should enrich the student work with more practical implementation of the concepts presented in this paper.

8. Acknowledgment

This work was supported in part by the NSF grant EPS-0903795-2010-702 awarded to South Carolina State University in 2009.

9. References

Ackovska N., Bozinovska L. & Bozinovski S (2010) Artificial chromosomes as genetic disks: A systems software metaphor for genetic engineering, *Proceedings of IEEE SouthestCon 2010*, pp. 324-327, 978-1-4244-5853-0, Charlotte, NC, March 18-21, 2010

Ackovska N. & Bozinovski S. (2008), Next Generation Operating Systems: A Biologically Inspired Future, *Proceedings of 2nd Annual IEEE Systems Conference*, pp. 1-7,978-1-4244-2150-3, Montreal, Canada, April 7-10, 2008

Ackovska N., Bozinovski S. & Jovancevski G. (2008a). Real-Time Systems – Biologically Inspired Future, *Journal of Computers*, Vol. 3, No.3, (March 2008), pp. 56-63, 1796-203X, 2008.

Ackovska N., Bozinovski S. & Jovancevski G. (2008b). File system organization in minimal biological system, *Proceedings of the 6th International Conference on Informatics and Information Technology*, pp. 44-47, 978-9989-668-78-4, Bitola, Macedonia, February 10-14, 2008

Ackovska N., Bozinovski S. & Jovancevski G. (2007). A New Frontier for Real – Time systems – Lessons from Molecular Biology, Proceedings of IEEE SoutheastCon 2007, pp.224-228, 1-4244-1029-0, Richmond, VA, March 22-25, (2007)

Ananiev E., Wu C., Chamberlin M., Switashev S., Schwartz C., Gordon-Kamm W. & Tingey S. (1985). Artificial chromosome formation in maize, *Chromosoma*, Vol. 118, No. 2, pp. 157-177, ISSN : 0009-9515, 1985

Been M. (2006). Versatility of Self-Cleaving Ribozymes, *Science*, Vol. 313, pp. 1745-1747, ISSN: 0036-8075, 2006

Bozinovski S. (1985) Guest Editor's Introduction, *Automatika*, *Vol. 26 (3-4), Special Issue on Biocybernetics*, pp. 128, Zagreb, ISSN: 0005-1144, 1985

Bozinovski S. (1987) Flexible manufacturing systems: Biocybernetics approach (In Russian), *Problems in Manufacturing and Control*, Vol. 16, pp. 31-34, ISSN: 0234- 6206, 1987

Bozinovski S. & Bozinovska L. (1987), Flexible production lines in genetics: a model of protein biosynthesis process, *Proceedings of International Conference on Robotics*, pp.1-4, Dubrovnik, Yugoslavia, 1987

Bozinovski S., Mueller B. & diPrimio F. (2000). Biomimetic autonomous factories: Autonomous manufacturing systems and systems software, *GMD Report*, No. 115, German National Research Center for Information Technology, Bonn, ISSN: 1435-2702, 2000

Bozinovski S. & Bozinovska L. (2001), Manufacturing science and protein biosynthesis, In N. Calaos, W. Badawy, S. Bozinovski (eds.) *Proceedings of SCI Conference 2001, Vol. XV*, pp. 59-64, ISBN : 980-07-7555-2, Orlando, FL, 2001

Bozinovski S., Jovancevski G. & Bozinovska N. (2001). DNA as a real time, database operating system, In N. Calaos, W. Badawy, S. Bozinovski (eds.) *Proceedings of SCI Conference 2001*, VOL XV : pp. 65-70, ISBN: 980-07-7555-2, Orlando, FL, 2001

Bozinovski S. & Bozinovska L. (2011). Human Body Parts Making: Educational Challenges for Engineered Physiology, Biorobotics, and Biofabrication, *Proceedings of IEEE SoutheastCon 2011*, pp. 307-308, ISBN: 978-1-61284-737-5 , Nashville, TN, 2011

Brown T. A. (2001). *Genetics, A Molecular Approach*, Nelson Thomas, ISBN 0-7487-4370-7, 2001

Brown T.A. (2002), *Genomes*, 2-nd Ed., Willey-Liss, ISBN: 0-471-31681-0, 2002

Burke D., Carle M. & Olson M. (1987). Cloning of large segments of exogenous DNA into yeast by means of artificial chromosome vectors, *Science*, No. 236, pp. 806-812, ISSN : 0036-8075, 1987

Crick F. (1958). On protein synthesis, *Proceedings of Symposium on Society for Experimental Biology*, No.12, pp. 138-163, ISSN: 0081-1386, 1958

Demeester L. ; Eichler K & Loch C. H. (2004). Organic Production Systems: What the Biological Cell Can Teach Us About Manufacturing, *Manufacturing and Service Operation Management*, Vol. 6, No. 2, INFORMS, pp. 115-132, ISSN: 1523-4814, 2004

Danchin A. & Noria S. (2004). Genome structure, operating systems and the image of the machine in *Molecules in Time and Scpace: Bacterial Shape, Division, and Phylogeny*, Vicente M., Tamames J., Valencia A., Mingorance J. (eds.), pp. 195-208, Kluwer, ISBN: 0-306-4857-8, 2004

Gibson D., Glass J., Lartigue C., Noskov V., Chuang R., Algire M., Benders G., Montague M., Ma L., Moodie M., Merryman C., Vashee S., Krishnakumar R., Assad-Garcia N., Andrews-Pfannkoch C., Denisova E., Young L., Oi Z-O., Segall-Shapiro T., Calvey C., Parmar P., Hutchison C., Smith H. & Venter C. (2010). Creation of a bacterial cell controlled by a chemically synthesized genome, *Science*, Vol. 329 (5987): 52–56. 2010

Grimes B. & Cooke H. (1998) Engineering mammalian chromosomes, *Human Molecular Genetics*, Vol. 7, No.10, pp. 1635-1640, ISSN: 0964-6906, 1998

Jackson D., Symons R. & Berg P. (1972) Biochemical method for inserting new genetic information into DNA of Simian Virus 40: Circular SV40 DNA molecules containing lambda phage genes and the galactose operon of Escherichia coli. *Proceedings of National Academy of Sciences*, 69 (10): 2904–2909, 1972.

Kruger K., Grabowski P. J., Zaug A. J., Sands J., Gottschling D. E. & Cech T. R. (1982). Self-splicing RNA: Autoexcision and autocyclization of the ribosomal RNA intervening sequence of tetrahymena, *Cell*, Vol.31, pp. 147-157 , ISSN: 0092-8674, 1982

Kanehisa M. (2000). *Post-genome Informatics*, Oxford University Press, ISBN: 0-19-850326-1, 2000

Larin Z. & Mejia . (2002). Advances in human artificial chromosome technology, *Trends in Genetics*, Vol. 18, No.6, pp. 313-319, ISSN: 0168-9525, 2002

Madevska-Bogdanova A. & Ackovska N. (2009), Different Approach to Information Technology – Teaching the Intelligent Systems Course in Technology, Education and Development, Lazinica A. & Calafate C. (eds), pp. 357-366, 978-953-307-007-0, In-Teh, Vukovar, Croatia, 2009

Nicholl D. (2008). *An Introduction to Genetic Engineering*, Cambridge University Press, ISBN: 051-80867-7, 2008

Nutt G. (1992). *Centralized and Distributed Operating Systems*, Prentice Hall, ISBN 0-13-122326-7, 1992

Osoegawa K., Mammoser A., Wu C., Frengen E., Zeng C., Catanese J., & de Jong P. (2001). A Bacterial Artificial Chromosome Library for Sequencing the Complete Human Genome, *Genome Research*, No. 11, pp.483–96, ISSN: 1088-9051, 2001

Pirim H. (2005). Biological Cell's production system, *Proceedings of 35th International Conference on Computers and Industrial Engineering*, pp. 1571-1575, Istabul, Turkey, 2005

Ratner V. (1975). Control Systems in Molecular Genetics, (In Russian) *Nauka*, Novosibirsk, 1975

Rebersek M. & Miklavcic D. (2011) Advantages and disadvantages of different concepts of electroporation pulse generation. *Automatika 42(1):, Special Issue on Recent Advances in Biomedical Engineering*, pp. 12-19, ISSN 0005-1114, 2011

Ren X., Tihimic C., Katoh M., Kurimasa A., Inoue T. & Oshimura M. (2006). Human artificial chromosome vectors meet stem cells, *Stem Cells Reviews and Reports*, Vol. 2, No.1, pp. 43-50, ISSN: 1558-6804, 2006

Tanenbaum A. (1995). *Distributed Operating Systems*, Prentice Hall, ISBN 10: 0321-99084, 1994

Tanenbaum A. & Van Steen M. (2007). *Distributed Systems; Principles and Paradigms.* ISBN 0-13-239227-5, Prentice Hall, 2007

Watson J. & Crick F. (1953). Molecular structure of nucleic acids: a structure of deoxyribose nucleic acid, *Nature*, No. 171, pp. 737-738, ISSN: 0028-0836, 1953

Williamson J. (2002) Dragon's island and other stories. Five Star, ISBN-10: 0786243147

Xiong J.(2006). *Essential Bioinformatics*, Cambridge University Press, ISBN-10 0-521-60082-0, 2006

Zaibak F., Kozlovski J., Vadolas J., Sarsero J., Williamson R. & Howden S. (2009). Integration of functional bacterial artificial chromosomes into human cord blood-derived multipotent stem cells, *Gene Therapy*, Vol. 16, No.3, pp. 404-414, ISSN: 0969-7128, 2009

10

Design of a PC-Based Electrocardiogram (ECG) Recorder as - Internet Appliance

Mahmud Hasan

Computer Science Program, Faculty of Science, University Brunei Darussalam
Brunei

1. Introduction

The advancement of medical science contributes towards lengthening the human life expectancy. Therefore the proportion of the elder people in the society is increasing. And taking proper care of these people should, most certainly, be one of the prime responsibilities of the society. ECG is the only way to check the heart condition, and if ECG can be done frequently, then it is easier for the physician to identify a problem from the ECG's history of a person (it has long been preferred by physicians).

The importance of ECG as a medical tool has led to the development of various types of ECG recording system. The developed systems vary from a simple ECG recorder that can only monitor the ECG signal to a sophisticated system with computer analysis and database. With the use of the computer, patients can record and save their reliable ECG data by themselves at home. Moreover, the Internet services like e-mail and File Transfer Protocol (FTP) can be used as a communication tool to send the recorded ECG data to the medical center. What they need is an ECG data acquisition system at home, which is easy for recording, viewing and sending the ECG data to the medical center with reliable accuracy. To send the ECG data over Internet, the recorded ECG data should be made into a form suitable for transmission; for example, the size and the fidelity of the ECG data should be taken into consideration.

This chapter will explain step by step development work to produce an ECG data acquisition system that is suitable for use at home and for transmitting ECG data over Internet. The system consists of an ECG recorder hardware that is used as a computer peripheral connected to the computer via USB port and software written to display the ECG data. The recorder hardware performs amplification, proper filtering and analog-to-digital conversion to the ECG signal. An 8-bit microcontroller is used to control the hardware and to communicate with the computer. The hardware is portable, battery-powered and its design emphasizes on low power consumption. The device should be isolated from the 240V power line; thus, working independently in a robust and reliable mode. The ECG signal was over sampled at 500 samples per second to improve its fidelity. The software can be written in Visual C++ or Java programming language. It will communicate with the hardware to control its recording process, monitor its battery status and display the ECG signal in real-time while recording is in process with the indication of bad or good ECG data. The software should provide a graphical user interface (GUI), which help general user to work with this system. Besides that, it detects the presence of the heartbeat, calculates the current heartbeat rate and beat-to-beat interval while recording the ECG data in memory. To make a reasonable file size for easy transmission over the Internet, a lossless compression can be performed to the ECG data before it is saved into a file.

Leveraging the growing impact of the Internet on healthcare, this device can be used for cardiac research, cardiac rehabilitation, cardiac activities follow-up (like Arrhythmia and Pacemaker), sports, emergency unit (wireless ECG transfer by wireless Internet) as well as regular cardiac care for improving health. This system can be expand according to the organizational need. A hospital database can be developed where the individual patient will upload their ECG information regularly from home for diagnostic purposes.

1.1 Objective

The objectives of this chapter are:

- To demonstrate a development process of a portable and reliable system that enables the recording of patient's ECG at home.
- To show how to recondition the recorded data that will be suitable for transmission over the Internet and suitable for use in the analysis and measurement of PQRST waves of the patient's ECG.
- To show how to develop software that must have the basic functions to get data from the hardware and send through the network.

2. Background study

2.1 The heart wiring

When the heart is working properly, it is a masterpiece of timed precision, with heart valves opening and closing on cue to prevent backward blood flow. Heart valves, chambers, electrical impulses, coronary arteries and veins all of these must be in perfect working order for the heart to function at its best.

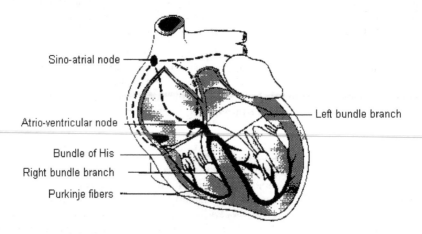

Fig. 1. Anatomy of the heart

2.2 The surface electrodes

The electrodes for surface recording of biopotential are generally made of silver-silver chloride (Ag-AgCl). This disposable foam-pad consists of an Ag-AgCl metal contact button at the top of a hollow column that is filled with a conductive gel. Most bioelectric measurements an interference level of 1 to $10\mu V$ peak-to-peak (pp) or less than 1% of the pp value of an ECG, is acceptable.

2.3 Interference

The capacitances between the patient, the power lines and ground cause a small interference current to flow through the body. These capacitances cause an interference current of $0.5\mu A_{pp}$ to flow from the power supply lines (220V, 50Hz) through the body to ground [9]. A typical situation with a mean current of $10nA_{pp}$ in the wires, mean electrode impedance of $20K\Omega$ and a relative difference electrode impedance of 50%, leads to an unacceptable high interference level of $200\mu V_{pp}$. There are two ways by which a high common mode voltage may cause interference. The first, obvious way is when the common mode rejection ratio of the amplifier is limited. Second, and much more important way a high common mode voltage may cause interference is when there are differences in electrode impedance and input impedance whish convert common mode voltage into a differential input voltage. There are three situations.

- The amplifier common is connected to ground. The amount of interference current Z_{rl} is determined mainly by the capacitance between patient's body and the main power supply (C_{pow}). In a typical situation (C_{pow} = 3pF, Z_{rl} = 20KΩ) the common mode voltage is an acceptable $10mV_{pp}$.
- The amplifier common is not connected to ground. The resulting current through the impedance of the electrode-skin interface of the right leg electrode depends on the values $C_{pow,}$ capacitance between body and ground (C_{body}), capacitance between amplifier common and main power supply (C_{sup}) and between amplifier common and ground (C_{iso}). It can be calculated that under typical conditions (Cbody = 300pF, Cpow = 3pF, Zrl = 20KΩ), the common mode voltage will be small, 1mVpp.
- In all cases, the common mode voltage can be largely reduced if a driven right leg circuit is added. An extra amplifier drives the patient to the same voltages, as the voltage of the amplifier common mode voltage can be made much smaller this way than the voltage across Z_{rl}.

2.4 Data acquisition system

With the analog low-pass filter, high frequency noise and interference can be removed from the signal path prior to the analog-to-digital conversion. Using a sampling frequency (f_s), typically called the Nyquist rate. If there is a portion of the input signal that resides in the frequency domain above $f_s/2$, that portion will fold back into the bandwidth of interest with the amplitude preserved [15]. This frequency folding phenomena can be eliminated or significantly reduced by using analog low-pass filter prior to the ADC input. Consequently, this signal will not be aliased into the final sampled output. There are two regions of the analog low-pass filter illustrated in Figure 2. The region to the left is within the bandwidth of DC to $f_s/2$. The second region, which is shaded, illustrates the transition band of the filter. Since this region is greater than $f_s/2$, signals within this frequency band will be aliased into the output of the sampling system. The effects of this error can be minimized by moving the corner frequency of the filter lower than $f_s/2$ or increasing the order of the filter. In both cases, the minimum gain of the filter at $f_s/2$ should be less than the signal-to-noise ratio (SNR) of the sampling system.

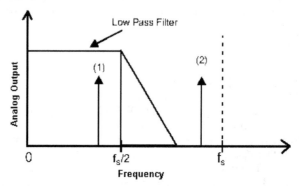

Fig. 2. Frequency Response of Low Pass Filter

2.5 Passive filters

Passive, low pass filters are realized with resistors and capacitors. The realization of single and double pole low-pass filters are shown in Figure 3. This value of resistor could create an undesirable voltage drop or make impedance matching difficult. Consequently, passive filters are typically used to implement a single pole.

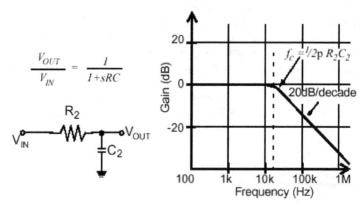

Fig. 3. Passive, Low-pass Filter

2.6 Active filters

The active filter offers the advantage of providing isolation between stages [11]. This is possible by taking advantage of the high input impedance and low output impedance of the operational amplifier. In all cases, the order of the filter is determined by the number of capacitors at the input and in the feedback loop of the amplifier. The Double Pole, Voltage Controlled Voltage Source is better known as the Sallen-Key filter realization. This filter is configured so the DC gain is positive. In the Sallen-Key Filter realization shown in Figure 4, the DC gain is greater than one. In the realization shown in Figure 5, the DC gain is equal to one. In both cases, the order of the filters is equal to two. The poles of these filters are determined by the resistors and capacitors values of R1, R2, C1 and C2.

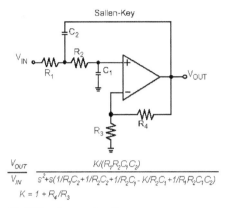

$$\frac{V_{OUT}}{V_{IN}} \quad \frac{K/(R_1 R_2 C_1 C_2)}{s^2 + s(1/R_1 C_2 + 1/R_2 C_2 + 1/R_2 C_1 - K/R_2 C_1 + 1/R_1 R_2 C_1 C_2)}$$

$$K = 1 + R_4/R_3$$

Fig. 4. Sallen-Key filter with DC gain greater than one

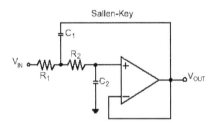

Fig. 5. Sallen-Key filter with DC gain equal to one.

2.7 Low power design

It is possible to have the system to manage its own power consumption by software control utilizing a microcontroller. Some microcontroller like the PICmicro™ family from Microchip can source up to 20mA of current and provides power to other components. In this case, one can simply connect the Vdd pin of an external component to an input-output (IO) pin of the microcontroller. Currently, most of the op-amps, analog-to-digital converter (ADC), and other devices manufactured today are low powered and therefore can be powered by this technique. Many techniques are used to reduce power consumption in the microcontroller. The most commonly used methods are SLEEP mode and external events. The microcontroller can periodically wake-up from sleep using the Watchdog Timer or external interrupt, execute code and then go back into SLEEP Mode.

2.8 Data compression

In these applications, the loss of even a single bit could be catastrophic. In the case of ECG file, it is important to preserve the actual recorded ECG data when transmitting to the physician for later analysis. Thus, the lossless compression technique is more suitable to be used for the compression of ECG data. Huffman codes have the unique prefix attribute, which means they can be correctly decoded despite being variable length.

2.9 Heartbeat detection techniques

The recognition of almost all ECG parameters is based on a fixed point identifiable at each cycle. R-peak is suitable for use as the datum point, because it has the largest amplitude and sharpest waveform that can be extracted from ECG. For the detection of R-peak, the amplitude level triggering method is based on the square of the derivative. The high frequency components of an original signal are indicated by differentiating the filtered ECG signal. The reliability of the identification is increased by squaring this differentiated value, which will highlight the high frequency components. After this processing, the R-peak can be detected by determining the known amplitude level as a triggering level. The triggering level can also be adaptively determined. Thus, the value of the triggering level is dependent on the upper values of the ECG signal. The R-wave triggering threshold value can be determined by try and error to find out the most suitable value [19]. The value of the triggering threshold can adaptively change with the base line value of the ECG signal. The R-peak can be recognized by determining the squared value that is greater than the R-wave triggering value. The current RR-interval can be calculated by differentiating the previous R-wave time with the current R-wave time and the heartbeat can be calculated by the following formula.

$$RR\,int\,erval = Xms$$

$$Beat\,/\,Second = \frac{1}{X} \times 1000$$

$$\therefore Beat\,/\,Minute = \frac{1}{X} \times 1000 \times 60 = \frac{60000}{X}$$

(1)

3. Hardware design

3.1 Block diagram of the hardware design

The proposed block diagram of the ECG recorder is shown in Figure 6. Three electrodes were used in the ECG recorder; two electrodes are used to sense the ECG signal and the other one is for noise reduction. The weak ECG signal acquired from electrodes is first amplified by the preamplifier with a gain of about 500 before it is filtered by a 96Hz low-pass filter to eliminate most of the electromagnetic noise. The offset null circuit is to eliminate the DC voltage between the two electrodes. The filtered signal is further amplified to about 1V and the offset of the signal is shifted to 1.25V by another stage of amplifier. It is necessary to shift the signal offset to 1.25V so that all the negative voltage is shifted to positive because the ADC accepts pseudo-differential inputs ranging from 0V to reference voltage of 2.50V. The amplified signal is digitized by a 12-bit ADC sampled at 500 samples per second before it is sent to the microcontroller via 3-wire serial interface. In the block diagram, there is a power supply unit that produces +5V and –5V needed by the analog circuits. The power supply for analog circuit can be shutoff by a control line connected to the microcontroller. Besides that, this unit can detect the battery level and signal the microcontroller in case of battery low. The recorder has an 8-bit microcontroller to control its operations like initializing the ECG recording circle, stop recording, shutdown the circuits when there is no activity and monitor the battery status. It also implements the communication protocol between recorder and the computer. The ECG recorder is physically isolated from the 240V power line using opto-isolator for the safety of the patient.

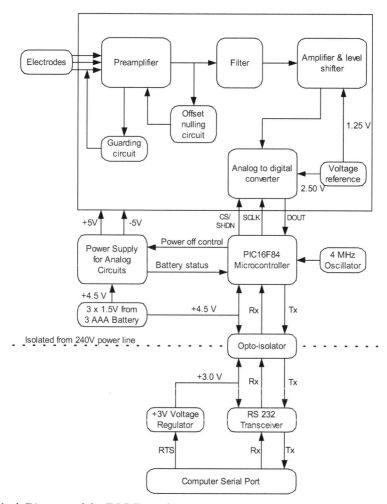

Fig. 6. Block Diagram of the ECG Recorder

3.2 Amplifier design

The amplification of ECG signal can be divided into two stages, the preamplification stage has a gain of about 500 to improve Common Mode Rejection Ratio (CMRR) and the second stage of amplification has a gain of two to make the overall gain of 1000 that will amplify the signal voltage to about 1.0V so that it is high enough to input to the ADC. The 50 Hz line interference can be sufficiently reduced by having a third electrode connected to the amplifier's common ground, which put the patient's body voltage potential to the amplifier's common voltage. The ECG recorder is optically isolated from the main power line and the amplifier is powered by battery, which delivers very clear voltage supply. DC input voltage of up to 200mV should not result in saturation of the amplifier. The INA118 uses a single external resistor to set gain from 1 to 10000. It is laser trimmed for very low offset voltage (50µV) and high CMRR (110dB at 1000 gain). It operates with power supplies as low as ±1.35V and quiescent current is only 350µA, which is very suitable for battery operation. The INA118 has eight pins which are the positive input pin ($V_{in}{}^{+}$), negative input pin ($V_{in}{}^{-}$), positive voltage supply pin (V^{+}), negative voltage supply pin (V^{-}), output reference pin (Ref), two pins for gain setting (R_G) and output pin (V_o). To set G to 500, In the recorder design, two 51Ω resistors are connected in series between the two R_G pins to set the amplifier gain to 491.

$$500 = 1 + \frac{50K\Omega}{R_G} \tag{2}$$

$$R_G = \frac{50K\Omega}{500-1} = 100.2\Omega$$

The LMC6464 quad Operational Amplifier (OA) from National Semiconductor is used in the second stage of signal amplification. The filtered ECG signal is input to the inverting input of the OA through a 10KΩ resistor and the output is feedback to the inverting input pin by a 20KΩ resistor to set the inverting amplifier gain to two. The offset of the ECG signal is shifted to +1.25V by connecting the non-inverting input of the OA to a +1.25V voltage reference through a 10KΩ resistor. A 20KΩ resistor is then connected between the ground and the non-inverting input to make the amplifier balance.

3.3 Filter design

The filtering system for the ECG recorder can be divided into two parts, the low-pass filter and the high-pass filter. The low-pass filter is a second order Butterworth Filter designed to give a cutoff frequency of 96Hz. At this cutoff frequency, most of the electromagnetic noise will be filtered out. The LMC6464 OA is used to build the filter with two resistors and two capacitors connected to form a Sallen-Key configuration as shown in Figure 9. The values of the two resistors were 8.2KΩ and 15KΩ and the values of the two capacitors are 0.1µF and 0.22µF. The calculation of cutoff frequency (f_c) is as follows:

$$f_c = \frac{1}{2\pi\sqrt{R_1 R_2 C_1 C_2}}$$

$$= \frac{1}{2\pi\sqrt{(8.2K)(15K)(0.22\mu)(0.1\mu)}} \tag{3}$$

$$= 96.8Hz$$

The high-pass filter has a cutoff frequency of 0.3Hz to eliminate the DC signal. It was built by using a LMC6464 OA, a 1MΩ resistor and a 0.47µF capacitor to form an integrator circuit. The output of the preamplifier was integrated and connected back to the Ref pin of the INA118. Cutoff frequency calculation for the high-pass filter is given by equation 4.

$$f_c = \frac{1}{2\pi(1M)(0.47\mu)}$$

(4)

$$= 0.3 Hz$$

3.4 Electrodes, wires and guarding circuit

The electrode that has been selected for the ECG recorder is the disposable foam electrode. Two electrodes are used to form a single lead pair where one will be placed on the left arm and the other will be placed on the right arm. Another electrode, which is used for noise reduction will be placed on the right leg or hand. The lead wires have to be made short, preferably less than one meter to reduce the electromagnetic interference. The length of the lead wires must be equal so that the difference of impedance in the two lead wires can be minimized. The difference of impedance in the lead wires can cause the potential divider effect, which converts the common mode voltage to differential voltage. The wires used in the ECG recorder were shielded with a single guarding circuit to drive the shield back to common mode voltage. The guarding circuit consists of the LMC6464 OA connected as a uni-gain amplifier. The common mode voltage from the INA118 was input to the non-inverting input of the OA and the output was feedback to the inverting input. The high input impedance from the non-inverting input of the OA will isolate the guarding circuit from the IA so that the operation of the IA will not be interrupted.

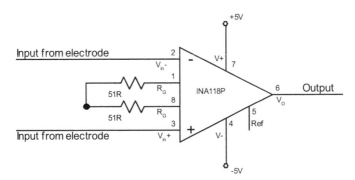

Fig. 7. Circuit Diagram for the Instrumentation Amplifier

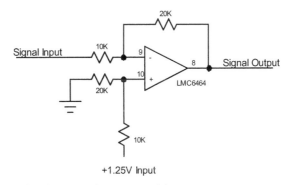

Fig. 8. Circuit Diagram for the SecondStage Amplifier

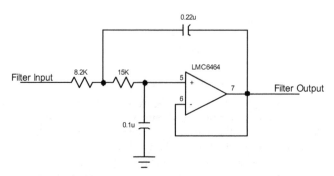

Fig. 9. Circuit Diagram for the Low Pass Filter

3.5 Analog to digital converter

The MAX145 ADC from MAXIM Integrated Product was selected in the development of the ECG recorder. It is a serial 12-bit ADC with low power consumption, automatic shutdown and fast wake-up time. It has an internal track-and-hold circuitry eliminating the need for external sample-and-hold amplifier. The reference voltage of 2.5V was used in the ECG recorder to limit the Quantization Noise to $1mV_{pp}$. A serial ADC has more complicated data transfer process compared to parallel ADC. However, a serial ADC has been chosen to minimize number of pin used for data transfer because the PIC16F84 microcontroller that is used to control the ADC has limited number of IO pin. In addition, a serial ADC will use up less space to make the recorder compact. The MAX145 pin 2 and pin 3 are pseudo-differential inputs which only accept input signal ranging from zero to a reference voltage set at pin 5, which is 2.5V in the ECG recorder design. The pin 2 of MAX145 is connected to the ground and the analog ECG signal is input to the pin 3 of MAX145. Because MAX145 only accept positive input voltage, the offset of the ECG signal is shifted to 1.25V before it is input to the pin 3 of MAX145.

Only three pins need to be connected to the microcontroller from MAX145, via the pin 7 (DOUT), pin 8 (SCLK) and pin 6 (active low chip select/active high shutdown). Pin 7 and 8 is used for reading out the digitized ECG sample serially and pin 6 is a control pin to activate the analog-to-digital conversion process. The serial data output format for MAX145 is a stream of 16-bit data stream. The first bit must be logic high to indicate the end of conversion. The next three bits must clock out high followed by the 12 bits of data in MSB-first format. After the last bit has been read out, additional serial clock pulses will clock out trailing zeros. DOUT changes on the falling edge of SCLK. That means the DOUT pin should be sampled when SCLK is high.

3.6 Power supply unit

The basic function of the power supply unit is to produce +5V and –5V from a single +4.5V supply from the battery. The MAX681 voltage converter is used to convert +4.5V to both +9V and –9V. It is a monolithic, dual charge-pump converter that provides positive and negative outputs of two times a positive input voltage. The +9V is regulated to +5V by using the MAX666 voltage regulator and the –9V is regulated to –5V by using the MAX664 voltage regulator. With this configuration, the input voltage from the battery can drop to +3V without affecting the output voltage of the power supply unit. This is because MAX681 voltage converter can convert the +3V to +6V for the MAX666 voltage regulator and –6V for the MAX664 voltage regulator.

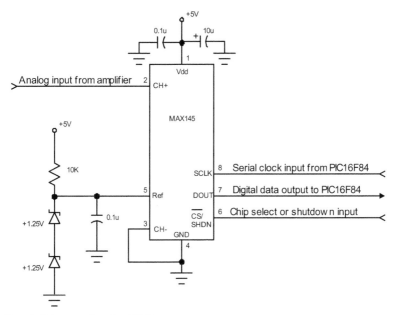

Fig. 10. Circuit Diagram for the ADC

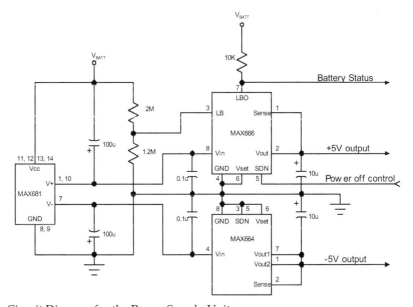

Fig. 11. Circuit Diagram for the Power Supply Unit

The MAX666 contains on-chip circuitry for low power detection. If the voltage at pin 3 of MAX666 falls below the regulator's internal reference of 1.3V, then pin 7 or the Low Battery Output pin goes low. The threshold for low battery indication can be set by using two external resistors with the values of 2MΩ and 1.2MΩ to form a voltage divider to the supply voltage. The divided voltage is then input to pin 3 of MAX666. The MAX666 can be shutdown by logic input at pin 5. When in Shutdown State, the maximum drain current is limited to 12µA, which is desired for low power design.

3.7 Microcontroller

The microcontroller acts as an intelligent unit to perform all the operation of the ECG recorder such as power management, ECG samples data acquisition and implementation of the computer interfacing protocol. Microcontroller selected for the ECG recorder is the PIC16F84 from Microchip Technology Inc. It is a low cost 8-bit microcontroller operating at the speed of 4MHz. The microcontroller has pins connected to the ADC, power supply unit and the opto-isolator. Pin 18 (RA1), pin 1 (RA2) and pin 2 (RA3) of the PIC16F84 microcontroller are connected to the MAX145 ADC. RA1 is the clock output to clock out serial data from the ADC. It is connected to the SCLK pin of the ADC. RA2 is the serial data input pin to receive the serial data from the ADC. It is connected to the DOUT pin of the ADC. RA3 is the chip-select output pin, which will power up the ADC when RA3 is high and shutdown the ADC when RA3 is low. It is connected to the pin 6 of the MAX145 ADC. Pin 3 (RA4) and pin 17 (RA0) are connected to the power supply unit. RA4 is an output pin to power off the analog circuitry and RA0 is the input pin for battery status. Pin 6 (RB0) and pin 7 (RB1) are used for serial data transfer with the computer. RB0 is for receiving data and RB1 is for transmitting data.

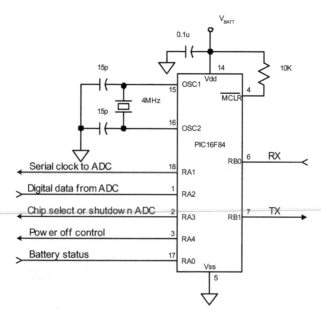

Fig. 12. Circuit Diagram for the Microcontroller

The flowchart for the microcontroller firmware is shown in Figure 17. Once the microcontroller power is up or being reset, it sets its IO pins direction by setting the internal registers, TRIS A and TRIS B [20]. RB0, RA0 and RA2 are set as input pins whereas RB1, RA1, RA3 and RA4 are set as output pins. The microcontroller then set all the output pins' initial state to logic high.

After initialization, it enters the Ready State. The Ready State puts the microcontroller in SLEEP mode and puts the ECG recorder to the lowest power consumption while waiting for the present of data from the computer serial port. Before entering the SLEEP mode, the microcontroller will enable RB0 pin interrupt.

When in SLEEP mode, the microcontroller will eventually wake up by the watchdog timer and go back to SLEEP mode. When a start-bit is detected at RB0 pin, the microcontroller will wake up, send back an "OK" string to the computer and wait for the command string from the computer. It then parses the command to determine whether the computer wants to start recording ECG or wants to check battery status. If the command is not valid, the microcontroller will reset.

In the check battery routine, the microcontroller will check the status of RA0 pin and send back the battery status to the computer. The routine then goes back to Ready State. In the record ECG routine, the microcontroller initializes its internal timer to start the analog-to-digital conversion of the ECG signal every two milliseconds. It sends one ECG sample together with the battery status to the computer every two milliseconds. The record ECG routine is designed such that the computer has to send back an acknowledgement byte to the microcontroller before 255 samples of ECG data have been sent to the computer. This is to prevent the recording process to run continuously without knowing whether the computer is receiving the ECG data. The recording process will stop and go back to SLEEP mode if there is no acknowledgement received from the computer after 255 samples have been sent to the computer.

3.8 Opto-isolator

The H11L1 opto-isolator is used to isolate the ECG recorder from the 240V power line. It has an infrared emitting diode optically coupled to a high-speed integrated detector with Schmitt trigger output. The H11L1 can isolate maximum peak voltage of 7500V before the isolation breaks down. For the ECG recorder, two H11L1 opto-isolators are used to isolate the transmit data and receive data lines for the serial port. With the use of opto-isolator, there will be no electrical path connecting the patient and the computer, which is then connected to the 240V power line.

3.9 RS-232 transceiver to USB

The RS-232 transceiver is used to convert the TTL level logic at the ECG recorder to RS-232 level logic at the serial comport and vice versa. The MAX3232 low-power RS-232 transceiver is used in the ECG recorder. It requires four 0.1μF external capacitors to operate and typical supply current of only 300μA. The low power consumption of this component makes it possible to be powered by the RTS line of the computer serial port. This part cannot be powered by the battery because it is not isolated from the 240V power line by the opto-isolator.

3.10 Communication protocol between computer and ECG recorder

The ECG recorder is interfaced to the computer via RS-232 port with the setting of 19200 baud, 8 bit data, no parity bit and one stop bit. To activate the ECG recorder, the computer has to send a set of commands to the ECG recorder. The communication starts with a function call from the computer software either to get the battery status of the ECG recorder or to start a recording process.

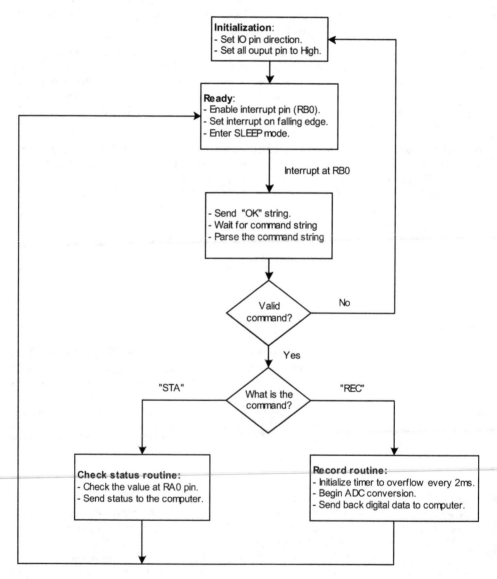

Fig. 13. Flow Chart for Microcontroller Firmware

To get the battery status of the ECG recorder, the computer will first send a one-byte data with hexadecimal value of '0xAA' to wakeup the ECG recorder. Upon receiving the one byte data, the ECG recorder will wakeup from sleep, power up the analog circuit and send back an "OK" string to indicate that it is now ready to receive a command from the computer. The command format begins with one byte of 'SOH' ASCII character and then followed by three letters string. To get the battery status, the letter string is "STA". When the ECG recorder receives a 'SOH' followed by a "STA" string, it will check the battery status and send back the battery status to the computer. It first sends a "SOH" byte followed by the status byte, which is an ASCII letter 'L' if battery level is low and an ASCII letter 'H' if battery level is normal. After sending the status byte to the computer, the ECG recorder goes back to sleep mode.

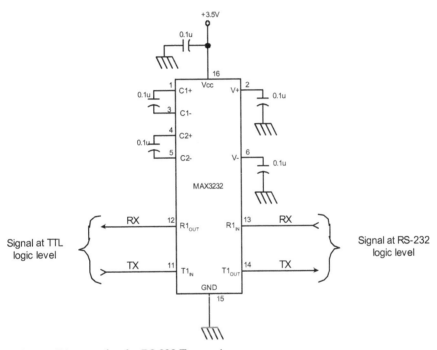

Fig. 14. Circuit Diagram for the RS-232 Transceiver

To start the ECG recording process, the computer sends a byte with value '0xAA' to the recorder. Then the ECG recorder wakes up from sleep, power up the analog circuit and sends back an "OK" string to the computer. The computer then sends a 'SOH' byte followed by a "REC" string command to the ECG recorder. When the ECG recorder receive the "REC" string command, it will initiate the ECG recording process and send back a 'SOH' byte to the computer to signal that the ECG samples are coming. Then it sends a sequence of ECG sample to the computer, two bytes of data per sample. Before the ECG recorder has sent 255 samples, the computer must send a byte with hexadecimal value '0xff' to the ECG recorder to continue the recording process. Otherwise, the ECG recorder will stop recording and go back to sleep mode. The communication protocol is illustrated by Figure 15.

4. Software design

The software design part involves the design of multiple functioning objects and the design of data flow between them [21, 22]. The functions and data structures of several main objects in the ECG software are described in the following sections. The algorithms used in the compression of ECG data and heartbeat detection are described in the Data Compression Unit and the Heartbeat Detection Unit.

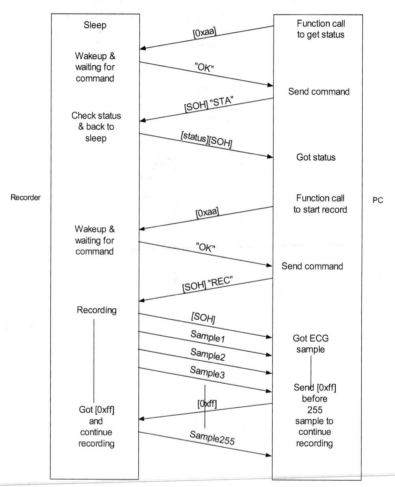

Fig. 15. Communication Protocol Between the Recorder and Computer

4.1 Data flow diagram

The data flow diagram in Figure 20 shows the flow of information between different objects and variables for the ECG recorder software. The object can be hardware, Graphic User Interface (GUI), file on disk, variable in the program memory or running thread. The information can be in the form of command or variable passed from one object to another.

The GUI objects are objects that interact with the user. They are the objects that the user can see in the main window or as a dialog box. The GUI objects in the data flow diagram together with their description are listed in the Table 1.

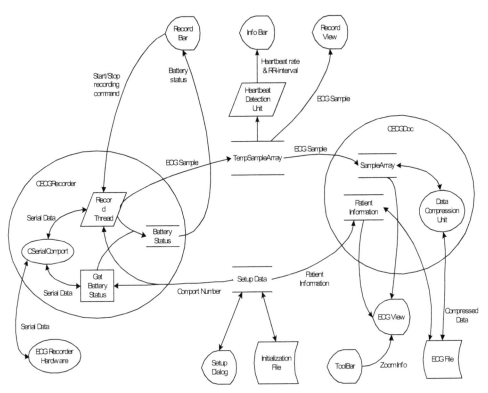

Fig. 16. Data Flow Diagram for the ECG Software

The recording process is initiated by the user from the Record Bar. When a Start Recording command is received by the CECGRecorder object, it will start the Record Thread. The Record Thread then activates the appropriate serial port and start sending and receiving serial data from the ECG recorder hardware. The Record Thread will decode the serial data into digital samples of ECG signal and the battery status. The ECG samples are then appended to the TempSampleArray, an instance of CWordArr that represents ECG signal in the memory. The battery status will be stored in a variable that later will be used to update the battery status display at the Record Bar. While the TempSampleArray represents the recording ECG signal in the memory, Record View displays it in graphical form. The Record View constantly reads the TempSampleArray and updates its display so that the user will see the real-time signal. Meanwhile, another thread, the Heartbeat Detection Unit will examine the recording signal to check whether a heartbeat is present in the signal. If a heartbeat is detected, the thread then calculates the current heartbeat rate and updates the Info Bar.

GUI object	Description
Record Bar	This is the control panel for the ECG recorder to start or stop the ECG recording process. It displays the battery status of the ECG recorder, the start time and the time elapsed.
Info Bar	Displays the heartbeat rate and the RR-interval. It also displays animation when a heartbeat is detected by the program.
Record View	Displays the real-time graph of the ECG signal when the recording process is running.
ECG View	Displays the recorded ECG signal or the ECG document in a graph. This object enables the user to scroll through the whole ECG document by a Scroll Bar. It also enables the ECG signal to be zoomed to 5 different sizes.
Setup Dialog	This dialog box is for the user to set the name of serial port where the ECG recorder hardware has been connected. Besides that, the user can also set the patient's information here.

Table 1. Description for the GUI Objects

When the recording process is stopped by the user, a new CECGDoc object will be created. The TempSampleArray will be copied to another instance of CWordArr class, which is the SampleArray and the patient information from the setup data will be copied to the member variable of the CECGDoc. The CECGDoc object represents the recorded ECG signal in the memory. It includes every detail of the recorded ECG such as the record time, the length of the recorded ECG and information of the patient, to whom the ECG belongs. The CECGDoc is graphically displayed by the CECGView. In the CECGView, one can see everything in the CECGDoc. It includes five types of zoom view to view the recorded ECG signal in different sizes. The zoom size of the CECGView can be controlled by the user in Tool Bar or in program menu. The CECGDoc can be saved on a disk as an ECG file. Actually, the patient information and the SampleArray are saved to the file when a CECGDoc is saved. The patient information is saved directly to the ECG file while the SampleArray will be compressed first before it is saved to the ECG file. The Data Compression Unit is responsible to compress and decompress the SampleArray.

4.2 Main frame window

The main frame window is the main display for the ECG software that is showed when the program starts. Like other standard window program, it has a main menu, toolbar, status bar, minimize box, maximize box and close box. It is a Multiple Document Interface (MDI) frame, which may contain several child frames so that two or more ECG files can be viewed at the same time. If the main frame window is closed, the program will also be terminated.

4.3 CECGRecorder class and the recording thread

CECGRecorder is a class that abstractly represents the ECG recorder in the ECG software. This class provides all the functions to control the ECG recorder hardware. All the low-level protocols to communicate with the hardware are automatically handled here. To use the CECGRecorder class, one must call the Activate function first and pass two arguments with it. One argument is strings to specify the name of serial port that connects to the recorder hardware, and the other argument is a pointer to the data structure to store the ECG samples when recording ECG. The member functions of this class are listed in the Table 2.

Member Function	Description
BatteryStatus	Method to return the battery status of the ECG recorder hardware.
StartRecord	Method to start the record thread.
StopRecord	Method to stop the record thread.
IsRecording	Method to check whether the record thread is running.
Activate	Method to activate the CECGRecorder.
DeActivate	Method to deactivate the CECGRecorder.
IsActivated	Method to check whether the CECGRecorder is activated.
HardwareAvailable	Method to check whether the recorder hardware is still connecting to the activated comport.

Table 2. Member Function for CECGRecorder Class

4.4 Data structure of the ECG signal

CWordArr Class represents the ECG signal in the program memory. It is a thread-safe array, which enables multiple threads to add data to it and get data from it at the same time without blocking each other. Each element of the CWordArr consists of 16-bit word, which manages to hold a 12-bit ECG sample. An ECG sample can only be added to the top of the CWordArr so that the position of ECG samples in the CWordArr will follow exactly like the ECG signal. When reading from the CWordArr, ECG sample in the CWordArr can be read out at any position and will not be deleted from the CWordArr. It is designed to be like this so that multiple threads can read out ECG sample in the CWordArr at different position simultaneously. Beside adding and reading ECG sample, the CWordArr class includes method to get the length of the ECG sample.

4.5 Data compression unit

The Data Compression Unit compresses the SampleArray into a compressed data stream before it is saved to a file. The compression process is divided into two steps. First, the ECG samples stored in the 16-bit array are converted to a stream of 12-bit ECG samples stored in CByteArray. In this step, no compression is actually done except to eliminate the redundant bits in the SampleArray. Then the CByteArray data stream is compressed by using Huffman algorithm. To implement the Huffman algorithm, the frequency of occurrence of each byte in the CByteArray is counted and the result is stored in the Counts array. The Counts array is then scaled down so that they fit in a byte and then stored as initial weights in the NODE array. From the NODE array, a Huffman tree is built. The Huffman tree is then converted to Huffman codes by a function that recursively walks through the tree, adding the child bits to each code until it gets to a leaf. When it gets to a leaf, it stores the code value in the CODE element and return. The Counts array, which has been scaled down, is saved in the ECG file so that the Huffman tree can be rebuilt in the data expansion process. When the Huffman codes are available, the CByteArray can be compressed by saving the Huffman code correspondence to each byte in the CByteArray into the ECG file. The Data Compression Unit is also responsible to expand the compressed data to SampleArray. The expansion process is also divided to two steps. First, it converts the Huffman codes to CByteArray and then converts the CByteArray to SampleArray.

4.6 Heartbeat detection unit

The Heartbeat Detection Unit is a thread that detects the present of heartbeat in the ECG signal. The algorithm used to detect the heartbeat is to find the presence of R-wave in the signal by amplitude triggering method. In this method, the average value of the signal is calculated by taking the latest 1000 samples into calculation. The triggering threshold is set to 600 units higher than the calculated average value. This value is determined by trial and error to find the most suitable value. Then, the ECG sample is read in one by one. An R-wave is considered detected when it can fulfill the follows conditions.

- The ECG sample value is higher than the triggering threshold.
- The differential value between the current ECG sample and the previous one is greater than three units.
- Four consecutive decrease of sample value followed by four consecutive increase of sample value is detected.

5. Discussion

The ECG recorder has been constructed prototype on a board and tested for its electrical characteristics. The microcontroller firmware is tested by software simulation before it is transferred to the PIC16F84 microcontroller. The microcontroller can successfully get the ECG samples from the ADC and send it to the computer. To test for the stability of the recorder, it is put in record mode for long duration and it is able to continuously sending data to the computer for more than 3 hours. When the recorder is in recording mode, it automatically goes back to power saving mode if the ECG recorder is disconnected from the computer. At battery low condition, the recorder can still work until the battery voltage drops to less than +3V. The battery low indicator will turn on when the battery level drop below +3.5V and left some time for the user to stop the recorder before it stop working by itself.

5.1 Testing for the amplifier gain

The voltage gain for every stage of the amplifier was measured. By using a sinusoidal signal with frequency of 20Hz, the peak-to-peak voltage at the input and output is measured by an oscilloscope. The measurement result for the preamplifier and second stage amplifier are shown in the Figure 17 and Figure 18. From the Figure 17 and Figure 18, the voltage gain for the preamplifier is 497.2 and the voltage gain for the second stage amplifier is 2.0. Thus, the total voltage gain for the two amplifiers is 994.4.

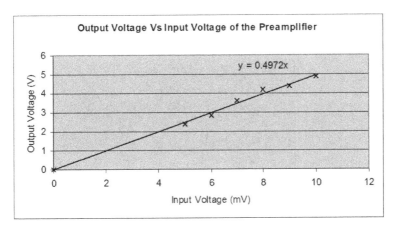

Fig. 17. Gain Measurement for the Preamplifier

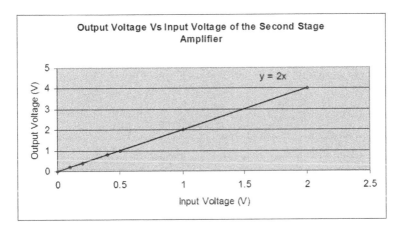

Fig. 18. Gain Measurement for the Second Stage Amplifier

5.2 Testing for the filter

To measure the frequency response for the analog circuit, a sinusoidal signal with different frequency was applied to the differential input of the preamplifier. The input signal is set to voltage of 25mV and then the output voltage for every frequency at the final stage amplifier was measured by an oscilloscope. From the measured result, a graph was plotted to find the cutoff frequency for the high pass filter and the low pass filter. Figure 19 shows the measurement result for frequency response. From the Figure 19, the cutoff frequency for high pass filter was 0.23Hz and the cutoff frequency for low pass filter was 79Hz.

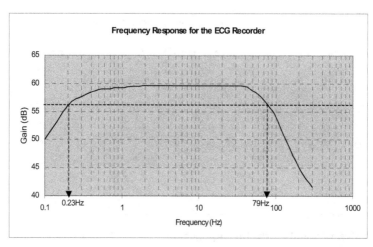

Fig. 19. Frequency Response for the Analog Circuit

5.3 Total current consumption

The total current consumption for the ECG recorder was measured with ammeter. Measurements have been made when the recorder was in power saving mode and active mode. The measurement results are shown in Table 3. As we can see in Table 6, the ECG recorder consumes very little power with only 1.85mA in power saving mode and 6.50mA in active mode.

Recorder Operation State	Total Current Consumption (mA)
Power Saving Mode	1.85
Active Mode (Recording ECG)	6.50

Table 3. Measurement for the Total Current Consumption

5.4 The ECG software

The ECG software can communicate with the recorder hardware and get data from the recorder. The recorded ECG signal can be displayed at five different scales, via 5mm/mV, 10mm/mV, 20mm/mV, 40mm/mV, and 80mm/mV. The P, Q, R, S and T waves of the ECG signal can be clearly displayed at the scale of 10mm/mV, 20mm/mV, 40mm/mV, and 80mm/m. The recorded ECG signal has noise level of about 0.12mV and can be noticed when displayed at the scale of 40mm/mv and 80mm/mV. The heartbeat detection unit can detect the heartbeat quite well when there is no movement artifact present. Therefore, the patient should prevent large movement when the recording is in progress to reduce the movement artifact noise

Fig. 20. The actual PC-based Electrocardiogram (ECG) recorder as - Internet appliance

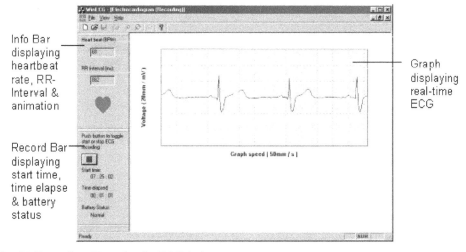

Fig. 21. Recording View for the ECG Software

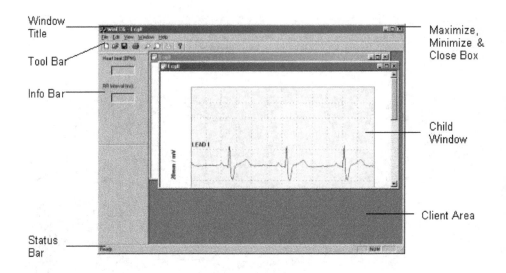

Fig. 22. Main Window for the ECG Software

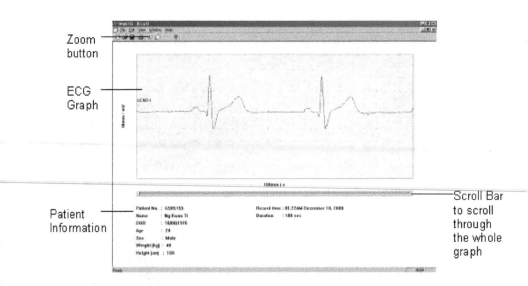

Fig. 23. Document View for the ECG Software

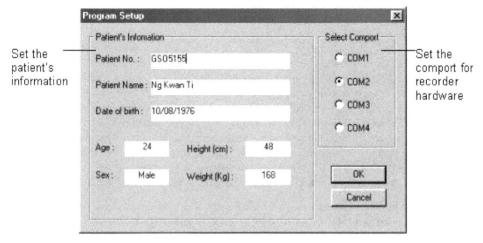

Fig. 24. Setup Dialog for the ECG Software

5.5 ECG file size

The graph in Figure 25 shows the size of the ECG files and its record duration. The increasing rate for the ECG file is approximately 0.5236 kilobyte per second. If the result is compared with the increasing rate without compression, which is 0.7324 kilobyte per second, the compression ratio achieved is 39.9%.

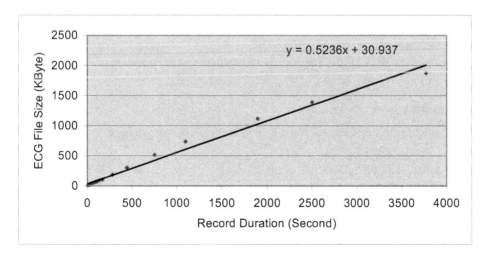

Fig. 25. ECG File Size versus Record Duration Graph

6. Conclusion

Electrocardiogram (ECG) has long been used as an important medical tool in monitoring patient's heart activities. In some cases, the patient has to go to the medical center very often to record their ECG for the diagnostic of the physician. This chapter describes the design and development of an electrocardiogram (ECG) recorder for single-lead recording that enables the recording of ECG at home and development of software to receive the ECG data from the recorder. The ECG data is saved in a compressed format for easy transmission over the Internet. The ECG recorder is a battery-powered device and its design emphasises on low power consumption. The ECG recorder is used as a peripheral connected to the computer via RS-232 port as well as USB simulator. The software was can be developed using Visual C++ or Java programming language. This design has been tested with the real patient and the ECG output has been verified by the medical doctors. The software can display the ECG data and clearly show the P, Q, R, S and T waves for diagnosis. The system consists of ECG recorder hardware and a software program. The recorder hardware has the following characteristics:

- It is battery powered and small in size. Therefore, it is very portable and easy to use.
- It provides amplification, noise filtering, analog-to-digital conversion and data acquisition of ECG signal. In addition, it has the ability to monitor its battery status and manage its power usage.
- It samples the ECG signal at high resolution of 12-bit and with the sampling rate of 500 samples per second. Therefore, the digitized ECG signal is suitable for use in the analysis and measurement of PQRST waves of the patient's ECG.
- This battery operated portable device has safety medical standards that comply with IEC601-1, AAMI EC11 (protective class).

The ECG software has the following characteristics:

- It can control the recorder hardware and get the digitized ECG data from the hardware via serial port.
- It provides the display of the real time ECG signal, patient's heartbeat rate and the battery status when recording ECG. When viewing the recorded ECG signal, the user can scroll through the signal by a scroll bar. Besides that, the ECG signal can be zoomed to five different sizes.
- It compresses the ECG data before saving the data to a file and expands it when the file is loaded. This makes the ECG file small and easy to send through the Internet.

6.1 Suggestion for future expansion

The following improvements can be made to make the system more powerful and useful.

- The program can be expanded to be a client-server program so that the patient's ECG file can update directly to the medical center's database and thus providing a more secure transmission.
- The security part of the program can be further improved by providing some data encryption to the ECG file so that only the intended person can view the patient's ECG data.
- Besides that, the program can be developed so that it includes digital filter to further reduce the noise level.

- The prototype can be expanded to make it work with handheld devices such as palm top to make the recorder mobile. The ECG recorded at the handheld device can then be downloaded on the computer.
- To make the recorder a wireless device by adding the radio communication circuitry to the recorder hardware and computer.
- To view of more flexibility, this device can be modified further to work with direct telephone line interface whereby the communication and ECG data will be transferred to the server automatically. A modem and a communication protocol need to be embedded with this device.

7. References

Application of Embedded Technology on a Portable ECG Recorder, Posted on April 8, 2010 by China Papers, http://mt.china-papers.com/2/?p=116829

Bonnie C. Baker, Anti-aliasing, analog filters for data acquisition systems, Microchip Technology Inc., 1999.

Claydokjun S and Chitsakul K, "Real Time Electrocardiogram Compression Technique Using Wavelet transform On MCS-51," 16th Biennial International Eurasip Conference Biosignal 2002.

Claydokjun S and Chitsakul K, "Real Time Electrocardiogram Compression Technique Using Wavelet transform On MCS-51," 16th Biennial International Eurasip Conference Biosignal 2002.

Daskalov, I. K., I. I. Christov, Electrocardiogram Signal Preprocessing for Automatic Detectioin of QRS boundaries, *Medical Engineering & Physics* 21, 1999, pp 37–44.

Edmond Zahedi, Wong Meng Meng, Chye Jun Lee, Development of a Portable Electrocardiogram Event Recorder with Digital Memory, Department of Electrical, Electronic and System Engineering, Universiti Kebangsaan Malaysia, 1999.

Farah Magrabi, Nigel H. Lovell, Branko G. Celler, (1999) A web-based approach for electrocardiogram monitoring in the home, *International Journal of Medical Informatics* 54, 1999, pp 145–153.

Ho, C.S.; Chiang, T.K.; Lin, C.H.; Lin, P.Y.; Cheng, J.L.; Ho, S.H.; , "Design of portable ECG recorder with USB storage. Electron Devices and Solid-State Circuits, 2007. EDSSC 2007. IEEE Conference.

ITU SG-2 Q14/2 Document 2/195 (Rev.1)-E "Telemedicine Directory". by Rapporteur for Question 14/2; 6 July 2001.

James W.Grier.Comparison of three handheld 1-lead ECG / EKG recorders. NDSUEDU.2008.

John D. Halamka, Carsten Osterland, Charles Safran, CareWeb™, a Web-based Medical Record for an Integrated Health Care Delivery System, *International Journal of Medical Informatics* 54, 1999, pp 1–8.

K. Janda, K. Chitsakul, Portable ECG monitor/record with wireless data transmission, The 3rd International Symposium on Biomedical Engineering (ISBME 2008).

Kohki Shinozaki Risk management of QT-prolonging Drugs by community pharmacists by using a mobile electrocardiography.The Pharmaceutical Society of Japan.2010.

Metting , A. C. Van Rijn, A. Peper, C. A. Grimbergen, High Quality Recording of Bioelectric Events: Interference Reduction, Theory and Practice, Academic Medical Center, Netherlands, 1998.

Metting, A. C. Van Rijn, A. Peper, C. A. Grimbergen, Amplifiers for Bioelectric Events: a Design with a Minimal Number of Parts, Academic Medical Center, 1998.

Metting, A. C. Van Rijn, A. Peper, C. A. Grimbergen, High Quality Recording of Bioelectric Events II: A Low-noise, Low-power Multichannel Amplifier Design, Academic Medical Center, Netherlands, 1998.

Tamas Hornos, Wireless ECG/EEG with the MSP430 Microcontroller,Dept. Of Electronics and Electrical Engineering, University of Glasgow, Thesis for Master of Science, 2009

Tikkanen, P. E., L. C. Sellin, H. O. Kinnunen, H. V. Huikuri, Using simulated noise to define optimal QT intervals for computer analysis of ambulatory ECG, *Medical Engineering & Physic* 21, 1998, pp 15–25.

11

Residual Stresses and Cracking in Dental Restorations due to Resin Contraction Considering In-Depth Young's Modulus Variation

Estevam Barbosa de Las Casas, João Batista Novaes Jr.,
Elissa Talma, Willian Henrique Vasconcelos,
Tulimar P. Machado Cornacchia, Iracema Maria Utsch Braga,
Carlos Alberto Cimini Jr. and Rodrigo Guerra Peixoto
Universidade Federal de Minas Gerais
Universidade de Campinas
Brazil

1. Introduction

Composite resins have been increasingly used as restorative material, both in anterior and posterior teeth, where they replace metal restorations. Its aesthetic characteristics, coupled with improved physical properties have made their use extend from just anterior teeth to also include posterior teeth. The use of such material in oral regions subjected to higher loading makes it important to account for the effect of residual stresses arising during polymerization, induced by resin contraction (Ausiello, Apicilla and Davidson 2002). Different reports can be found in the literature focused in this aspect and using different approaches, such as X-ray micro-computed tomography (Sun, Eidelman and Gibson 2009), 3D evaluation of the marginal adaptation (Kakaboura et al 2007) and 3 D deformation analysis from MCT images (Chiang et al 2010). These authors agree in the critical role played by resin contraction in restoration success. The Finite Element Method can therefore be a useful tool to investigate the cracking of interfaces, stress concentrations in the internal angles and effects of variations in the mechanical properties in the overall behavior of the resin-based dental restorations.

Some composite resins for dentistry use are the result of the interaction of an organic matrix, a coupling agent and an inorganic material, since it was first devised by Bowen et al. (1983). In the last decades, industry has altered significantly the inorganic part of this composition. As a result, composite hybrid resins, with the inclusion of nanoparticles have not only improved strength and abrasion resistance but also presented greater polishing characteristics. Nevertheless, their physicochemical properties are still dependent of the polymerization of the resin matrix, induced by light penetration with 400 to 500 nm wave lengths, the peak absorption of camphorquinone (Braga et al., 2005).

The beginning of the polymerization process of the resinous matrix happens when it is irradiated with an external source, and stops when the maximum conversion of the monomers is reached, which varies with time of exposure to light, quality of the polymerizing agent and depth of light penetration (Jandt et al., 2000; Felix et al., 2006; Jong et al., 2007; Ceballos et al., 2009; Ferracane, 2005). The degree of conversion of the resin matrix can then be associated directly to the physical properties of the restoration, as well as its strength under dental loading (Peutzfeldt et al., 2000; Sakaguchiet al., 1991).

When a composed resin is applied directly to dental tissues, after the hybridization by the adhesive agent of the cavity walls, the requirement is to rigidly connect enamel and dentine to the restoring material, in the attempt to generate a solid and continuous body as a mechanically ideal restoration. The ideal restoration would be totally impermeable to the infiltration of fluids and resistant to the opening of gaps in the tooth-restoration interface.

The use of the composed resin as restorative material implies in the polymerization of the resinous matrix and in its fixation to the cavity walls. However, during the polymerization the resinous matrix suffers contraction, which can surpass the adhesive rupture resistance, leading to imperfections in the restoration as marginal gaps, cracking, hypersensitivity and infiltration. The reduction of the generated residual stresses can be obtained in different ways during polymerization: (i) when a controlled fluid flow is allowed during the process (Braga et al.,2005); (ii) by the use of devices with the *soft-start* technique (Yoshikawa et al., 2001); (iii) reducing the associated factor C in the layers deposition (Ceballos et al., 2009; Petrovic et al., 2010); (iv) through the application of resinous material with low modulus of elasticity that can deform without producing high residual stresses (Dietschi et al., 2002); and (v) using a method of improving marginal adaptation by eliminating stress concentration points during resin photo-polymerization (Petrovic et al., 2010).

After polymerization, the composite resin exposition to the humid oral environment leads to water penetration to the inner regions of the restored cavity, causing volumetric hygroscopic expansion. Due to the penetration of water inside the hydrophilic organic matrix, a gradual expansion occurs, until a balanced value is eventually reached.

The aim of this study is to quantify the effect of the use of two different irradiation sources (halogen and second-generation LED) on the elastic modulus and the degree cure during the polymerization process of a nanofilled composite restorative material. The effect of the variation of the elastic modulus on the polymerization residual stresses is then accessed by means of3-D finite element simulations, in order to understand the mechanical response of a cylindrical Class I restoration under polymerization contraction.

2. Material and methods

2.1 Evaluation of composite resin mechanical properties

A composite resin (Filtek Z250, 3M Co., St Paul., MN, USA) and two different irradiation sources, XL3000 (halogen - 3M) and Elipar Freeligth 2 (LED - 3M ESPE), were used to evaluate, *in vitro*, the degree of conversion and the Young's modulus (also modulus of elasticity or elastic modulus) distribution along sample height. This last parameter was then used to examine, through a computer simulation, the effect of through-height stiffness variation in expected polymerization residual stresses.

2.1.1 Evaluation of the Degree of Conversion

To measure the Degree of Conversion (DC), test samples were prepared in standardized plastic molds with diameter of 4 mm and height of 2 mm and 4 mm. Four groups of samples were prepared, each one with three samples (n=3). Three measurements were performed for each sample. The groups were classified as: Group 1 (G1) – XL 3000, 2 mm height; Group 2 (G2) – Elipar, 2 mm height; Group 3 (G3)– XL3000, 4 mm height; Group 4 (G4) – Elipar, 4 mm height. The moulds were filled to the top in one step, and irradiated during a period of 40 seconds. After that, they were stored in a dark environment. The Degree of Conversion (DC) was measured at the base of the samples. Raman spectra were obtained on a JobinYvon/Horiba LABRAM-HR 800 spectrograph equipped with a He-Ne laser (632.8nm). The power in each sample was around 7mW. The Raman signal was collected by an Olympus BHX microscope provided with objectives (10x, 50x and 100x). The detector used was a N_2 liquid cooled CCD (Charge-Couple Device) of the Spectrum One, back illuminated. Depending on the sample background fluorescence, the acquisition time ranged from 10 to 30 seconds. To reduce signal/noise ratio, spectra were acquired 5 times after a 5 minutes photobleaching. Collected Raman spectra were analyzed and optimized with Labspec 1.1 and PeakFit v4. Spectra collected were averaged. The background was corrected and if necessary, normalized and peak deconvoluted.

The samples were excited by laser in the region between 2000 to 1000 cm^{-1} after five 60 second accumulations in order to evaluate the conversion proportion of the vinylic function in aliphatic function, comparing the residual non-polymerized methacrylate band C=C (1640cm^{-1}) to the aromatic 1610cm^{-1} C=C stretching band used as reference. The DC value was defined as given in Eq. (1),

$$DC(\%) – 100x [1 –Rp/ Rnp] \qquad (1)$$

where Rp and Rnp are the signals for the polymerized resin and non-polymerized resin. The degree of conversion was determined at three locations and the mean value calculated for each of the three specimens. From these three averages a new mean value was calculated.

2.1.2 Young's modulus measurement

Three specimens were prepared for each condition in order to measure the modulus of elasticity (Young's modulus) and were stored dry at room temperature for at least 24 hours before testing. All these specimens were prepared with 4 mm height, and the radiation source was kept at a distance between 1 and 2 mm. Therefore, there were also four groups of specimens for Young's modulus measurement: Group 1 (G1) – XL 3000, light source at 1 mm distance; Group 2 (G2) – XL3000, light source at 2 mm distance; Group 3 (G3)–Elipar, light source at 1 mm distance; Group 4 (G4) – Elipar, light source at 2 mm distance. For each group, three specimens were made (n=3) and, for each specimen, three indentations were done at four different depths (0.5, 1.5, 2.5 and 3.5 mm). The indentations were performed using a micro-durometer Shimadzu, at a speed of 2.6 mN/sand kept for 5 seconds. The values of Young's modulus and hardness were obtained for each depth in order to detect through-depth changes in the mechanical properties.

2.2 Finite element simulation

To simulate the restoration, a cylindrical model with a cavity was created to represent the dentine. On the cavity's inner walls an adhesive layer was applied, followed by the resin addition and then polymerization is performed. Despite being an axisymmetric problem, a three dimensional model was necessary due to software limitations to represent the adhesive material. Previous works in the literature describe numerical analysis of resin expansion, but disregarded the stiffness variation due to the light activation of the resin (Rüttermann et al., 2007; Carvalho Filho, 2005; Ausiello et al., 2002).

To simulate both the dentine and the resin regions, ten-nodded tetrahedral elements were used, with three degrees of freedom at each node. Program Ansys, release 10, was used in the analyses. Figure 1 represents the geometry of the model. A special element with a constitutive model developed for brittle materials was used for the adhesive layer in order to simulate cracking. This material constitutive model has a cutoff that imposes a limit to normal and tangential stresses, after which fracture occurs.

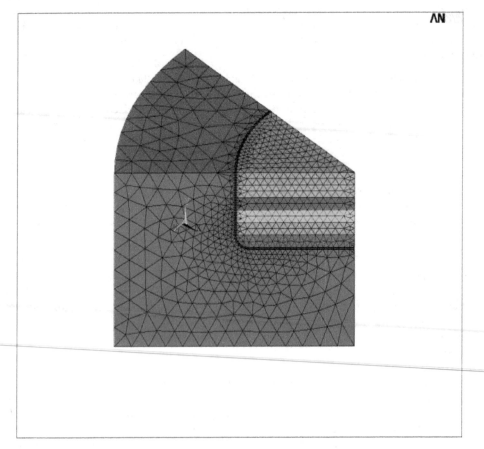

Fig. 1. Geometric model

Boundary conditions were imposed so as to restrict displacements at the base and non-vertical displacements in the faces, consistent with the restoration axisymmetry. Figure 2 shows dentine with the cavity, and Table 1 its material properties according to Ausiello et al. (2002). It is assumed that the dentine is under small displacements, with elastic behavior, and there is no stiffness variation. It is also considered homogeneous and isotropic. The analysis is then linear for this component.

The adhesive is represented by a thin layer with low traction strength. This material model introduces a nonlinear effect to the model, besides demanding a three dimensional geometric model. The traction at the interface was limited to 45 MPa, as indicated in the literature (Ausielo et al, 2002). In the process of solving the equations, the program verifies if this limit is exceeded at each iteration. If so, a crack is generated. Later, if compression at a normal direction to the crack's plane occurs, the crack is closed.

Fig. 2. Discrete model

Material	Young's Modulus	Poisson ratio
Dentine	18000 MPa	0.31
Adhesive	1000 MPa	0.30

Table 1. Material properties for dentine and adhesive (Ausielo et al, 2002)

Figure 3 shows the adhesive layer, and Table 1 presents its material mechanical properties (Ausielo et al, 2002).

The geometry of the resin was divided in six layers, each one with a half millimeter height. To these layers different values for the Young's modulus were assigned, which were measured from the laboratory experiments, as described in item 2.1.2.

In order to impose a volumetric contraction, a thermal expansion coefficient was introduced, and set to 0.01 °C⁻¹. Then, a temperature variation was applied to the resin in the model, being negative to contraction and positive to expansion, calculated so as to produce the desired resin contraction.

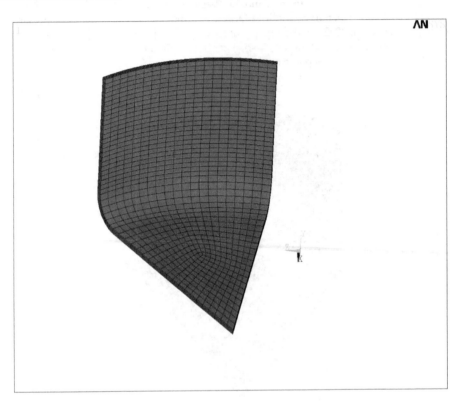

Fig. 3. Interface layer model

Two different clinical situations were studied. First, the cavity complete filling in a single step (here called "*one increment technique*"), consisting in filling the whole cavity with resin and then polymerizing it in one step. Then, the cavity filling in steps (here called "*horizontal incremental technique*") consisting in filling the cavity with resin one layer at a time, polymerizing it and moving to the next, outer layer. A new layer of resin is added, and the process is repeated until the resin reaches the top of the cavity. This way, the layers will be thin, causing less stiffness variation, as indicated by the experimental results. For this study, the cavity was considered filled with three increments of equal thickness.

In the first method (one increment technique) the Young's modulus variation is significant, and its value will be smaller on the top of the restoration. The polymerization is simulated

Residual Stresses and Cracking in Dental Restorations due to Resin Contraction Considering
In-Depth Young's Modulus Variation

233

using the *Birth and Death* resource available in the software, which reduces the "dead" element's stiffness to a very low value, practically eliminating the contribution of that region's stiffness from the analysis. This resource allows the user to activate only part of the domain in one step, and then consider previously "dead" regions for further load steps.

To simulate the horizontal incremental technique, three layers were created. The two lowest were initially "killed" and a temperature variation applied at the top layer. The numerical system was solved. Then, the middle layer was "born" or activated, the temperature variation was applied, and the new system was solved. The same process was followed by the third layer.

This technique of simulation was similar to the one increment technique, but now the layers were polymerized from the bottom to the top, with the layer activation sequence following a downward sequence. An incremental iterative Newton-Raphson algorithm with a force based convergence criteria was used. The stopping criterion is given in Eq. (2), with $\beta = 0.001$ (F_{int} is the internal force and F_{ext} is the external force):

$$\left\| F_{int} - F_{ext} \right\| < \beta \cdot \left\| F_{ext} \right\| \tag{2}$$

3. Results

3.1 Degree of Conversion (DC)

The Raman test results for XL3000 (G1 and G3) and Elipar (G2 and G4) groups are respectively presented in Figures 4 and 5.

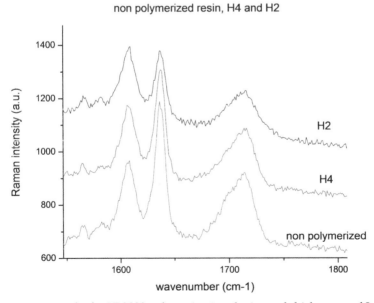

Fig. 4. Raman test results for XL3000 polymerization device and thicknesses of 2 mm (H2-G1) and 4mm (H4-G3)

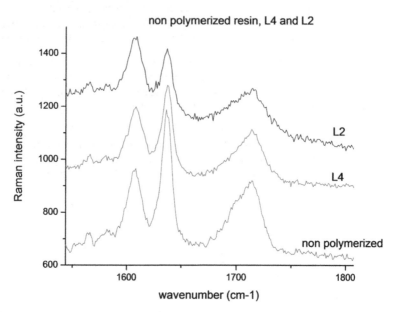

Fig. 5. Raman test results for Elipar polymerization device and thicknesses of 2 mm (L2-G2) and 4mm (L4-G4)

The measured average degrees of conversion values are listed in Table 2 for the four groups.

Sample height	XL3000 - QTH		Elipar Freeligth 2 - LED	
2mm	55 %	±3 %	59%	± 3 %
4mm	26%	± 1 %	32%	± 2 %

Table 2. Degree of conversion (in %) for Z350 resin measured using the Raman technique

The results for DC fit into a normal distribution, and three-way ANOVA model was used to analyze the influence of mode of cure sample height. Comparisons of the QTH (XL3000) and LED (Elipar Freeligth 2) curing methods showed significant influence of initial height in the degree of conversion ($p < 0.05$). For higher initial height, it was observed a sharp reduction in DC values. As for different irradiation techniques, it did not produce significant difference between the groups with the same initial height (2 or 4mm).

3.2 Young's modulus

The Young's modulus (E) showed a decrease for points located farther away from the light sources, as it is shown in Table 3, for both curing methods, regardless of the distance between the sample and the light source.

| Group-depth | E (GPa) | | W | p |
	Average	Std Dev		
G1-0.5 mm	20.094	1.4360	0.678	0.0089
G1-1.5 mm	19.375	0.6461	0.9526	0.6901
G1-2.5 mm	17.720	0.8739	0.9351	0.9351
G1-3.5 mm	12.288	1.7453	0.9194	0.4180
G2-0.5 mm	19.004	1.4362	0.8287	0.0497
G2-1.5 mm	18.071	1.1987	0.838	0.0649
G2-2.5 mm	15.947	0.7665	0.9426	0.5816
G2-3.5 mm	10.755	0.6570	0.9247	0.4457
G3-0.5 mm	19.288	2.0709	0.9658	0.8331
G3-1.5 mm	18.221	1.4476	0.9018	0.3252
G3-2.5 mm	17.146	1.3244	0.9489	0.6501
G3-3.5 mm	15.226	1.2777	0.8505	0.0858
G4-0.5 mm	19.876	1.5748	0.7793	0.0157
G4-1.5 mm	19.056	1.5425	0.9095	0.3659
G4-2.5 mm	18.147	1.2606	0.9711	0.8905
G4-3.5 mm	16.016	1.3564	0.9736	0.9132

Table 3. Young's modulus (in GPa) variations with depth for different curing devices

Comparisons (Dunn method)	Differences	z calculed	z critical	p
G1-0.5mm x G1-1.5mm	2.3333	0.4698	2.638	ns
G1-0.5mm x G1-2.5mm	13.3333	2.6846	2.638	< 0.05
G1-0.5mm x G1-3.5mm	23.2222	4.6757	2.638	< 0.05
G1-1.5mm x G1-2.5mm	11	2.2148	2.638	ns
G1-1.5mm x G1-3.5mm	20.8889	4.2059	2.638	< 0.05
G1-2.5mm x G1-3.5mm	9.8889	1.9911	2.638	ns

Table 4. Young's modulus (in GPa). Variations with depth for group G1, for different depth
values, XL3000at distance of 1 mm

Comparisons (Dunn method)	Differences	z calculed	z critical	p
G2-0.5mm x G2-1.5mm	4.5556	0.9172	2.638	ns
G2-0.5mm x G2-2.5mm	15.4444	3.1097	2.638	< 0.05
G2-0.5mm x G2-3.5mm	24.6667	4.9666	2.638	< 0.05
G2-1.5mm x G2-2.5mm	10.8889	2.1924	2.638	ns
G2-1.5mm x G2-3.5mm	20.1111	4.0493	2.638	< 0.05
G2-2.5mm x G2-3.5mm	9.2222	1.8569	2.638	ns

Table 5. Young's modulus (in GPa). Variations with depth for group G2, for different depth
values, XL3000 for distance of 2 mm

After verifying that the results do not follow a normal distribution, Kruskal-Wallis analysis was used to statistically determine the significance of the variation in the obtained data. The test did not indicate significant differences in elastic modulus between the depths of 0.5 and 1.5 mm for the QTH device, not even by increasing the position of the polymerization tip by 1 mm. At each 2 mm in depth a statistically significant variation was observed for the elastic modulus.

Comparisons (Dunn method)	Differences	z calculed	z critical	p
G3-0.5mm x G3-1.5mm	3.7222	0.7495	2.638	ns
G3-0.5mm x G3-2.5mm	10.5556	2.1253	2.638	ns
G3-0.5mm x G3-3.5mm	20.1667	4.0605	2.638	< 0.05
G3-1.5mm x G3-2.5mm	6.8333	1.3759	2.638	ns
G3-1.5mm x G3-3.5mm	16.4444	3.311	2.638	< 0.05
G3-2.5mm x G3-3.5mm	9.6111	1.9352	2.638	ns

Table 6. Young's modulus (in GPa). Variations with depth for group G3, for different depth values, Elipar distance of 1mm

Comparisons (Dunn method)	Differences	z calculed	z critical	p
G4-0.5mm x G4-1.5mm	5.3889	1.085	2.638	ns
G4-0.5mm x G4-2.5mm	10	2.0135	2.638	ns
G4-0.5mm x G4-3.5mm	21.5	4.329	2.638	< 0.05
G4-1.5mm x G4 -2.5mm	4.6111	0.9284	2.638	ns
G4-1.5mm x G4-3.5mm	16.1111	3.2439	2.638	< 0.05
G4-2.5mm x G4-3.5mm	11.5	2.3155	2.638	ns

Table 7. Young's modulus (in GPa). Variations with depth for group G4, for different depth values, Elipar distance of 2 mm

For LED polymerization, the results also failed the normality assumption. Kruskal-Wallis analysis did not indicate statistically significant differences in elastic modulus between the depths of 0.5, 1.5 and 2.5 mm for the LED device, not even by increasing the position of the polymerization tip by 1 mm. As for the comparison between 0.5 mm depth and 3.5, p<0.05 was obtained, as well as with the distances of 1.5 and 3.5 mm. The LED device showed an improved polymerization performance at 2.5 mm depth, but results showed that at each 2 mm in depth a statistically significant variation has occurred for the elastic modulus.

The variation in the polymerization device`s tip from the sample did not affect the obtained stiffness results, as shown by a comparison between tables 4 and 5 and 6 and 7.

3.3 Numerical results

The mechanical properties for the resin used in the finite element model were the Young's modulus measured at the four different depths presented on Table 8 and the Poisson's ratio equal to 0.3 throughout the resin model. For the finite element simulation it was used the lowest values measured for the Young's modulus (Group 2 – XL3000, light source at 2 mm distance). In this group it was also verified the highest variations on the Young's modulus with respect to the depth (Tables 3 to 7).

| Depth (mm) | Young's Modulus (GPa) for the resin | |
	Test results for Group 2 (G2) – XL3000, light source at 2 mm distance	Properties used in the Finite Element simulations
0.5	19.004	19.0
1.5	18.071	18.1
2.5	15.947	15.9
3.5	10.755	10.8

Table 8. Young's modulus used in the finite element model (E) for the resin according to depth

The following figures show the first principal stress (σ_1), in MPa, after the polymerization contraction. Figures 6 to 8 show the simulation results for resin contraction in the three-layers horizontal incremental technique. The first principal stresses (σ_1) representing maximum tension are shown, as all the materials have a reduced strength for traction.

Figure 9 depicts the stress values after filling the restoration in a single step (one layer technique). Then, in Fig. 10, the simulation was repeated keeping the Young's modulus for the resin constant, that is, neglecting the variation due to incomplete polymerization. The results are given for both filling techniques.

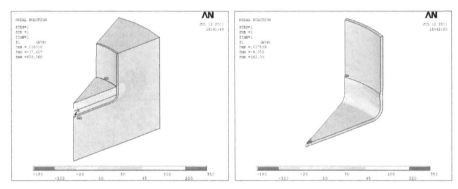

Fig. 6. First principal stresses (σ_1- MPa) for first layer in three-layers horizontal increment technique – restoration on the left and interface on the right

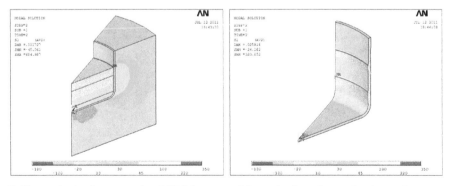

Fig. 7. First principal stresses (σ_1- MPa) for second layer in three-layers horizontal incremental technique – restoration on the left and interface on the right

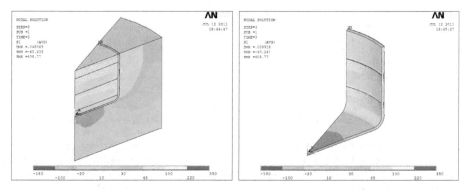

Fig. 8. Final first principal stresses (σ₁- MPa) in three-layers horizontal incremental technique – restoration on the left and interface on the right

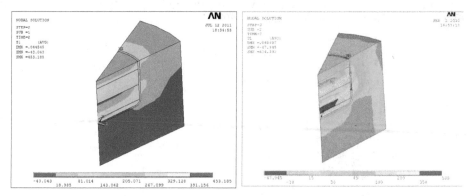

Fig. 9. Final first principal stresses (σ₁- MPa) horizontal incremental technique (2 layers) and one increment (or bulk) filling

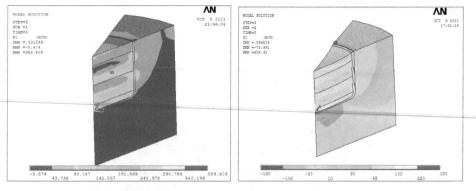

Fig. 10. Final first principal stresses (σ₁- MPa) three and one increment technique with constant E – restoration and dentine

4. Discussion

The study of residual stresses in dental restorations using a finite element simulation tries to reproduce *in silico* a phenomenon frequently observed in dental clinic. By including in the simulation the sequence of cavity filling, the variation in modulus of elasticity resulting from light penetration and the stress relief due to cracking the computational model can represent the process with the inclusion of key aspects of the restoration.

In order to consider the change in properties along the depth of the layers, experimental evaluation of this mechanical property was made, for different light sources and tip positions, as well as for hardness and degree of conversion based on micro-hardness.

Many authors use the degree of conversion associated to harness to verify the effectiveness of the polymerization of composite resins (Kawaguchi et al., 1989; Chung & Greener, 1990). The evaluation of micro-hardness is influenced not only by its degree of conversion, but also by the percentage of filler and presence or absence of oxygen (Peutzfeld, 1997; Yap et al., 2002; Halvorson et al., 2002). As the composite's viscous deformation is a time-dependent event, slower curing rates may provide extended periods where the material is able to yield to contraction forces before acquiring higher elastic modulus (Feilzer et al., 1993). In fact, reducing polymerization rates in composites has been shown to lower stress levels significantly (Bouschlicher & Rueggeberg, 2000; Lim et al., 2002).

There is a direct relationship between the degree of conversion of a resin and its modulus of elasticity. The maximum polymerization of a given resin implies in higher modulus of elasticity and higher contractive stresses in the tooth-resin restoration interface (Davdison-Kaban, 1997). As a consequence, higher mechanical properties for the composite resin results in a corresponding higher value for residual stresses at the adhesive interface (Braem et al., 1987; Silikas et al., 2000).

The degree of conversion was obtained by the Raman method. As for the Young's modulus associated to micro-hardness, the measurements were performed changing the distance from the resin to the tip of the device from 1 to 2 mm. Results showed DC results ranging from 55 and 60% for 2 mm depth, unchanged regardless of the polymerization technique or distance to the tip. For 4 mm depth, DC felt to between 26 and 32% for the two devices, showing a reduction with depth, which can lead to deterioration in its mechanical properties and a reduction in polymerization contraction (Ceballos et al., 2009).

The measurement of the modulus of elasticity showed, for all tested groups, a statistically relevant reduction at each 2 mm depth increase, regardless of the polymerization technology used.

The results obtained for Young's modulus presented a statistically significant reduction for each 2 mm variation in depth, regardless of the curing method. These results reinforce the idea of avoiding to apply layers thicker than 2 mm, which will assure a minimum of non-homogeneity in the material's stiffness, for any restoration volume or configuration factor (Chiang et al., 2010).

According to Dewaele et al. (2009), who investigated different polymerization protocols, the reduction in contraction is directly related to the reduction in the degree of cure. Cho et al. (2011), in an analysis of the particles movement in resin restorations, observed that it depends on the degree of conversion and the contraction value, and that a mathematical equation can be derived for the particles' displacements, especially for those below a 3.5 mm depth.

The computer simulations show a clear stress concentration between the cavity walls and the restoration material, especially in the regions with more abrupt geometry changes corresponding to restoration internal angles, as well as interface rupture at different points of the wall. Stresses are higher at the interface due to the change in stiffness at that point.. These results agree with the conclusions of Chiang et al. (2010), which also verified gap formation in cylindrical restorations with and without the use of an adhesive layer, both numerically and experimentally, through the use of micro-CT images. Their results pointed to higher stresses for adhesives with smaller modulus of elasticity, sometimes higher than the adhesive limit strength.

Most other regions of the restoration stay under stresses lower than 200 MPa. Stress concentration patterns, that is, final stresses after the polymerization process, are dependent of the curing technique. Stress levels indicate that, without subsequent relief, restoration can be easily damaged by additional loading such as temperature variation or biting. During the polymerization process, as the sequence in Figures 6, 7 and 8 shows, stresses redistribute and vary in the different parts of the restoration, reaching intermediate values that can lead to fracture in the interface and damage at the resin. The limit in stresses at the cement provide a buffer for dentine, as very high stresses are not transmitted through the interface layer, that cracks and avoid compromising the tissue.

As for a comparison between the three polymerization techniques, one, two and three increments, it is evident that higher stresses occur for the last option. Figures 8, 9 and 10 show higher peak stresses at dentine and resin for less incremental filling techniques (two and one layer, respectively).

In Figure 9 different filling techniques are explored. First, on the left, instead of three layers, only two 1.5 mm layers were considered. In the right side, bulk or one increment technique was used. In this case, the effect of the variation of Young's modulus with depth is expected to be more important.

In order to examine the influence of the variation in stiffness on the final results, two models were developed with constant resin modulus of elasticity. The results are shown in Fig. 10, both for the complete restoration and for dentine, and show a reduction in stress levels. The real, variable distribution of Young's modulus through depth leads to an increase in stresses in the restoration. This effect is more pronounced for the one increment technique, as expected, as the stiffness variation is larger, and are in agreement with the work of Chiang et al., 2010.

When the influence of the cavity size and incremental restoration technique were analyzed in Class I restorations, it was verified that the adhesive was damaged in several locations, and cracks were occurring in the interface.

He et al. (2007) observed that the size of the restoration and the restoration technique affected its overall behavior. Incremental filling of the cavity reduced the stress levels in large cavities, but did not affect the resulting tractions for small restorations. The authors suggested that incremental technique be used in large restorations with a high C factor.

Sun et al. (2009) noted the occurrence of contraction at the top and base of the restorations, fact also observed in the present finite element simulation. Gaps formation is directly connected to the C factor and volume of composite resin, eventually leading to micro-leakages. The authors concluded that neither C-factor nor sample volume affects polymerization shrinkage at a constant Degree Conversion. They observed in an analysis with a micro-CT that resin contraction occurred independently of the geometry. Using either μCt or infiltration with dyes, if was possible to verify the occurrence of infiltration and its dependence on the C factor and resin volume.

As for Young's modulus and DC values for composite resin, it was expected that a stiffness reduction along the axial direction would lead to a reduction in residual stress levels. This was verified by Oliveira et al. (2010), using resins with low modulus of elasticity. The resulting gaps, even though reduced after hygroscopic expansion, were permanent, being observed even 4 years after treatment (Krämer et al., 2009). The complex behavior of light activated resins and the several parameters involved in its use, as discussed, reinforce the need and importance of further study in the area and the need to consider different aspects in the definition of the clinical restoration procedure (Kakaboura et al., 2007).

5. Concluding remarks

The results described in the previous sections point to the following conclusions:

- The polymerization device, when comparing Elipar and laser-driven devices, used to cure the resin, did not affect significantly the degree of cure (DC) of the examined samples.
- As for the thickness of the sample, for both polymerization devices, the results proved a high correlation between depth and DC value.
- The stiffness of the composite resin was affected by the thickness of the sample. This effect was more pronounced for depths larger than 2 mm
- The numerical analysis showed that layered techniques lead to improved stress distribution and reduced residual stresses in cylindrical restorations.
- The variation in elastic modulus in layers with thickness smaller than 2 mm is not pronounced. As for 3 mm layers, the resulting stress distribution is significantly different when compared with the results for a homogeneous stiffness through restoration thickness.
- The deterioration of the modulus of elasticity in thick layers led to a reduction in stress concentration in the model. Other relevant information, such as reduction in strength, change in contraction and quality of adhesion of the interface should be also investigated for a more complete picture of this case.
- Current recommendations for restricting layer thickness to 2 mm are in accordance with the obtained results.

6. Future directions

The reported results show that a refinement in finite element analysis can deepen the knowledge in the mechanical behavior of resin restorations, providing a tool to improve clinical success. Future work directions should include, among other aspects:

- examine the effect of hygroscopic expansion on the residual stress distribution;
- include inclined layers as an alternative to horizontal layers in the models;
- simulate drop-shaped cavities and restorations;
- study the consequences of the residual stresses and cracks in the performance of the restoration.

7. Acknowledgments

The authors kindly acknowledge the support of the Brazilian funding agencies FAPEMIG and CNPq in the development of this work.

8. References

Ansys Release 10.0.

Ausiello, P.; Apicella, A. & Davidson, C.L. (2002). Effect of adhesive layer properties on stress distribution in composite restorations - a 3D finite element analysis. *Dental Materials*, Vol.18, No.4, pp. 295-303, ISSN 0109-5641

Bouschlicher, M.R. & Rueggeberg, F.A. (2000). Effect of ramped light intensity on polymerization force and conversion in a photoactivated composite. *Journal of Esthetic Dentistry*, Vol.12, No.6, pp. 328-339, ISSN 1040-1466

Bowen R.L.; Nemoto, K. & Rapson, J.E. (1983). Adhesive bonding of various materials to hard tooth tissues: forces developing in composite materials during hardening. *Journal of the American Dental Association*, Vol.106, No.4, pp.475–477, ISSN 0002-8177

Braem, M.; Lambrechts, P.; Vanherle, G. & Davidson, C.L. (1987). Stiffness increase during the setting of dental composite resins. *Journal of Dental Research*, Vol.66, No.12, pp. 1713–1716, ISSN 0022-0345

Braga, R.R.; Ballestera, R.Y. & Ferracane, J.L. (2005). Factors involved in the development of polymerization shrinkage stress in resin-composites: A systematic Review. *Dental Materials*, Vol. 21, pp. 962–970, ISSN 0109-5641

Carvalho Filho, F. (2005). Modelagem via método dos elementos finitos da contração de resinas odontológicas fotopolimerizáveis, M. Sc. thesis, Structural Engineering, Engineering School, Federal University of Minas Gerais, UFMG (in Portuguese)

Ceballos, L.; Fuentes, M.V.; Tafalla, H.; Martínez, A.; Flores, J. & Rodríguez, J. (2009). Curing effectiveness of resin composites at different exposure times using LED and halogen units. *Journal of Clinical and Experimental Dentistry*, Vol.1, No.1, pp. 8-13, ISSN 1989-5488

Chiang, Y.C.; Rösch, P.; Dabanoglu, A.; Linc, C.P.; Hickel, R. & Kunzelmann, K.H. (2010). Polymerization composite shrinkage evaluation with 3D deformation analysis from μCT images. *Dental Materials*, Vol.26, No.3, pp. 223–231, ISSN 0109-5641

Cho, E.; Sadr, A.; Inai, N. & Tagami, J. (2011). Evaluation of resin composite polymerization by three dimensional micro-CT imaging and nanoindentation. *Dental Materials*, Vol.27, No.11, pp. 1070-1078, ISSN 0109-5641

Chung, K.H. & Greener, E.H. (1990). Correlation between degree of conversion filler concentration and mechanical properties of posterior composite resins. *Journal of Oral Rehabilitation*, Vol.17, No.5, pp. 487–494, ISSN 0305-182X

Dewaele, M.; Asmussen, E.; Peutzfeldt, A.; Munksgaard, E.C.; Benetti, A.R.; Finné, G.; Leloup, G. & Devaux, J. (2009). Influence of curing protocol on selected properties of light-curing polymers: degree of conversion, volume contraction, elastic modulus, and glass transition temperature. *Dental Materials*, Vol.25, No.12, pp. 1576-1584, ISSN 0109-5641

Dietschi, D.; Monasevic, M.; Krejci, I. & Davidson, C. (2002). Marginal and internal adaptation of class II restorations after immediate or delayed composite placement. *Journal of Dentistry*, Vol.30, Nos.5-6, pp. 259–269, ISSN 0300-5712

Feilzer, A.J.; De Gee, A.J. & Davidson C.L. (1993). Setting stresses in composites for two different curing modes. *Dental Materials*, Vol.9, No.1, pp. 2–5, ISSN 0109-5641

Felix, C.A.; Price, R.B.T. & Andreou, P. (2006). Effect of reduced exposure times on the microhardness of 10 composites cured by high power LED and QTH curing lights. *Journal of the Canadian Dental Association*, Vol.72, No.2, pp. 147a–147g, ISSN 0008-3372

Ferracane, J.L. (2005). Developing a more complete understanding of stresses produced in dental composites during polymerization. *Dental Materials*, Vol.21, No.1, pp. 36–42, ISSN 0109-5641

Halvorson, R.H.; Erickson, R.L. & Davidson, C. L. (2002). Energy dependent polymerization of resin-based composite. *Dental Materials*, Vol.18, No.6, pp. 463–469, ISSN 0109-5641

He, Z.; Shimada, Y. & Tagami, J. (2007). The effects of cavity size and incremental technique on micro-tensile bond strength of resin composite in Class I cavities. *Dental Materials*, Vol.23, No.5, pp. 533–538, ISSN 0109-5641

Jandt, K.D.; Mills, R.W.; Blackwell, G.B. & Ashworth, S.H. (2000). Depth of cure and compressive strength of dental composites cured with blue light emitting diodes (LEDs). *Dental Materials*, Vol.16, No.1, pp. 41–47, ISSN 0109-5641

Jong, L.C.G.; Opdam, N.J.M.; Bronkhorst, E. M.; Roeters, J.J.M.; Wolke, J.G.C. & Geitenbeek, B. (2007). The effectiveness of different polymerization protocols for class II composite resin restorations. *Journal of Dentistry*, Vol.35, No.6, pp. 513-520, ISSN 0300-5712

Kakaboura, A.; Rahiotis, C.; Watts, D.; Silikas, N. & Eliades, G. (2007). 3D-marginal adaptation versus setting shrinkage in light-cured microhybrid resin composites. *Dental Materials*, Vol.23, No.3, pp. 272–278, ISSN 0109-5641

Kawaguchi, M.; Fukushima, T. & Horibe, T. (1989). Effect of monomer structure on the mechanical properties of light cured composite resins. *Dental Materials Journal*, Vol.8, No.1, pp. 40–45, ISSN 0287-4547

Krämer, N.; Reinelt, C.; Richter, G.; Petschelt, A. & Frankenberger, R. (2009). Nanohybrid vs. fine hybrid composite in Class II cavities: Clinical results and margin analysis after four years. *Dental Materials*, Vol.25, No.6, pp. 750–759, ISSN 0109-5641

Lim, B.-S.; Ferracane, J.L.; Sakaguchi, R.L. & Condon, J.R. (2002). Reduction of polymerization contraction stress for dental composites by two-step light-activation. *Dental Materials*, Vol.18, No.6, pp. 436–444, ISSN 0109-5641

Oliveira, L.C.A.; Duarte Jr., S.; Araujo, C.A. & Abrahão, A. (2010). Effect of low-elastic modulus liner and base as stress-absorbing layer in composite resin restorations. *Dental Materials*, Vol.26, No.3, pp. e159–e169, ISSN 0109-5641

Petrovic, L.M.; Drobac, M.R.; Stojanac, I.L. & Atanackovic, T.M. (2010). A method of improving marginal adaptation by elimination of singular stress point in composite restorations during resin photo-polymerization. *Dental Materials*; Vol.26, No.5, pp. 449–455, ISSN 0109-5641

Peutzfeld, A. (1997). Resin composites in dentistry: The monomer systems. *European Journal of Oral Sciences*, Vol.105, No.2, pp. 97–116, ISSN 1600-0722

Peutzfeldt, A.; Sahafi, A. & Asmussen, E. (2000). Characterization of resin composites polymerized with plasma arc curing units. *Dental Materials*, Vol.16, No.5, pp. 330–336, ISSN 0109-5641

Rüttermann, S.; Krüger, S.; Raab, W. & Janda, R. (2007). Polymerization shrinkage and hygroscopic expansion of contemporary posterior resin-based filling materials — A comparative study. *Journal of Dentistry*, Vol. 35, No.10, pp. 806-813, ISSN 0300-5712

Sakaguchi, R. L.; Sasik, C. T.; Bunczak, M.A. & Douglas, W.H. (1991). Strain gauge method for measuring polymerization contraction of resin composite restoratives. *Journal of Dentistry*, Vol.19, No.5, pp. 312–316, ISSN 0300-5712

Silikas, N.; Eliades, G. & Watts, D.C. (2000). Light intensity effects on resin-composite degree of conversion and shrinkage strain. *Dental Materials*, Vol.16, No.4, pp. 292–296, ISSN 0109-5641

Sun, J.; Eidelman, N. & Gibson, S.L. (2009). 3D mapping of polymerization shrinkage using X-ray micro-computed tomography to predict microleakage. *Dental Materials*, Vol.25, No.3, pp. 314–320, ISSN 0109-5641

Yap, A.U.; Soh M.S. & Siow, K. S (2002). Effectiveness of composite cure with pulse activation and soft-start polymerization. *Operative Dentistry*, Vol.27, No.1, pp. 44–49, ISSN 0361-7734

Yoshikawa, T.; Burrow, M.F. & Tagami, J. (2001). A light curing method for improving marginal sealing and cavity wall adaptation of resin composite restorations. *Dental Materials*, Vol.17, No,4, pp. 359-366, ISSN 0109-5641

12

Implications of Corporate Yoga: A Review

Rudra B. Bhandari[1], Churna B. Bhandari[2], Balkrishna Acharya[3],
Pranav Pandya[4], Kartar Singh[5], Vinod K. Katiyar[6] and Ganesh D. Sharma[7]
[1]University of Patanjali, Haridwar, Uttarakhand,
[2]Department of Physics, Case Western University, Ohio,
[3]University of Patanjali, Haridwar, Uttarakhand,
[4]Dev Sanskriti Vishwavidyalaya, Uttarakhand,
[5]University of Patanjali, Haridwar, Uttarakhand,
[6]Department of Mathematics, Indian Institute of Technology, Roorkee, Uttarakhand,
[7]Department of Yogic Sciences, University of Patanjali, Haridwar, Uttarakhand,
[1,3,4,5,6,7]India
[2]USA

1. Introduction

Yoga is an art of life management and a universal means for self realization. Health benefits and improvement of human intelligence are inseparable byproducts of yoga practices that can be achieved by every practitioner. Aurobindo (1999) defines yoga as "a practical discipline incorporating a wide variety of practices whose goal is the development of a state of mental and physical health, well-being, inner harmony and ultimately a union of the human individual with the universal and transcendent existence". Yoga is an ancient discipline designed to bring balance and health to the physical, mental, emotional, and spiritual dimensions of the individual (Iyengar, 1976). In contemporary scenario, a part of oriental wisdom, yoga has been widely known even in western countries and a substantial number of people have been practicing it for different purposes such as physical fitness, flexibility, stress management, psychological well being, emotional rectification, good habits cultivation and disease management as adjunct therapy. Only USA invests 5.7 billion US dollars annually for yoga classes and yoga products (Macy, 2008). A substantial number of women have been found practicing yoga in UK and other countries. The emergence of many more yoga studios in Europe and South Asia and research studies made pertaining diverse efficacies of yoga portray its ascending popularity and scientific validation and standardization by scientific community.

At present, there are number of scientific researches that substantiate preventive, rehabilitative, therapeutic and excelling powers of yoga at individual and corporate levels (Becker, 2000; Jacobs, 2001; Khalsa, 2004; Ornish, 2009). One of the most exciting developments in the last few decades is the cross fertilization of western science with ideas from Eastern wisdom system such as yoga. With increasing precision, scientists are able to look at the body, mind and spirit and detect the sometime subtle changes than practitioners of yoga and meditation undergo. A scientific interpretation of yogic effects has been made on the basis of bio-psycho-socio-spiritual research model (Evans et al., 2009).

On the other hand, a flood of chronic diseases (cardiac problems, diabetes, cancer, lower back pain, obesity, depression etc.) (World Economic Forum [WEF], 2010, p. 9), organizational misbehaviors (work place incivility, insidious and insulting behaviors, social undermining, theft of company assets, acts of destructiveness, vandalism and sabotage, substance abuse and misconduct perpetrated against fellow employees) (Fox & Spector, 2005), interpersonal conflict/work-life conflict and dearth of spiritual leadership have been met as the precursors for global recessions and workplace disharmony. Therefore, most of the today's successful companies of the world have prioritized workplace yoga/spirituality as an emerging avenue for corporate- wellness (CW) and excellence (CE).

"The business of business and business of life are one. The reason for living and working is to act and the reason to act is to seek excellence in everything that you do" (Sinclair, n.d, as cited in Pruzen & Pruzan, 2001). This quotation from a CEO and chairman of a leading company (Tan Range Exploration, Ltd., USA/Tanzania) portrays the relevance of spiritual insight for business management and performance excellence. Persistent practice of yoga and allied disciplines as a part of corporate culture improves somatic, psychic, social and spiritual health and intelligence of an individual and organizational workforce.

Additionally, levels of four human intelligences- spiritual (SI), emotional (EI), creative (CI) and rational (RI), acquired by an individual govern his/her way of feeling, thinking and behavior and undoubtedly can be regarded as the determiners of human personality and human excellence too. Optimal health (physical, mental, social and spiritual) and the four elements of intelligence (SI, EI, CI and RI) that can be acquired and sustained by prolonged yoga practices underpin individual or CW and CE.

Health problems- stress or distress, obesity, low backache, respiratory disorders, cardiovascular problems, digestive disorders and genitourinary disorders are prevalent at corporate world and cause a huge decline in the corporate health and wealth. On the other hand, regular yoga practice is found more helpful for total health promotion, disease prevention and rehabilitation as well. Particularly, yoga has been found effective to manage work related stress, respiratory disorders (asthma, pulmonary tuberculosis, pleural effusion, obstructive pulmonary diseases, chronic bronchitis), cardiovascular disorders (ischemic heart disease, coronary artery disease, angina, chronic heart failure, hypertension), digestive disorders (irritable bowel syndrome, hyperacidity, colitis, indigestion, diabetes, gastroesophageal reflux disease, hepatitis, gall stones, celiac disease) and genitourinary problems (urinary stress incontinence, women sexuality, climacteric syndrome, premature ejaculation, pregnancy outcomes, labor pain and duration). Thus, this writing is contained with sub-headings that describe the efficacy of CY to manage aforesaid health problems.

Being a secular, global, holistic and cost effective tool for boosting holistic health and awakening four faculties of human intelligence, yoga needs to include as a part of corporate culture along with scientific researches to substantiate its multidimensional efficacies. So the prime theme of this chapter is to highlight contemporary significance of corporate yoga (CY) to enrich health, happiness and harmony at workplace by promoting CW and CE. More specifically, the chapter is intended to argue-

1. The concept and contemporary significance of CY
2. Link among yoga, health care and four human intelligences
3. Efficacy of yoga for CW and CE
4. Preventive and therapeutic value of yoga relating to work related stress, respiratory disorders, cardiovascular disorders, digestive disorders and genitourinary disorders.

2. Contemporary significance of CY

Sages of yore have argued eternal significance of yoga for the welfare of entire mankind and global harmony. Same has been reinstated by contemporary enlightened masters such as Shriram Sharma Acharya, Swami Vivekananda, Maharishi Aurovindo, Swami Shivanand and Swami Rama so forth on the basis of their experiential and experimental knowledge. The father of scientific spirituality, Shriram Sharma Acharya defines yoga as "an art of living". This contemporized definition of yoga implies that yoga is nearer to life business or management. Correspondingly, business of business and business of life are one and the reason for living and working is to seek excellence in everything that we do (Sinclair, n.d, as cited in Pruzen & Pruzan, 2001). Interestingly, this indicates art of self business (yoga) as a foundation of sustainable corporate business and success. This substantiates the contemporary significance of CY for CW and CE. The contemporary significance of CY for corporate success can be concisely discussed in three sub heads- concept, popularity and health impacts of yoga for the ease of readers' comprehension.

2.1 Concept of Yoga

Yoga is a Sanskrit word that means union, to yoke or to unify; the merging of the microcosm of our existence in our body with the macrocosm. In other words, this also implies the fusion of embodied consciousness with cosmic consciousness (Chaoul & Cohen, 2010). The famous yoga exponent, sage Patanjali defines the yoga as "the inhibition of psychic modifications (Patanjali Yoga Sutra, 1:2)" that ultimately results in the fission of *Prakriti* (the equilibrium condition of three strands- *sat, raj* and *tam* that is eternal but changeable) and *Purusha* (pure consciousness that is immortal, eternal, omnipresent, omniscience and omnipotent). The next famous ancient text of yoga, Shrimad Bhagvat Geeta (SBG) defines the yoga as "a state of mental equanimity at each moment of the life" (SBG, 2:47). Subsequently, SBG also defines yoga from the behavioral perspective as "the excellence in action" (SBG, 2:48). In the West, yoga is often referred to as a mind-body technique from Asia, usually categorized as meditation (for those seated practices) and yoga (practices that include movement and the active participation of the body) (Chaoul & Cohen, 2010). Thus, yoga is perceived as an overarching category that includes all Asian mind-body practices, whether from India (Hatha yoga, etc.), Tibet (Tsa lung Trul khor [rTsa rlung 'Phrul 'khor]), China (T'ai chi, qi gong) or other Asian origin. Nonetheless, in Indian context, yoga is more than mind-body practices that also incorporates spiritual practices.

Basically, four major streams of yoga: *Karma Yoga* (The yogic path of undertaking selfless deeds by using attained wisdom, power and prosperity), *Bhakti Yoga* (The yogic path of devotion), *Jnana Yoga* (The yogic path that prioritizes rational thinking over knowledge), and *Raj Yoga* (The eightfold yogic path synthesized by sage Patanjali 5000 years ago) can be met in Indian classical texts. However, Raj Yoga as conceptualized by sage Patanjali is supposed to have synthesized all yogic paths as a garland. It has metaphorically comprised of eight subsequent limbs (a tree of eight limbs): *Yama* (universal ethics/social codes), *Niyama* (individual ethics), *Asana* (physical postures), *Pranayama* (breath control), *Pratyahara* (control of the senses), *Dharana* (concentration), *Dhyan* (meditation), and *Samadhi* (bliss). Indeed, this path of *Raj Yoga* is an integral form of *Karma Yoga, Bhakti Yoga,* and *Jnana Yoga* that can be adopted by any individual for total health and ascetic elevation (spiritual advancement). Correspondingly, Satyanand (2000) argued that from the perspective of yoga psychology *(Raj Yoga)*, human personality can be categorized into four types: dynamic, emotive, rational and volitional (p. 16).

Karma Yoga is preferred yogic path for an individual with active personality who can traverse inner journey of psychic refinement through selfless deeds. An individual with emotive personality may love *Ishwarparnidhan (Bhakti Yoga)* for psychic refinement and subsequently inhibition of psychic modifications. The path of *Jnana Yoga* is an optimal yogic way that prioritized by rational personalities. Eminent yoga scholar and seer, Patanjali put forth *Raj Yoga* (the royal path of yoga) that is equally applicable to each aspirant desired for perfect health, happiness, harmony, and ultimate bliss. Obviously, that is an integral and concise yogic way for all possible personalities.

2.2 Popularity of Yoga

Popularity of yoga practice in the West is in ascending order since 1960 and particularly in UK, where yoga classes are open to everyone although women tend to make up 70 to 90 per cent of the student base of most classes as well as the majority of yoga teachers (Newcombe, 2007). Moreover, perceived better physical health and emotional well-being by the yoga practice is an important reason for women's more participation in the classes. Additionally, yoga also served as an important support for women becoming more aware of feelings of alienation from traditional biomedical practitioners. "Only US invest \$5.7 billion dollars per year in yoga classes by involving 15.8 million people. Among US yoga practitioners, 72.2 percent are women who practice yoga to be slim, flexible, de-stressed and attractive (Macy, 2008). In a national population-based telephone survey (n = 2055), 3.8% of respondents reported using yoga in the previous year and cited wellness (64%) and specific health conditions (48%) as the motivation for doing yoga (Saper et al., 2004). In South Asian countries, everyone has craze for yoga and yoga has greater space in corporate circles too. Turnover of yoga business in Asia is more than 50 crore per year and a large number of corporate personnel are being trained in yogic ways of stress management and mind management in Pure Yoga Studio of the Hong Kong (Singh, 2009). Moreover, the rise of yoga masters like Swami Ramdev has promoted mass media communication of yoga worldwide.

2.3 Yoga versus health

A famous yoga exponent of contemporary time, Aurobindo (1999) defines yoga as "a practical discipline incorporating a wide variety of practices whose goal is the development of a state of mental and physical health, well-being, inner harmony and ultimately a union of the human individual with the universal and transcendent existence". Iyengar (1976) defines yoga as an ancient discipline designed to bring balance and health to the physical, mental, emotional, and spiritual dimensions of the individual. These two definitions of the yoga given by Aurobindo and Iyengar clearly hint its bio-psycho-socio-spiritual efficacy for attainment and maintenance of total health (physical, mental, social and spiritual health) as an elementary benefit and a byproduct if practiced persistently.

The *Hatha Yoga* is widely known in the present scenario, especially in the west, is supposed to be an elementary practice for the practice of *Raj Yoga* which includes body cleansing techniques *(Shatkarmas)*, postural exercises *(Asanas)*, gestures *(Mudras)*, psychic locks *(Bandhas)*, breath control *(Pranayama)*, concentration *(Dharana)* and meditation *(Dhyana)*.

Yoga practice consists of the five-principles including proper relaxation, proper exercise, proper breathing, proper diet, positive thinking and meditation (Chanavirut, Khaidjapho, Jaree & Pongnaratorn, 2006). Impacts of yoga practices can be better explained via bio-psycho-socio-spiritual model- at physical level it improves musculoskeletal functioning, cardiopulmonary status, autonomic nervous system (ANS) response and endocrine functioning; at psychosocial level, it enhances self-esteem, social support and positive mood; and at spiritual level it elevates compassionate understanding and mindfulness (Evans et al., 2009). Same hypothesis is supported as "well-rounded yoga practice may have benefits on structural, physiological, psycho-emotional and spiritual levels" (Herrick & Ainsworth, 2000).

Mechanisms underlying the modulating effects of yogic cognitive-behavioral practices (e.g., meditation, *Asanas, Pranayama*, caloric restriction) on human physiology can be classified into four transduction pathways: humoral factors, nervous system activity, cell trafficking, and bio-electromagnetism that shed light how yogic practices might optimize health, delay aging, and ameliorate chronic illness and stress from disability (illness and stress from disability) (Kuntsevich, Bushell, &Theise, 2010). Moreover, they provided standpoints for in-depth study of underlying mechanisms by postulating three possible hypotheses regarding mechanisms of yogic effects. Correspondingly, yogic practices may: 1) promote restoration of physiologic setpoints to normal after derangements secondary to disease or injury, 2) promote homeostatic negative feedback loops over non-homeostatic positive feedback loops in molecular and cellular interactions, and 3) quench abnormal "noise" in cellular and molecular signaling networks arising from environmental or internal stress. The detailed elaboration of the proposed hypotheses is beyond the scope of this writing unless it is quite intriguing and comprehensive that includes all possible modes of varied yogic effects (effects of *Asanas, Pranayams*, varieties of meditations and caloric restriction) till now.

It is claimed that these techniques bring an individual to a state of perfect health, stillness and heightened awareness by increasing the body's store of *prana*, or flow of vital energy (Kulkarni & Bera, 2009; Nayak & Shankar, 2004).Other claimed benefits of regular yoga practice leads to suppleness, muscular strength, feelings of well-being, reduction of sympathetic drive, pain control and longevity (Brown & Gerbarg, 2009; Garfinkel & Schumacher, 2000; Lipton, 2008). Yogic breathing exercises allegedly reduce muscular spasms and expand available lung capacity (Brown & Gerbarg, 2005).Yoga is thus advocated as a symptomatic treatment for a wide range of conditions, including anxiety, arthritis, back pain, cardiovascular problems, gastrointestinal complaints, headaches, insomnia, premenstrual syndrome, respiratory problems and stress (Ernst & Soo, 2010). However, substantial evidences from clinical trials also should be undertaken to generalize the therapeutic efficacy of yoga.

3. Yogic prescription for CW and CE

Yogic prescription (YP) is a sort of yogic capsule that is comprised of yogic practices from all major yogic streams (*Jnana, Bhakti, Karma and Raj*) and designed as per workplace problems met at individual and organizational level. YP presumes nine hurdles (physical or mental illness, dullness, doubt, procastination, laziness, over indulgence, delusion, inability and instability) behind inidvidual and corporate failure and basically targets their dissolution (Pandya, 2006, p. 118) via its prolonged practice. The components of YP may vary as per nature of participants and workplace. Nonetheless, YP considers four possbile types of human personality (rational, emotive, dyanmic, and volitional) and incorporates yogic practices from aforesaid four yogic streams accordingly. Dissolution of these hurdles via sustained YP practice leads to complete harmony (homogenity among feeling, thinking and action), harmony leads to talent, talent results in creativity and innovations, creativity results in perfection that further results in excellence as depicted in Figure 1.

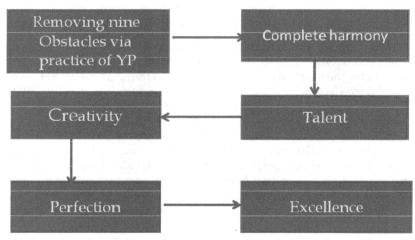

Fig. 1. Tentative model showing emergence of excellence via YP practice.

YPs have been found effective for health promotion and diseases management. Mind-body interventions derived from yoga (including breathing, meditation, postures, concentration and visualization) ameliorate stress-related mental and physical disorders—asthma, high blood pressure, cardiac illness, elevated cholesterol, IBS, cancer, insomnia, multiple sclerosis, and fibromyalgia (Becker 2000; Jacobs 2001). Curative effect of yoga has been seen in psychiatric problems, cardiovascular problems (CAD, hypertension), respiratory disorders (Bronchial asthma, OPD, pneumonia etc.), diabetes, neurological problems, musculoskeletal disorders, and others (Khalsa 2004). Ornish (2009) asserted that lifestyle changes (yogic way of living) could be considered not only as preventing chronic diseases but also reversing their progression—as an intensive non-surgical, non-pharmacological intervention. Moreover, the coronary heart disease, prostate and breast cancer, diabetes, and obesity account for 75% of health-care costs, yet the progression of these diseases can be stopped or even reversed with intensive lifestyle changes. Falus et al. (2010) highlighted the considerable connection between the length of telomeres and intensive changes in lifestyle and nutrition as well as behavioral and psychological factors. Epel et al. (2009) concluded that some forms of meditation might have salutary effects on telomere length by reducing cognitive stress and stress arousal and increasing positive states of mind and hormonal factors that might promote telomere maintenance. Between times one (before the Life Force Yoga program) and two (two weeks after learning it), participants reported 64% decrease in total mood disturbance, 53% decrease in average depression scores and overall mood disturbance continued to drop after two months (Bennett, Weintraub & Khalsa 2008).

Besides, the enhancement of SI, EI, CI and RI and their harmonious interplay by yogic practices is also substantiated by various scientific researches which are deemed essential for love and happiness at workplace: visionary leadership, sound management practices, creativity and innovations, and optimal work performance. Interestingly, levels of SI and EI are more about love and happiness at workplace. Moreover, happiness assists organization's members to be more productive, creative, fulfilled and with high morale that lead to outstanding performance and therefore, have a direct impact on organization's financial success (Claude & Zamor, 2003).

But the level of these four intelligences varies from person to person as per their personality types (dynamic, emotive, rational and mystic). Therefore, that holistic YP designed to promote CW and CE needs to include selected practices from *Ganan, Bhakti, Karma and Raj yoga*. Moreover, YP must include yogic practices of gross body, subtle body and causal body. As per author's self experience, pervious research findings and needs assessed in corporate companies, different tentative YPs can be designed that need to be tested to assess their effectiveness for promotion of four human intelligences and holistic health.

3.1 Yoga for CW

Yoga may be an integral part of worksite health promotion program (WHPP). "WHPP is an organized program in the worksite that is intended to assist employees and their family members (and/or retirees) in making voluntary behavior changes which reduce their health and injury risks, improve their health consumer skills and enhance their individual productivity and well-being whereas wellness is an intentional choice of a lifestyle characterized by personal responsibility, moderation, and maximum personal enhancement of physical, mental, emotional and spiritual health. Wellness programs typically begin by focusing on the reduction of health risks and then target issues that affect personal productivity, general well being, quality of work-life, personal growth, and other areas of interest" (Hunnicutt & Chapman, 2006, p. 4). On the other hand, CW is a good physical, mental, social and spiritual health of an individual and organizational workforce.

Royal path of yoga includes subsequent steps: *Yam, Niyam, Asana, Pranayama, Pratyahara, Dharana, Dhyan* and *Samadhi. Yama* (social codes) is the practice of improving social health and harmony that incorpates five vows: non-violence, truthfulness, non-stealing, non-possessiveness and celibacy. *Niyam* (moral codes) is the practice for creating homogeneity and harmony among feelings, thinking and action and comprised of five code of conducts-purity, contentment, austerity, self-study and complete surrender to God.

Asana (posture) is the practice for improving physical health, physical flexibility and fitness; overcoming conflicts, and maintaining steady posture for meditation. Yoga quietens the body and mind through vascular and muscular relaxation (Monro, 1995). Maintaining of posture was thought to lead strengthening and relaxation of voluntary muscles and eventually to control over the autonomic nervous system (ANS) (Vahia et al. 2004). In the same way, another study had reported that intensive practice of postural sequences as *Surya Namaskar* for longer than 10 minutes was associated with sufficiently elevated metabolic and heart response to improve cardio-respiratory fitness (Hagins et al., 2007). The continuous extension and flexion of muscles during yoga poses is associated with activation of antagonistic neuromuscular system as well as tendon-organ feedback resulting in increased range of motion and relaxation (Riley, 2004).

Pranayama is the fourth step in the *Ashtanga* yoga system of Patanjali. The control of the breath leads to the control of the life force or *prana* and mind. The ancient yogis have propounded many breathing techniques to maximize the benefits of *prana* at somatic and psychic level. The word *"Pranayama"* is made up of two words, *Prana* and *Ayama*. *Prana* stands for the capacity to keep body alive by air, i.e breath and *Ayama* means expansion, stretching or extension and control of breath. Thus, Pranayama means the art of controlling breath. *Pranayama* is basically undertaken for somatic and psychic purification, regulation of *prana* to each body organ and to optimize the cardio-pulmonary and autonomic functions. The ancient yogis of yore searched the intimate connection between breath and mind. Breath control has indirect influence over the mind thereby showing mind-body interplay. Breathing is an automatic process controlled by the autonomic nervous system. The science of bio-energy including the breathing movements is the practical yoga par excellence. One of the main texts of *Hatha Yoga*, *Hatha Yoga Pradeepika*(HP) advocates that unsteady flow of *prana* in body leads to unsteady mind and vice-versa (HP, 2:2). The ancient yoga texts state that *Pranayam* practiced properly can cure all diseases, but if practiced wrongly can onset diseases. Therefore, *Pranayama* needs to be learned under the supervision of an experienced teacher by taking needful precautions.

Breathing is the most important bodily function. Learning of breath control helps control body metabolism. There are generally 10 types of *Pranayama* (techniques of breath control) but five (*Bhastrika, Kapalbhati, Anulom-Vilom, Bharamari* and *Udgeeth Pranayama*) of them are found more in practice due to their prominent benefits. *Pranayam* (breathing mechanics for control and expansion of *prana*) is the practice for attaining a sound mental health by channeling *pranic* flow in subtle energy channels, expanding and controlling *pranic* energy. Its regular practice regulates secretions of endocrine hormones and neuro-hormones. The voluntary control of breath can modulate autonomic nervous system functions including cardiac vagal tone as measured by heart rate variability (Lehrer 1999; Sovik 2000), vigilance and attention (Fokkema 1999), chemoreceptor and baroreflex sensitivity (Bernardi 2001; Spicuzza 2000), as well as the level of central nervous excitation (Brown & Gerbarg 2005). *Pranayam* like *Ujjayi* breathing increases vagal tone, heart rate variability (HRV) (Telles and Desiraju 1992) and respiratory sinus arrhythmia (RSA) (Carney et al. 1995) by inducing parasympathetic activity through numerous mechanisms, including slow breath rate, contraction of the laryngeal musculature, inspiration against airway resistance and breath holds (Cappo & Holmes 1984). Furthermore, they emphasized that slow breathing with prolonged expiration was shown to reduce psychological and physiological arousal, anxiety, panic disorder, depression, IBS, early Alzheimer's disease and obesity (Friedman & Thayer 1998; Haug et al. 1994). Thus, *Pranayam* is the best practice of boosting morale, will power, self-confidence and mind-body health.

Deep yoga breathing exercises like *Bhastrika* reduce the work load on the heart in two ways. Firstly, deep breathing leads to more efficient lungs, which means more oxygen is brought into contact with blood sent to the lungs by the heart. So, the heart doesn't have to work hard to deliver oxygen to the tissues. Secondly, deep breathing leads to a greater pressure differential in the lungs, which leads to an increase in the circulation, thus resting the heart a little. Deep breathing improves the quality of the blood by promoting increased oxygenation in the lungs and thereby aiding the elimination of toxins and morbid matters; increases the digestion and assimilation of food by promoting enriched oxygen supply to stomach; and improves the health of the nervous system, including the brain, spinal cord, nerve centers and nerves by promoting supply of oxygenated blood accompanied by nutrients. This improves the health of the whole body, since the nervous system communicates to all parts of the body. Moreover, it rejuvenates glands, especially the pituitary and pineal glands. The brain has a special affinity for oxygen, requiring three times more oxygen than the rest of the body. This has far-reaching effects on well being. The movements of the diaphragm during the deep breathing exercise massage the abdominal organs - the stomach, small intestine, liver and pancreas. The upper movement of the diaphragm also massages the heart. This stimulates the blood circulation in these organs. The lungs become healthy and powerful, a good insurance against respiratory problems. Deep and slow yoga breathing reduces the work load for the heart. The result is more efficient and stronger heart that operates better and lasts longer. It also means controlled blood pressure and less chances of heart disease. Deep, slow breathing assists in weight control. If we are overweight, the extra oxygen burns up the excess fat more efficiently. If we are underweight, the extra oxygen feeds the starving tissues and glands. Slow, deep and rhythmic breathing causes a reflex stimulation of the parasympathetic nervous system which results in a reduction in the heart rate and relaxation of the muscles. These two factors cause a reflex relaxation of the mind, since the mind and body are very interdependent. In addition, oxygenation of the brain tends to normalize brain function and reduce excessive anxiety levels. The breathing exercises cause subsequent contractions of lung tissues thereby increasing elasticity of the lungs and rib cage. This creates an increased breathing capacity all day, not just during the actual exercise period.

Alternate Nostril Breathing (ANB) produces optimum functions to both sides of the brain-optimum creativity and optimum logical verbal activity. Regulated and harmonious rhythms of left and right nostrils calm the mind and the nervous system. Substantial studies have proven that the nasal cycle is associated with brain function. The electrical activity of the brain was found to be greater on the opposite side of decongested nostril. The right side of the brain controls creative activity while the left side controls logical verbal activity. The researches have shown that predominance of left nostril activates the right side of the brain thereby bettering creative performance. Similarly, the predominance of the right nostril activates the left side of the brain and betters verbal skills.

The concept of yoga therapy seems more advance and ancient compared to modern medical science. Yoga therapy advocates that manifestation of every disease accompanies with unrhythmic breathing and disturbed nasal cycles. Moreover, the onset of each disease can be perceived just by checking nasal cycle. Alternate nostril breathing technique is met efficacious to regulate alternate predominant flow of left and right nostril and hence activation of just two opposite sides of the brain. This clears blockage of the nasal passage and reestablishes the natural nasal cycle. For example, the yogis have known for a long time that prolonged breathing through the left nostril only (over a period of years) causes asthma. They also knew that this so-called incurable disease can be easily cured by teaching the patient to breathe through the right nostril and then to prevent its recurrence by practicing the alternate nostril breathing technique. The yogis also believed that diabetes is caused to a large extent by breathing mainly through the right nostril.

Pratyhara (senses withdrawal) is the practice of conserving energy or *prana* by diverting senses inward from their external objects to seal outward *pranic* flow. It is an introspective practice of increasing bio-immunity, psycho-immunity and spiritual immunity at large. The prevalent practice like *Yoga Nidra* comes under *Pratyahara* in which practitioner goes in relaxed meditative state and gets dissociated from wish to act. Kjaer et al. (2002) made a study to investigate whether endogenous dopamine release increased during loss of executive control in meditation (*Yoga Nidra*) and found a 65% increase in endogenous dopamine release, concomitant increase in theta activity, decreased desire for action and heightened sensory imagery.

Dharana (concentration) is the practice of hitting target by being pin pointed. i.e., hundred percent focused mental flow at a particular target. The practice like mindful awareness, mindful based stress reduction technique, guided imagery and advance stage of *Yoga Nidra* come under this. Siegel (2009) hypothesized that mindful awareness induced internal attunement thereby catalyzing the fundamental process of integration. Moreover, he asserted that integration — the linkage of differentiated elements of a system led to the flexible, adaptive, and coherent flow of energy and information in the brain, the mind and relationships. *Dharana* has shown remarkable effect on brain activity. The brain is an electrochemical organ that uses electromagnetic energy to function. Electrical activity emanating from the brain is displayed in the form of brainwaves. There are four categories of these brainwaves. They range from delta with high amplitude and low frequency to beta with the low amplitude and high frequency. Men, women and children of all ages experience the same characteristic brainwaves. They are consistent across cultures and country boundaries. During meditation brain waves alter. Emission of Beta waves (13-30 cycles per second) is an indication of awaking awareness, extroversion, concentration, logical thinking and active conversation. A debater would be in high beta. A person making a speech, or a teacher, or a talk show host would all be in beta when they are engaged in their work. Emission of Alpha (7-13 cycles per second) is associated with relaxation, non-arousal, meditation and hypnosis. Emission of Theta (4-7 cycles per second) is associated with the activities like dreaming, day-dreaming, creativity, meditation, paranormal phenomena, Extra Sensory Perception (ESP) and shamanic journeys. Emission of Delta (1.5-4 or less cycles per second) is an indicator of deep and dreamless sleep. Mindfulness meditation and related techniques are intended to train attention for the sake of provoking insight. It can be thought as the opposite of attention deficit disorder. A wider, more flexible attention span makes it easier to be aware of a situation, easier to be objective in emotionally or morally difficult situations and easier to achieve a state of responsive, creative awareness or 'flow'.

Dhyan (meditation) is the prolonged concentration on a particular target that culminates in self-realization and paranormal accomplishments. The subsequent practice of meditation is supportive for awakening ESP and reaching self-realization. Neuroimaging studies had shown that meditation resulted in activation of the prefrontal cortex, the thalamus and inhibitory thalamic reticular nucleus and a resultant functional deafferentation of the parietal lobe (Mohandas, 2008). He further asserted that neurochemicals' (GABA, endogenous dopamine, epinephrine, nor-epinephrine, encephalin, acetylcholine etc.) changes as a result of meditative practice involved all the major neurotransmitter systems that contributed to ameliorate anxiety, depressive symptomatology and psycotogenic property. Meditation works because of the relationship between the amygdala and the prefrontal cortex. Simply, the amygdala is the part of the brain that decides when to get angry or anxious (among other things) and the pre-frontal cortex is the part that makes us stop and think about things (it is also known as the inhibitory centre). Moreover, intuitive flashes and ESPs are very common when mind gets tranquilized and calm in deep meditative stage. In such condition there happens harmonious interplay among conscious, subconscious and unconscious minds thereby causing intuition and ESPs.

Samadhi (trance or super-consciousness) is fusion of embodied consciousness with cosmic consciousness; a steady feeling of holism and interconnectedness. As per yoga, *Samadhi* is supposed as the stage of total health where an aspirant gets freed from the effects of three strands — *Sat, Raj and Tam* and realizes one's real self. On other word, it is *Nirudha* stage of psyche that represents the total health.

3.2 Yoga for CE

Leadership is one of the most important components of CE and is more about putting first things first to translate vision into action. Prolonged yoga practice is deemed responsive to develop spiritual traits- self awareness, field independence, humility, tendency to ask fundamentals- why; ability to reframe, positive use of adversity and sense of vocation (Zohar, 2005) that are essential for effective leadership and translating holy organizational vision into action. Correspondingly, this sub head will advocate yogic efficacy for promoting organizational excellence thereby excelling leadership and four human intelligences (SI, EI, CI and RI).

CE is the function of four intelligences — SI (farsightedness, serenity, discriminative wisdom, personal meaning production, critical existential thinking, transcedental awareness and concious state expension), EI (affectionate and loving relationship with family and society; memorizing God's compassion is unbounded, transfer of privilege, career development, team building, empathy, sound leadership and civility), CI (creativity and innovations) and RI (good managerial capabilities, job placements and technical performances) born by an organization family. The optimal level of these intelligences among organization family members can be induced by inculcating yogic culture among them. On the basis of the ladder proposed by *Raj yoga*, an interesting model for achieving CE can be set to overcome the nine hurdles. Removing aforesaid hurdles by appropriate yogic practices induces inner harmony; inner harmony induces talent, talent leads to creativity and innovations; creativity and innovation lead to perfection; and perfection culminates in excellence. On the other hand, employee health and performance are closely linked to each other. Good workers' health leads to productivity at the work; productivity at the work leads to business competitiveness; business competitiveness leads to economic development and prosperity; economic prosperity leads to social well being and wealth; social well being and wealth again help maintain good employee health (Burton, 2010) as depicted in Figure 2.

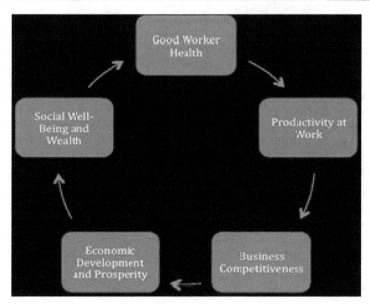

Fig. 2. Relationship between Health and Wealth.

This hierarchical relationship between health and wealth also displays the high possibility of achieving CW and CE via yoga practices. Therefore, total health and perfection need to be developed first at individual level for CW and CE by adopting persistent yoga practices. This may be plausiable by developing corporate yoga culture to provide equal chance of practicing yoga for each and every organizational family member.

4. Preventive and therapeutic value of Yoga

Preventive and therapeutic value of yoga has been argued as a side benefit of yogic practices especially associated with gross and subtle body and most of the yoga practitioners have been found to have the same concern. Substantial scientific studies undertaken have also attempted to substantiate the preventive and curative value of yoga practices like cleansing techniques (*Shatkriyas*), postures (*Asanas*), breathing techniques (*Pranayamas*), gestures (*Mudras*), psychic locks (*Bandhas*), concentration (*Dharana*) and meditation (*Dhyan*). As far as preventive and curative value of CY for CW is concerned, it is deemed contextual to discuss efficacy of CY in the connection of the most prevalent corporate health problems such as work related stress, respiratory problems, cardiovascular problems, digestive problems and genitourinary problems.

4.1 Yoga versus stress

Globalization, technological advancements, intermixing of work cultures, recessions and subsequent changes in the nature of work are in fast pace. Consequently, stress is found with everyone at workplace whether rich or poor, young or old, male or female; no one is immune from it. Stress may be the biggest single cause for illness or premature death. WHO has declared stress as worldwide epidemic and reported job stress as "the twentieth-century disease". The American Institute of Stress (AIS) states that stress related illness costs economy more than $ 100 billion per year. Additionally, AIS estimated in 2001 that stress costs organizations $ 300 billion in healthcare, workers compensation, absenteeism, and turnover. The productivity losses hover around $17 billion annually. Every health problem from simple headache to heart attack, from psychosomatic disorders to stroke can be linked to stress that is called the plague of the 21st century. Stress-related illness and injuries account for almost three-fourths of employee absenteeism.

A growing body of research evidence supports the belief that certain yoga techniques may improve physical and mental health through down-regulation of the hypothalamic–pituitary–adrenal (HPA) axis and the sympathetic nervous system (SNS) (Ross & Thomas, 2010). The HPA axis and SNS are triggered as a response to a physical or psychologic demand (stressor), leading to a cascade of physiologic, behavioral, and psychologic effects, primarily as a result of the release of cortisol and catecholamines (epinephrine and norepinephrine). This response leads to the mobilization of energy needed to combat the stressor through the classic "fight or flight"syndrome. Over time, the constant state of hypervigilence resulting from repeated firing of the HPA axis and SNS can lead to dysregulation of the system and ultimately diseases such as obesity, diabetes, autoimmune disorders, depression, substance abuse, and cardiovascular disease (Sterling, 2004; McEwen, 2000, as cited in Ross & Thomas, 2010). Conversely, substantial studies have shown yoga to have an immediate downregulating effect on both the SNS/HPA axis response to stress. Studies show that yoga decreases levels of salivary cortisol (Michalsen, 2005; West, 2004), blood glucose (Gokal & Shillito, 2007; Khatri et al. 2007) as well as plasma rennin levels, and 24-hour urine norepinephrine and epinephrine levels (Selvamurthy et al., 1998). Yoga significantly decreases heart rate and systolic and diastolic blood pressure (Damodaran et al., 2002; McCaffrey, Ruknui, Hatthakit & Kasetsomboon, 2005; Selvamurthy et al., 1998) (as cited in Ross & Thomas, 2010) . Studies suggest that yoga reverses the negative impact of stress on the immune system by increasing levels of immunoglobulin A (Stuck et al., 2003) as well as natural killer cells (Rao et al., 2008) (as cited in Ross & Thomas, 2010). Yoga has been found to decrease markers of inflammation such as high sensitivity C-reactive protein as well as inflammatory cytokines such as interleukin-6 (Pullen et al., 2008) and lymphocyte-1B (Schultz et al., 2007) (as cited in Ross & Thomas, 2010).

Aforementioned studies show that yoga has an immediate quieting effect on the SNS or HPA axis response to stress unless the precise mechanism of action has not been determined. The proposed hypotheses substantiate that yoga exercises cause a shift toward parasympathetic nervous system dominance, possibly via direct vagal stimulation (Innes, Bourguignon and Taylor, 2005); significant reductions in low-frequency heart rate variability (HRV) — a sign of sympathetic nervous system activation — in depressed patients following an 8-week yoga intervention (Shapiro et al., 2007); decrease in anxiety (Gupta et al., 2006; Michalsen, 2005; Telles et al., 2006; West, 2004) and increase in emotional, social, and spiritual well-being(Moadel et al., 2007) (as cited in Ross &Thomas, 2010).

4.2 Yoga versus respiratory problems

Yogic practices are always undertaken with synchronization of action, mental awareness and breathing pattern. Particularly, *Pranayam, Bandha, Mudra and Asana* incorporate systemic and subsequent inhalation, exhalation, inner breath retention and outer breath retentions; contraction and relaxation of lung tissues and chest wall thereby affecting the cardiopulmonary function (lung function, heart rate, breath rate, heart rate variability, oxygen consumption, and CO_2 expulsion), endocrine secretions and neural secretions and function of associated visceral organs. *Asanas* (postures) are basically somatic techniques for physical conditioning; *Pranayam* is a technique for breath control (inhalation, exhalation and retention) and expansion of *prana* (life energy) that strengthens respiratory muscles and better ventilation; a *Mudra* (gesture) is a sort of seal- a body movement to hold energy, or concentrate awareness; and a *Bandha* is an energy lock, using muscular constriction to focus awareness (Raub, 2002). Early studies (Joshi et al., 1992; Makwana et al., 1988) reported improvement in some, but not all, measures of ventilation after breath control exercises alone. For example, Joshi et al. (1992) followed lung function in 75 males and females with an average age of 18.5 years during yoga breath-control exercises. After 6 weeks of practice, they reported significant increases in forced vital capacity (FVC), forced expiratory volume in 1 second (FEV1), peak expiratory flow rate (PEFR), maximum voluntary ventilation (MVV), as well as a significant decrease in breathing frequency (fB), and prolongation of breath-holding time. Other studies reported similar improvement in lung function after practicing yoga postures alone or combined with other yoga techniques. Konar et al. (2000) reported that the practice of *Sarvangasana* (shoulder stand) twice daily for 2 weeks significantly reduced resting HR and left ventricular end-diastolic volume in 8 healthy male subjects. Birkel and Edgren (2000) reported that yoga postures, breath control, and relaxation techniques taught to 287 college students (89 men and 198 women) in two 50-minute class meetings for 15 weeks significantly improved FVC of the lungs measured by spirometry. In a similar study, 1 hour of yoga practice each day for 12 weeks significantly improved FVC, FEV1, and PEFR in 60 healthy young women having 17 to 28 years of age(Yadav & Das, 2001).

Moreover, yogic interventions are also found beneficial for improving ailments like asthma (Bhagwat, Soman, & Bhole, 1981; Bhole, 1967; Nagendra & Nagarathna, 1986; Nagendra & Nagarathna, 1985; Jain & Talukdar, 1993; Sabina et al., 2005; Singh, Wisniewski, Britton, & Tattersfield, 1990; Vedanthan et al., 1998), pulmonary tuberculosis (Milani, Valli & Bhole, 1992; Prakasamma & Bhaduri, 2004; Visweswaraiah & Telles, 2004), pleural effusion (Prakasamma & Bhaduri, 1984), rhinitis (Sim, 1981), sinusitis(Rabago et al., 2002), chronic obstructive pulmonary diseases (Behera, 1998; Donesky-Cuenco, Nguyen, Paul & Carrieri-Kohlman, 2009; Kamath & Chauhan, 1982; Pomidori, Camigotto, Amatya, Bernardi & Cogo, 2009; Tandon, 1978), chronic bronchitis (Behera, 1998) (as cited in McCall, 2009).

Controlled clinical studies have shown that an integrated approach of yoga therapy (consisted of yoga exercises and postures for 25 minutes; slow, deep breathing for 10 minutes; slow mental chanting for 15 minutes; and a devotional session as a daily practice) to be beneficial in the clinical management of asthma. A 65-minute daily practice of yoga for 2 weeks improved PEFR, medication use, and asthma attack frequency in 53 patients when compared to an age-, gender-, and clinically matched control group (Nagarathna & Nagendra, 1985). In a long-term, follow-up (3 to 54 months) prospective study (Nagendra & Nagarathna, 1986) made among 570 asthmatics showed overall significant improvement in PEFR after a similar yoga training program. The greatest improvement was found in patients who had practiced yoga most frequently and intensively thereby showing asthma medication reduction among approximately 70% of the practitioners. The effects of two

Pranayamas on lung function, airway reactivity, respiratory symptoms, and medication use were assessed in 18 patients with mild asthma in a randomized, double-blind, placebo-controlled, crossover trial (Singh et al., 1990). The subjects were taught *Pranayamas* by using a breathing device called the Pink City lung (PCL; Pulmotech, Jaipur, India) exerciser that could be used with a matched placebo breathing device.

After a baseline measurement, the subjects were undergone through the practice of slow deep breathing for 15 minutes, two times a day for consecutive 2-week periods, randomly alternating the breathing devices for each practice period. Measured lung function variables (FEV1, FVC, PEFR), symptom scores, and medication use improved with the PCL device with small and statistically insignificant changes. However, there was a statistically significant increase in the dose of histamine required to produce a 20% decrease in FEV1, a provocative airway test commonly used to assess lung responsiveness to nonspecific bronchoconstrictors. The findings indicate that *Pranyama* may lead to an overall clinical improvement in mild asthma. In a subsequent letter to the editor, Stanescu (1990) commented on possible autonomic mechanisms suggested by Singh et al. (1990) that might lead to reduced airway responsiveness. Studies previously conducted by Stanescu et al. (1981) on healthy subjects showed the efficacy of controlled yoga breathing techniques (i.e., slow, near VC maneuvers accompanied by apnea at end inspiration and end expiration) for significant lowering of ventilatory responsiveness to increased carbon dioxide. Two early studies (Behera & Jindal, 1990; Jain & Talukdar, 1993) reported on quality of life benefits provided by the effects of various yoga exercises among asthmatics. Behera & Jindal (1990) assessed the benefits of daily yoga exercises contained with breath control and postures, over a 6- to 8-week period in 41 asthmatics. Although the authors reported an overall subjective improvement in asthma symptoms, objective lung function measurements showed improvement in some, no change and reduced in some. Jain and Talukdar (1993) reported a similar overall effect of yoga therapy on exercise capacity in 46 asthmatics and reported improvement in a 12-minute walking test, a modified Harvard step test, and a more subjective index of exercise tolerance. However, it was unclear if the improvements were due, in part, to a placebo response. In the more recent literature (after 1995), breath-control and relaxation techniques in both children and adults with asthma have been reported to improve some, but not all, measures of lung function (e.g., PEFR, MVV, FEV1, and FVC), decrease usage of medication, and increase exercise tolerance (Blanc-Gras et al., 1996; Khanam et al., 1996; Manocha et al., 2002; Sathyaprabha et al., 2001; Vedanthan et al., 1998). Large variability in the subject population, questionable compliance in the yoga treatment groups, and potentially adverse outcomes in some subjects further complicates interpretation of the effects specific to a particular relaxation technique (Ritz, 2001). Further studies are warranted, therefore, to better understand the mechanisms of response to yogic intervention and to determine its clinical value for asthmatics.

Behera (1998) reported improvement in shortness of breath and some lung function parameters in patients (*n*= 15) with the history of chronic bronchitis and age range 48 to 75 years (58.9 6±11.1 years) who received yoga therapy that consisted of breath control and 8 types of *asanas* for a period of 4 weeks. The patients had baseline assessment of their history of chronic bronchitis, including spirometry, medication strategy, and exercise tolerance. They were instructed yogic postures (e.g., *Vajrasana, Simhasana, Sarvangasana, Chakrasana & Matsyasana*) and breathing techniques for 1 week and were encouraged to practice daily with follow-up yoga sessions each subsequent week along with medication. Reevaluated clinical status and pulmonary function revealed a significant improvement in FEV1 and PEFR after second week and significant improvements in VC and PEFR after fourth week excluding a patient's reporting of perceptual decrease in shortness of breath. No changes were noted in the amount of medication taken. This preliminary study (with poor research design) was made among few patients for short duration. Unfortunately, no other studies examining the possible benefits of yoga on chronic obstructive lung disease have been published yet and hence generalization of the outcomes needs further studies to eliminate its limitations.

4.3 Yoga versus cardiovascular problems

Cardiovascular disease continues to be a significant health issue, contributing to more deaths than any other disease in developed countries while becoming the leading cause of death in developing countries worldwide (Yach, Leeder, Bell, & Kistnasamy, 2005, as cited in deJong, 2009). Although the risk factors for cardiovascular disease are well known, they remain poorly controlled in the United States (Glover, Grerenlund, Ayala, & Croft, 2005) leading to increased costs for treatment (American Heart Association, 2005, as cited in deJong, 2009). Smoking, hypertension, diabetes mellitus, obesity, poor dietary patterns, physical inactivity, alcohol consumption, elevated blood apolipoprotein levels, and psychosocial factors are estimated nine risk factors that account for approximately 90% of population-attributed risk for cardiovascular disease (Yusuf et al., 2004, as cited in deJong, 2009). Besides, emotional stress is one more major cause in the pathogenesis of cardiac diseases like ischemic heart disease (IHD) (Eastwood &Trevelyan, 1971, as cited in Ornish et al., 1983). Some emotions and behaviors are associated with IHD include intense anxiety, depression, feelings of helplessness, and "type A behavior," characterized by ambitiousness, competitiveness, impatience, and a sense of time urgency (Hackett & Rosenbaum , 1980, as cited in Ornish et al, 1983). Significant lipid risk factors for CVD are increased levels of serum cholesterol and triglycerides, increased low-density lipoprotein (LDL) cholesterol, decreased high-density lipoprotein (HDL) cholesterol, and increased concentration of apoB-carrying lipoproteins (Raub, 2002). In fact, Chiuve, McCullough, Sacks and Rimm (2006) estimated that 62% of all coronary events could be avoided if all men adhered to a low-risk lifestyle that included smoking abstinence, regular exercise, healthy diet, moderate alcohol intake, and the maintenance of a healthy weight (as cited in deJong, 2009). Consistent citation was also made by Ornish et al. (1983) that concluded that bio-behavioral techniques such as yoga (meditation, pranayama, and progressive relaxation) may reduce some of the cardiovascular risk factors- BP (Bensen, 1977) and plasma cholesterol levels (Patel, 1976; Cooper & Aygen, 1979). Yoga's potential benefit to patients with CVD has been reported by limited literature (Raub, 2002). Adoption of a yoga lifestyle can significantly reduce many of the risk factors for CVD, including increased body weight, altered blood lipid profile, and elevated blood pressure (BP) (Mahajan et al., 1999; Manchanda et al., 2000; Schmidt et al., 1997, as cited in Raub, 2002). As cited in Field (2010), Yogendra et al. (2004) reported benefits of one year long yoga life style among the patients of advanced coronary artery disease. They found 23% reduction in cholesterol   in the yoga group as compared to 4% in the standard treatment control group. Besides, serum low density lipids also seen reduced more in the yoga group (26% versus 3% in the control group). In a similar study on coronary artery disease, a dietary change plus yoga group was compared to a group who only made dietary changes (Manchanda et al., 2000, as cited in Field, 2010). After one year of weekly sessions, the yoga group had fewer anginal episodes per week, improved exercise capacity, decreased body weight and lower serum total cholesterol levels. Low-density lipoprotein, cholesterol and triglyceride levels also decreased in the yoga group. Revascularization procedures (coronary angioplasty or bypass surgery) were less frequently required in the yoga group, and coronary angiography showed that more lesions regressed (20% versus 2%) and fewer lesions progressed (5% versus 37%) in the dietary change plus yoga group. Pullen et al. (2009) also reported improved cardiovascular endurance and decreased inflammatory markers (Interleukin-6 and C- reactive protein) in heart failure patients thereby showing similar effects of yoga as massage therapy. Schmidt et al. (1997) reported that a 3-month residential training program of yoga, meditation, and vegetarian nutrition decreased body mass, total serum and LDL cholesterol, fibrinogen, and BP (as cited in Raub, 2002). Mahajan et al. (1999) reported a similar reduction in risk factors for patients with coronary artery disease (CAD) and documented angina (chest pain) where subjects with risk factors for CAD were randomly assigned to a yoga intervention group ($n=5$ 52) or a control group ($n =5$ 41) (as cited in Raub, 2002). Both groups received lifestyle advice and the intervention group received additional yoga training. Serial evaluations at 4, 10, and 14 weeks showed a regular decrease in all lipid parameters, except for HDL, only in the patients with angina receiving yoga intervention. The most impressive of these studies was a 1-year prospective, randomized controlled trial of 42 men with angiographically documented CAD (Manchanda et al., 2000, as cited in Raub, 2002). A subgroup ($n =5$ 21) treated with an active program of risk factor and diet control along with yoga and moderate aerobic exercise showed significant reduction in angina, improved exercise capacity, and greater reductions in body weight, total cholesterol, LDL cholesterol, and triglyceride than the control group ($n = 5$ 21) treated conventionally with risk factor control and the American Heart Association (AHA) Step I diet.

Revascularization procedures also were less frequent in the yoga group and coronary angiography repeated at 1 year showed a significant regression of atherosclerotic lesions. The Lifestyle Heart Trial (Ornish et al., 1998) demonstrated that intensive lifestyle changes could lead to regression of CAD after only 1 year of a 5-year program. Forty-eight (48) patients with moderate-to-severe CAD were randomized to an intensive lifestyle change group or to a usual-care group. The lifestyle changes consisted of a 10% fat whole–food vegetarian diet, aerobic exercise, stress management training (yoga and meditation), smoking cessation, and group psychologic support. Clinical status was followed by quantitative coronary angiography and frequency of cardiac events. Of the 35 patients completing the 5-year follow-up, 20 in the experimental group showed a 4.5% relative improvement in cardiovascular status after 1 year and a 7.9% relative improvement after 5 years. The control group had a relative worsening of cardiovascular status after 1 and 5 years (5.4% and 27.7%, respectively), and more than twice as many cardiac events. Intensive lifestyle changes/yogic life style, therefore, can cause a regression of CAD. Nonetheless, the sparse of randomized controlled studies for assessing the efficacy of yogic intervention on CVD, especially in comparison to the conventional practice of Western medicine, has made it difficult to assess the direct benefits of an integrated yoga practice on patients with CAD.

Another most prevalent risk factor for cardiac problems is hypertension. Early studies on yoga intervention for hypertension investigated the value of total body relaxation postures, primarily *Savasana* (Chaudhary et al., 1988; Mogra and Singh, 1986, as cited in Raub, 2002). The authors reported reductions in BP that were similar to control by drug therapy or biofeedback;

however, small numbers of subjects were utilized in the studies and there were no controls. I hour daily yoga practice of three month decreased blood pressure, blood glucose, cholesterol and triglycerides and improved subjective well-being and quality of life among mild to moderate hypertensive participants (Damodaran et al., 2002, as cited in Field, 2010).

Yoga exercises twice a day for 11 weeks were found to be as effective as standard medical treatment in controlling measured variables of hypertension (Murugesan et al., 2000). In a randomized study, 33 hypertensive patients with 35 to 65 years of age, were assigned into three groups receiving yoga therapy, physician-provided medication, and no treatment (control group). Preanalysis/postanalysis regarding systolic and diastolic blood pressure, pulse rate, and body weight revealed that both the treatment groups (i.e., yoga and drug) were effective in controlling hypertension. Twenty male patients with essential hypertension (EH) were treated for 3 weeks with postural tilt stimulus (tilt table) or with postural yoga *asanas* to restore normal baroreflex sensitivity (Selvamurthy et al., 1998). Progressive autonomic changes were assessed by cardiovascular responses to head-up tilt and cold pressor stimulus, electroencephalographic indices, blood catecholamines, and plasma rennin activity. There was a significant reduction in blood pressure after 3 weeks in both treatment groups, indicating a gradual improvement in baroreflex sensitivity. A similar improvement in baroreflex sensitivity, and significant reductions in systolic and diastolic blood pressure, were seen in 81 patients (58-61 years of age) with stable chronic heart failure (CHF) who practiced slow and deep breathing (Bernardi et al., 2002). The same authors (Bernardi et al., 1998) previously reported that a slow rate of breathing in patients with CHF increases resting oxygen saturation, improves ventilation/perfusion mismatching, and improves exercise tolerance. These changes were obtained by simply modifying the breathing pattern, from a resting, spontaneous ventilation of approximately 15 breaths per minute to 6 breaths per minute, which seems to cause a relative increase in vagal activity and a decrease in sympathetic activity. The effects on baroreflex sensitivity were similar to those obtained with captopril treatment in patients with CHF (Osterziel et al., 1988). Captopril belongs to a group of drugs called angiotensin-converting enzyme (ACE) inhibitors that help to lower blood pressure and make the heart beat stronger. This medication is used to treat hypertension (high blood pressure) and heart failure.

4.4 Yoga versus digestive problems

Hectic and unnatural corporate life style (fast food, materialistic relationship, hectic schedule, distanced natural environment, odd duty hours, smoking, alcoholism, poor intake of mental and emotional diet etc.) has been a sufficient condition to trigger and progress digestive problems like hyper acidity, irritable bowel syndrome, gastritis, pancreatitis, flatulence, ulcerative colitis, diabetes, inflammatory bowel disease, constipation, indigestion, hiccups, gastroesophageal reflux disease (GERD), hepatitis, gall stones, celiac disease among corporate workforces. Additionally, because of being a major system of nutrients absorption and morbid matter elimination, disordered digestive system is sufficient to disturb homeostasis and to trigger varieties of somatic and psychological problems. Sages of yore have rightly spoken that good digestion is a key to radiant health and is the function of psychological well being. Yoga views the digestive system as a very sensitive mirror of the mind and encourages examining overall lifestyle choices, emotions and other mental components in the diagnosis and healing of the digestive problems (Butera, 2010, p. 14).

Within a few seconds of "flight or fight" response happened in the central nervous system due to distress, most of the blood in the body gets shunted out from the digestive system and into the major muscle groups. This negatively impacts on the contractions of the digestive muscles that help move food through the body as well as the fluids and secretions that are needed for healthy digestion. Consequently, persistent mental distress results in esophagus spasms, indigestion, nausea, diarrhea, constipation, stomach ulcers, celiacs disease, irritable bowel syndrome as well as other more severe digestive ailments.

Yogic life style (regular practice of a complete yogic approach including selected- cleansing techniques, postures, gestures, psychic locks, concentration, meditation, natural diet as per body constitution and season, exercising charter of righteousness, and observance of social and moral codes) is found beneficial for the proper digestion and elimination, and healing of various digestive disorders like hyper acidity, irritable bowel syndrome, gastritis, pancreatitis, flatulence, ulcerative colitis, diabetes, inflammatory bowel disease, constipation, indigestion, hiccups, Gastroesophageal reflux disease (GERD), Hepatitis, Gall stones, Celiac disease etc.

Practice of yogic postures causes sponge like squeezing in the soft tissues of the digestive organs, and encourages stale and waste-bearing fluids to be out of the tissues thereby facilitating the elimination of the morbid matters and subsequently supply of essential nutrients to these areas. Subsequent opening and stretching of digestive organs during the practice of yogic postures regulates the Peristalsis movement that is a key involuntary process for the proper digestion and elimination. Besides, yogic breathing exercises send oxygen deep into the cells of the body and help it to absorb nutrients and excrete morbid matters thoroughly. On the other hand, efficacy of Yoga for stress management, rebalance of the autonomic nervous system to create deep relaxation and dominate parasympathetic system is well documented.

Langhorst et al. (2007) analyzed the effects of a comprehensive lifestyle modification program (a structured 60-hour training program over a period of 10 weeks which included stress management training, psycho-educational elements, and self-care strategies) on health-related quality-of-life, psychological distress, and clinical parameters in 60 patients with ulcerative colitis (UC). The 60 patients were randomly assigned to an intervention group or a usual-care control group. Comparison of the measurements taken at baseline, after 3 and 12 months showed significant improvement in the quality of life and emotional well-being of the participants as compared to controls.

4.5 Yoga versus genitourinary disorders

Most of the corporate workforces have been competing for material prosperity and focused on sensual indulgences. The most fashionable one is over and multi-partner romantic relationship. Prevalent dress codes, especially girls', unnatural food at cafeteria, pornographic literatures and audio-visual aids and western corporate culture are serving as the best catalysts to provoke more lust and engage in frequent sexual activities. The incestuous workplaces are in ascending order. Consequently, corporate workforces are extremely prone to the genitourinary problems and seriously loosing their health and wealth. Hence, it seems quite contextual to address significance of CY for managing the genitourinary problems. As yoga argued, over sexual activities result in suppressed immunity, low self-esteem, low morale and different health problems, especially genitourinary problems such as urinary stress incontinence, poor sexual orgasm, HIV AIDS, syphilis, infertility, miscarriage, premature ejaculation (PE), climacteric syndrome and pregnancy problems.

The efficacy of mind-body intervention like yoga was supposed effective means for bettering genitourinary health for long. Yoga has been found effective to promote genitourinary health and heal the concerned problems like urinary stress incontinence (Milani, Valli & Bhole, 1992), women sexuality (Brotto, Krychman & Jacobson, 2008; Dhikav et al., 2007), climacteric syndrome (Chattha et al., 2008), PE (Dhikav et al., 2007), pregnancy outcomes (birth weight, preterm labor, and IUGR either in isolation or associated with PIH) (Narendran et al., 2005), labor pain and duration (Chuntharapat, Petpichetchian & Hatthakit, 2008). Interestingly, Yoga appeared as a non-pharmacological measure for improving female sexual functions (Dhikav et al., 2007). Dhikav et al. (2010) also reported that after the completion of yoga sessions; the female sexual functions scores were significantly improved ($P< 0.0001$). The improvement occurred in all six domains of female sexual function index (FSFI) (i.e., desire, arousal, lubrication, orgasm, satisfaction, and pain) was more in older women (age > 45 years) compared with younger women (age < 45 years) thereby proving yoga as an effective method of improving all domains of sexual functions in women. Considering widespread acceptability of yoga, non-pharmacological nature, and apparent beneficial effects in the present study, this modality deserves further study. Chattha et al. (2008) studied the effect of 8 week-integrated yoga program consisted of breathing practices, sun salutation and cyclic meditation on cognitive functions in climacteric syndrome by sampling 120 premenopausal women between 40 and 55 years with follicle-stimulating hormone level equal to 15miu/ml. Sample was randomly assigned in two groups as participants and controls and participants were allowed to practice yoga module one hour per day, 5 day per week for 8 weeks. In yoga group they reported improvement on flushes and night sweats; and cognitive functions such as remote memory, mental balance, attention and concentration, delayed and immediate recall, verbal retention and recognition tests after 8 week. Dhikav et al. (2007) conducted another study to know if yoga could be tried as a treatment option in PE and to compare it with fluoxetine. For the same, they sampled 68 patients (38 yoga group; 30 fluoxetine group) attending the outpatient department of psychiatry of a tertiary care hospital and employed both subjective and objective assessment tools to evaluate the efficacy of the yoga and fluoxetine in PE and found that all 38 patients (25–65.7% = good, 13–34.2% = fair) belonging to yoga and 25 out of 30 of thefluoxetine group (82.3%) had statistically significant improvement in PE thereby showing yoga as a feasible, safe, effective and acceptable non-pharmacological option for PE. Nonetheless, more studies involving larger patients could be carried out to establish its utility in this condition. Vaze and Joshi (2010) also concluded that Yoga as a free-of-cost noninvasive method that is fairly effective and is strongly recommended to all women of menopausal age. Another study conducted by Brotto et al. (2008) reported that Eastern techniques like acupuncture, yoga, mindfulness and other forms of spiritual practice might offer a unique approach to enhance women's sexuality. However, it needs the development of sound theory and controlled studies; they might be the key for improving women's lack of sexual satisfaction. Narendran et al. (2005) conducted a study to assess the effect of yoga on pregnancy outcomes and recruited 335 women with 18-20 weeks of pregnancy. Yoga program including selected postures, breathing techniques and meditation was given to 169 women of yoga group one hour daily until delivery whereas 166 women of control group were suggested to go for 30 minutes walk twice a day during the study. Intervened integrated yoga approach during pregnancy was found safe thereby showing improvement in birth weight, preterm labor, and IUGR either in isolation or associated with PIH, with no increased complications in yoga group as compared to control group. Chuntharapat et al.'s (2008) findings suggested that 30 min of yoga practice at least three times per week for 10 weeks is an effective complementary means for facilitating maternal comfort, decreasing pain during labor and 2 hour post delivery, and shortening the length of labor that highlighted yoga as an alternative nursing intervention to improve the quality of maternal and child health care.

5. Summary

In 21st century the corporate world is associated with the most tension giving elements such as competition, deadlines, market conditions and above all the desire to reach high on the corporate ladder. These four elements are ultimately responsible to impair the harmonious interplay of body, mind and spirit thereby leading to various health problems among corporate workforces. On the other hand, yoga seems as an emerging avenue for the worldwide corporate health and wealth. The packaging of CY for corporate life style is the best preventive and therapeutic measure to optimize organizational health and culture as well. Persistent practice of CY by a corporate executive makes him/her healthy and wealthy. CY will prove to be an art and science of life for a corporate executive.

Yoga is an ultimate attempt for the fusion of embodied consciousness with supreme consciousness that subsequently proceeds from the practice of social adjustment (*Yam*), moral observance (*Niyam*), postures (*Asana*), breathing mechanics (*Pranayama*), senses withdrawal (*Prathyara*), concentration (*Dharana*), meditation (*Dhyan*) and super-consciousness (*Samadhi*). Regular practice of yoga is supposed to empower corporate health, happiness and harmony and hence wealth too.

The relevance and popularity of yoga is ascending in the entire West and Europe. Particularly, in UK and USA, yoga has become more popular where the women form 70-90% of the student population. Nonetheless, it is difficult to say in numerical figures exactly how many people are benefiting from yoga around the world. But one can easily notice that a huge number of people especially the women have been practicing yoga daily. The participation of a huge number of women itself signifies how important the yoga is for the health and the happiness. This trend also recommends that yoga must be taken seriously into the consideration as a part of workplace curriculum or culture to promote CW, CE and corporate social responsibility (CSR).

The CY will consists of normal subtle yogic warming up exercises, postures, yogic breathing exercises(*pranayama*), gestures, psychic locks, concentration, meditation, and spiritual counseling. CY can be taught collectively and practiced individually in order to gain its wholesome effects. Impacts of CY practices can be better explained via bio-psycho-socio-spiritual model- at physical level it improves musculoskeletal functioning, cardiopulmonary status, ANS response and endocrine functioning; at psychosocial level, it enhances self-esteem, social support and positive mood; and at spiritual level it elevates compassionate understanding and mindfulness.

The progress of CY practice will positively produce prolonged physical, mental, emotional, social and spiritual effects on executives. This will also help in producing effective leadership in corporate world. Nurturing effective leaders is one of the most important functions of the corporate excellence. This aspect also can be achieved by persistent practice of CY. There is no other method better than yoga that can make a corporate executive physically fit, mentally alert, emotionally rectified, socially adapted, rationally positive, completely self analytic and spiritually elevated.

Particularly, work related stress, respiratory problems, cardiac problems, digestive problems and genitourinary problems are seen improved by the specific and regular yoga practice. Mechanisms underlying the modulating effects of yogic cognitive-behavioral practices (eg, meditation, *Asanas*, *Pranayama*, caloric restriction) on human physiology can be classified into four transduction pathways: humoral factors, nervous system activity, cell trafficking, and bio-electromagnetism that shed light how yogic practices might optimize health, delay aging, and ameliorate chronic illness and stress from disability (illness and stress from disability). That implies that yoga is a cost effective and common avenue to minimize medical expenditure and maximize corporate performance and productivity.

Promotion of total health, happiness, harmony and four human intelligences- rational intelligence, creative intelligence, emotional intelligence and spiritual intelligence are side benefits of CY practice. Scientific validation and standardization of the effects of yoga practices at individual and corporate level follows bio-psycho-socio-spiritual research model and substantiate efficacy of CY for CW and CE. The general mechanism of yogic effects and efficacy of yoga for managing work stress, and improving health problems related to stress, respiratory, cardiopulmonary, digestive and genitourinary systems in organizational family is portrayed on the basis of concerned research findings. Regular practice of yoga or CY culture is directly linked to wellness and optimal intelligences of organizational family. Good employees' health leads to productivity at work, productivity at work leads to business competitiveness, business competitiveness leads to economic prosperity and well-being which is again associated with employees' good health.

Cognitive intelligence that can be slightly enhanced and maintained through yoga practice is helpful for sound managerial capabilities and technical skill empowerment. Creative intelligence and concentration promoted by yoga practice is supportive for the generation of creative and innovative ideas. Emotional intelligence that can be remarkably increased by yoga practice is the key for galvanizing leadership, team building, optimal interpersonal relationship and harmony. Spiritual intelligence that can be increased subsequently via yoga practices is near to the corporate social responsibility, holism, empathy, farsightedness, compassion and universal love. Thus, it can be concluded that CY is a cost-effective, eternal and universal means for workplace wellness and excellence that needs to be included as an indispensable part of corporate culture.

6. References

American Institute of Stress. (2011). *Effects of stress.* Retrieved May 31, 2011 from
        http://www.stress.org/topic-effects.htm
Aurobindo, S. (1999). *The Synthesis of Yoga* (5th ed.). Pondicherry, India: Sri Aurobindo
        Ashram Publication Department.

Becker, I. (2000). *Use of yoga in psychiatry and medicine.* In P. R. Muskin (Ed), *Complementary and alternative medicine and psychiatry* (pp. 107- 145). Washington, D.C.: American Psychiatric Press, Inc.

Behera, D. (1998). Yoga therapy in chronic bronchitis. *Journal of the Associations of Physicians of India, 46,* 207–208.

Behera, D. & Jindal, S. K. (1990). Effect of yogic exercises in bronchial asthma. *Lung India, 8*(4), 187-189.

Bennett, S. M., Weintraub, A. & Khalsa, S. B. (2008). Initial evaluation of the LifeForce Yoga® Program as a therapeutic intervention for depression. *International Journal of Yoga Therapy, 18,* 49-56. Retrieved from http://www.yogafordepression.com/IJYT-2008-Bennett.pdf

Bernadi, L., Gabutti, A., Porta, C. & Spicuzza, L. (2001). Slow breathing reduces chemoreflex response to hypoxia and hypercapnia, and increases baroreflex sensitivity. *Journal of hypertension, 19,* 2221- 2229.

Bernardi, L., Porta, C., Spicuzza, L., Bellwon, J., Spadacini, G., Frey, A. W., ...Tramarin, R. (2002). Slow breathing increases arterial baroreflex sensitivity in patients with chronic heart failure. *Circulation, 105,* 143–145.

Bernardi, L., Spadacini, G., Bellwon, J., Hajric, R., Roskamm, H., & Frey, A. W. (1998). Effect of breathing rate on oxygen saturation and exercise performance in chronic heart failure.Lancet, 351, 1308–1311.*British Medical Journal (Clin Res Ed), 291,* 1077 . doi: 10.1136/bmj.291.6502.1077

Birkel, D. A. & Edgren, L. (2000). Hatha Yoga: Improved vital capacity of college students. *Alternative Therapy and Health Medicine, 6*(6), 55-63.

Blanc-Gras, N., Benchetrit, G. & Gellego, J. (1996). Voluntary control of breathing pattern in asthmatic children. *Perceptual Motor Skills, 83* (3 Pt 2), 1384-1386.

Brotto, L. A., Krychman, M., & Jacobson, P. (2008). Eastern approaches for enhancing women's sexuality: Mindfulness, acupuncture, and yoga. *Journal of Sex Medicine, 5,* 2741-2748.

Brown, R. P., & Gerbarg, P. L. (2009). Yoga breathing, meditation, and longevity. *Annals of the New York Academy of Sciences, 1172,* 54–62.

Brown, R. P. & Gerbarg, P. L. (2005). Sudarshan Kriya Yoga Breathing in the treatment of stress, anxiety, and depression: Part I – Neurophysiological model. *Journal of Alternative and Complementary Medicine, 11,* 189–201.

Brown, R. P. & Gerbarg, P. L. (2005). Sudarshan KriyaYoga breathing in the treatment of stress, anxiety, and depression: Part II: Clinical applications and guidelines. *Journal of Alternative and Complementary Medicine, 11,* 711-717.

Burton, J. (2010, February). *WHO Healthy Workplace Framework and Model, Background and Supporting Literature and Practices.* Submitted to Evelyn Kortum WHO Headquarters, Geneva, Switzerland. Retrieved at www.who.int/entity/occupational../healthy_workplace_framework.

Butera, K. (2010). Yoga therapy for the digestive health. *Yoga Living, xii*(ii), 14. Retrieved from http://yogalivingmagazine.com/wp-content/issues/2010/sept/YogaWebFall10%201_16.pdf.

Cappo, B. M. & Holmes, D. S. (1984). The utility of prolonged respiratory exhalation for reducing physiological and psychological arousal in non-threatening and threatening situations. *Journal of Psychosomatic Research, 28,* 265–273.

Carney, R. M., Saunders R. D., Freedland, K. E., Stein, P., Rich, M. W., & Jaffe, A. S. (1995). Association of depression with reduced heart rate variability in coronary artery disease. *American Journal of Cardiology, 76,* 562–564.

Chaoul, M. A. & Cohen, L. (2010). Rethinkging Yoga and the Application of Yoga in Modern Medicine. *Crosscurrents, 60*(2), 144-167. doi:10.1111/j.1939-3881.2010.00117.x

Chattha, R., Nagarathna, R., Padmalatha, V., & Nagendra, H. (2008). Effect of yoga on cognitive functions in climacteric syndrome: a randomized control study. *BJOG, 115,* 991-1000. doi: 10.1111/j.1471-0528.2008.01749.x

Chuntharapat, S., Petpichetchian, W., & Hatthakit, U. (2008). Yoga during pregnancy: Effects on maternal comfort, labor pain and birth outcomes. *Complementary Therapies in Clinical Practice, 14,* 105–115.

Claude, J., & Zamor, G. (2003). Workplace spirituality and organizational performance. *Public administration review, 63*(3): 355-362.

Damodaran, A., Malathi, A., Patil, N., Shah, N., Suryavansihi & Marathe, S. (2002). Therapeutic potential of yoga practices in modifying cardiovascular risk profile in middle aged men and women. *Journal of the Association of Physicians of India, 50*(5), 633-640.

deJong, A. (2009). Cardiovascular disease: Using a polypill, lifestyle modification, or a combined approach to reducing overall risk. *ACSM's Health & Fitness Journal, 13*(6), 38-40.

Dhikav, V., Karmarkar, G., Gupta, M., & Anand, K. S. (2007). Yoga in premature ejaculation: A comparative trial with fluoxetine. *Journal of Sex Medicine, 4,* 1726–1732.

Dhikav, V., Karmarkar, G., Gupta, R., Verma, M., Gupta, R., Gupta, S., & Anand, K. S. (2010). Yoga in female sexual functions. *Journal of Sex Medicine, 7,* 964–970.

Epel, E., Daubenmier, J., Moskowitz, J. T., Folkman, S., & Blackburn, E. (2009). Can meditation slow rate of cellular aging? Cognitive stress, mindfulness, and telomeres. Can meditation slow rate of cellular aging? Cognitive stress, mindfulness, and telomeres. *Annals of New York Academy of Science, 1172,* 34-53.

Ernst, E., & Soo, M. L. (2010). How effective is yoga? A concise overview of systematic reviews. *Complementary Therapies, 15*(4), 274–279. doi:10.1111/j.2042-7166.2010.01049.x

Evans, S., Tsao, J. C. I., Sternlieb, B., & Zeltzer, L. K. (2009). Using the biopsycosocial model to understand the health benefits of yoga. *Journal of complementary and integrative medicine, 6*(1): Article 15. doi: 10.2202/1553-3840.1183

Falus, A., Marton, I., Borbényi, E., Tahy, A., Karádi, P., Aradi, J., ...Kopp, M. (2010). The 2009 Nobel Prize in Medicine and its surprising message: style is associated with telomerase activity. *Orv Hetil., 13,* 151(24), 965-970.

Field, T. (2010). Yoga clinical research review. *Complementary Therapies in Clinical Practice xxx,* 1-8.

Fokkema, D. S. (1999). The psychobiology of strained breathing and its cardiovascular implications: A functional system review. *Psychophysiology, 36*(2), 164-175.

Fox, S., & Spector, P. E. (2005). *Counterproductive work behavior: Investigations of actors and targets.* Washington, DC: American Psychological Association.

Friedman, B. H. & Thayer, J. F. (1998). Autonomic balance revisited: panic anxiety and heart rate variability. *Journal of Psychosomatic Research, 44*, 133–151. from http://www.who.int/chp/chronic_disease_report/en/.

Garfinkel, M., & Schumacher, H. R. Jr. (2000).Yoga. *Rheumatic Disease Clinics of North America, 26*, 123– 32.

Glover, M. J., Grerenlund, K. J., Ayala, C. & Croft, J. B. (2005). Racial/ethnic disparities in prevalence, treatment, and control of hypertension V United States, 1999-2002. *Morbidity and Mortality Weekly Report, 54*, 7-9.

Gokal, R., & Shillito, L. (2007). Positive impact of yoga and pranayam on obesity, hypertension, blood sugar, and cholesterol: A pilot assessment. *Journal of Alternative and Complementary Medicine, 13*, 1056–1057.

Hagins, M., Moore, W. & Rundle, A. (2007). Does practicing hatha yoga satisfy recommendations for intensity of physical activity which improves and maintains health and cardiovascular fitness? *BMC Complementary and Alternative Medicine, 30*, 7-40

Haug, T. T., Svebak, S., Hausken, T., Wilhelmsen, I., Berstad, A. & Vrsin, H. (1994). Low vagal activity as mediating mechanism for the relationship between personality factors and gastric symptoms in functional dyspepsia. *Psychosomatic Medicine, 56*, 181–186.

Herrick, C. M., & Ainsworth, A. D. (2000). Yoga as a Self-Care Strategy. *Nursing Forum, 35* (2), 32-36.

Hunnicutt, D. & Chapman, L. S. (2006). Planning wellness getting off a good start. *Absolute Advantage, 5*(4), 4. Retrieved from http://www.welcoa.org/freeresources/pdf/financial_wellness.pdf

Iyengar, B. K. S. (1976). *Light on Yoga* (2nd ed.). New York: Schocken Books.

Jacobs, G. D. (2001). Clinical applications of the relaxation response and mind-body interventions. *Journal of Alternative and Complementary Medicine, 7* (Suppl 1), S93-S101.

Jain, S. C. & Talukdar, B. (1993). Evaluation of Yoga Training Programmae for Patients of bronchial asthma. *Singapore Medical Journal, 34*(4), 306-308.

Joshi, L. N., Joshi, V. D. & Gokhale, L. V. (1992). Effect of short term Pranayama practice on breathing rate and ventilatory functions of lung. *Indian Journal of Physiology and Pharmacology, 36*, 105-108.

Khalsa, S. (2004). Bibilometirc study on therapeutic interventions of Yoga. *Indian Journal of Physiology and Pharmacology, 48*(3), 269-285. Retrieved from http://www.ijpp.com/vol48_3/vol48_no3_spl_invt_art.pdf

Khanam, A. A., Sachdeva, U., Guleria, R., & Deepak, K. K. (1996). Study of pulmonary and autonomic functions of asthma patients after yoga training. *Indian Journal of Physiology and Pharmacology, 40*(4), 318-324.

Khatri, K., Goyal, A. K., Gupta, P. N., Mishra, N., & Vyas, S. P. (2008). Plasmid DNA loaded chitosan nanoparticles for nasal mucosal immunization against hepatitis B. *International Jouranl of Pharmacology, 354*(1-2), 235-41.

King, D. B. (2009). A Viable Model and Self-Report Measure of Spiritual Intelligence. *International Journal of Transpersonal Studies, 28*, 68-85.

Kjaer, T. W., Bertelsen, C., Piccini, P., Brooks, D., Alving, J., & Lou, H. C. (2002). Increased dopamine tone during meditation-induced change of consciousness. *Cognitive Brain Research, 13*, 255-259.

Konar, D., Latha, R., & Bhuvaneswaran, J. S. (2000). Cardiovascular responses to head-down-body-up postural exercise (Sarvangasana). *Indian Journal of Physiology and Pharmacology*, 44(4), 392-400.

Kulkarni, D. D., & Bera, T. K. (2009). Yogic exercises and health – a psycho-neuro immunological approach. *Indian Journal of Physiology and Pharmacology*, 53, 3–15.

Kuntsevich, V., Bushell, W. C., & Theise, N. D. (2010).Mechanisms of Yogic Practices in Health, Aging, and Disease. *Mount Sinai Journal of Medicine*, 77, 559–569

Langhorst, J., Mueller, T., Luedtke, R., Franken, U., Paul, A., & Scand, J. (2007). Effects of a comprehensive lifestyle modification program on quality-of-life in patients with ulcerative colitis: a twelve-month follow-up. *Gastroenterology*, 42(6), 734-45.

Lehrer, P., Sasaki, S. & Saito, Y. (1999). Zazen and cardiac variability. *Psychosomatic medicine*, 61, 812-821.

Lipton, L. (2008). Using yoga to treat disease: an evidence based review. *Journal of the American Academy of Physician Assistants*, 21, 38–41.

Macy, D. (2008). 'Yoga in America' market study practitioner spending grows to nearly $6 billion a year. *Yoga Journal: press release*. Retrieved May 6, 2011 from http://www.yogajournal.com/advertise/press_releases/10

Makwana, K., Khirwadkar, N., & Gupta, H. C. (1988). Effect of short term yoga practice on ventilatory function tests. *Indian Journal of Physiology and Pharmacology*, 32(3), 202-8.

Manocha, R., Marks, G. B., Kenchington, P., Peters, D., & Salome, C. M. (2002). Sahaja yoga in the management of moderate to severe asthma: a randomised controlled trial. *Thorax.*, 57(2), 110-5.

McCaffrey, R., Ruknui, P., Hatthakit, U., & Kasetsomboon, P. (2005). The effects of yoga n hypertensive persons in Thailand. *Holistic Nursing Practice*, 19, 173–180.

McCall, T. (2009).*Yoga as Medicine: The Yogic Prescription for Health and Healing*: Bantam. Retrieved from www.DrMcCall.com

Michalsen, A., Grossman, P., Acil, A., Langhorst, J., Lüdtke, R., Esch T, … Dobos, G. J. (2005). Rapid stress reduction and anxiolysis among distressed women as a consequence of a three month intensive yoga program. *Medical Science Monitor*, 11, 555–561.

Moadel, A. B., Shaw, C., Wylie-Rossett, J., Harris, M. S., Patel, S. R., Hall, C. B. & Sparano, J. A. (2007). Randomized controlled trial of yoga among a multiethnic sample of breast cancer patients: Effects on quality of life. *Journal of Clinical Oncology*, 25(28), 4387-4395.

Modeling the impact of a comprehensive wellness program. (2010). *Enhancing corporate performance by tackling chronic diseases*. Retrieved May 4, 2011 from at https://members.weforum.org/pdf/Wellness/BCG-Report.pdf

Mohandas, E. (2008). Neurobiology of spirituality. In A. R. Sing and S. A. Singh (Eds), Medicine, mental health, science, religion, and well being, MSM, 6 Jan- Dec 2008.

Monro, R., Nagarathna, R. & Nagendra, H. R. (1995). *Yoga for common ailments*. New York/London: Simon & Schuster.

Murugesan, R., Govindarajulu, N., & Bera, T. K. (2000). Effect of selected yogic practices on the management of hypertension. *Indian Journal of Physiology and Pharmacology*, 44(2):207-10.

Nagarathna, R., & Nagendra, H. R. (1985). Yoga for bronchial asthma: a controlled study. *British Medical Journal*, 291(6502), 1077-1079.

Nagendra, H. R., & Nagarathna, R. (1986). An integrated approach of yoga therapy for bronchial asthma: A 3-54-month prospective study. *Journal of Asthma, 23*(3), 123-37.

Narendran, S., Nagarathna, R., Narendran, V., Gunasheela, S. & Nagendra, H. R. (2005). Efficacy of yoga on pregnancy outcome. *The Journal of Alternative and Complementary Medicine, 11*(2), 237–244

Nayak, N. N., & Shankar, K. (2004). Yoga: a therapeutic approach. *Physical Medicine & Rehabilitation Clinics of North America, 15*, 783–98.

Newcombe, S. (2007). Stretching for health and well-being: Yoga and women in Britain, 1960-1980. *Asian Medicine, 3*, 37-63.

Ornish, D. (2009). Intensive life style changes and health reform. *The Lancet Oncology, 10,* 198-199.

Ornish, D., Scherwitz, L. W., Billings, J. H., Brown, S. E., Gould, K. L., Merritt, T. A., ...Brand, R. J. (1998). Intensive lifestyle changes for reversal of coronary heart disease. *JAMA 280,* 2001–2007.

Ornish, D., Scherwitz, L. W., Doody, R. S., Kesten, D., McLanahan, S. M., Brown, S. E., & Gotto, A. M. (1983). Effects of Stress Management Training and Dietary Changes in Treating Ischemic Heart Disease. *JAMA, 249*(1), 54-59.

Osterziel, K. J., Rohring, N., Dietz, R., Manthey, J., Hecht, J., & Kubler, W. (1988). Influence of captopril on the arterial baroreceptor reflex in patients with heart failure. *European Heart Journal, 9,* 1137–1145.

Pandya, P. (2006). *Antarjagat Ke Yatra Ka Jnana-Vijnana* (Science of Inner Journey). Haridwar, India: Vedmata Gayatri Trust.

Pruzan, P. & Pruzan, M. K. (2001). *Leading with wisdom: Spiritual based leadership in business.* New Delhi: Response Books from SAGE.

Pullen, P. R., Thompson, W. R., Benardot, D., Brandon, L. J., Mehta, P. K., Vadnais, D. S., ...Khan, B. V. (2010). Benefits of yoga for African American heart failure patients. *Medicine and Science in Sports and Exercises, 42*(4):651-7.

Ramsay, H., Scholarios, D. & Harley, B. (2002). Employee and high-performance work system: testing inside the black box'. *British Journal of Industrial Relations, Worker.* Ithaca, NY: Cornell University Press.

Raub, J. A. (2002). Psychophysiologic effects of Hatha Yoga on musculoskeletal and cardiopulmonary function: a literature review. *Journal of Alternative and Complementary Medicine, 8*(6), 797-812.

Riley, D. (2004). Hatha yoga and the treatment of illness. *Alternative Therapy and Health Medicine, 10*(2), 20-1.

Ritz, T. (2001). Relaxation Therapy in adult asthma. Is there new evidence for its effectiveness? *Behavior Modification, 25,* 640-666,

Ross, A. & Thomas, S. (2010). The Health Benefits of Yoga and Exercise: A Review of Comparison Studies. *The Journal of Alternative and Complementary Medicine, 16*(1), 3–12. doi: 10.1089=Acm.2009.0044

Saper, R., Eisenberg, D., Davis, R., Culpepper, L., & Phillips, R. (2004). Prevalence and patterns of adult yoga use in the United States: Results of a national survey. *Alternative Therapy & Health Medicine, 10,* 44–48.

Sathyaprabha, T. N., Murthy, H., & Murthy, B. T. (2001). Efficacy of naturopathy and yoga in bronchial asthma--a self controlled matched scientific study. *Indian Journal of Physiology and Pharmacology, 45*(1), 80-86.

Satyananda, S. (2002). *The Four Chapters on Freedom*. Bihar, India: The Yoga Publication Trust.

Selvamurthy, W., Sridharan, K., Ray, U. S., Tiwary, R. S., Hegde, K. S., Radhakrishan, U., & Sinha, K. C. (1998). A new physiological approach to control essential hypertension. *Indian Journal of Physiology and Pharmacology, 42,* 205–213.

Siegel, D. J. (2009). Mindful awareness, mindsight and neural integration. *The Humanistic Psychologists, 37,* 137- 158.

Singh, N. (2009, October). Yog Ne Napi Puri Dharti. *Kadambini, 12 (49),* 18-23.

Singh, S. M., Longmire, W. P. Jr., & Reber, H. A. (1990). Surgical palliation for pancreatic cancer. The UCLA experience. *Annals of Surgery, 212, 132-139.*

Sivananda, S. (2003). *The Bhagavad Gita* (11th ed.). Uttarakhand, India: The Divine Life Society.

Sovik, R. (2000). The science of breathing- the yogic view. *Progressive Brain Research, 122,* 491-505.

Spicuzza, L., Gabutti, A., Porta, C., Montano, N., & Bernardi, L. (2000). Yoga and chemoreflex response to hypoxia and hypercapnia. *Lancet, 356*(9240), 1495–1496.

Stanescu, D. (1990). Yoga breathing exercises and bronchial asthma. *Lancet, 336*(8724), 1192.

Stanescu, D. C., Nemery, B., Veriter, C., & Marechal, C. (1981). Pattern of breathing and ventilator response to CO_2 in subjects practicing hatha-yoga. *Journal of Applied Physiology, 51,* 1625-1629.

Sterling, P. (2004). Principles of allostasis: Optimal design, predictive regulation, pathophysiology, and rational therapeutics. In Schulkin, J. (Ed.), *Allostasis, Homeostasis, and the Costs of Physiological Adaptation* (pp. 17–64). Cambridge: Cambridge University Press.

Stück, M., Meyer, K., Rigotti, T., Bauer, K., & Sack, U. (2003). Evaluation of a yoga based stress management training for teachers: Effects on immunoglobulin A secretion and subjective relaxation. *Journal for Meditation and Meditation Research, 3,* 59-68.

Taimni, I. K. (1961). *The science of yoga*. Madaras, India: The Theosophical Publishing House.

Vahia, N. S., Vinekar, S. L. & Doongaji, D. R. (1966). Some ancient Indian concepts in the treatment of psychiatric disorders. *British Journal of Psychiatry, 112*(492), 1089-1096.

Vaze, N. & Joshi, S. (2010). Yoga and Menopausal Transitiion. *Journal of Mid-life Health, 1*(2), 56-58. doi: 10.4103/0976-7800.76212

Vedanthan, P. K., Kesavalu, L. N., Murthy, K. C., Duvall, K., Hall, M. J., Baker, S., Nagarathna, S. (1998). Clinical study of yoga techniques in university students with asthma: a controlled study. *Allergy and Asthma Proceedings, 19*(1), 3-9.

West, J., Otte, C,. Geher, K., Johnson, J. & Mohr, D. C. (2004). Effects of Hatha yoga and African dance on perceived stress, affect, and salivary cortisol. *Annals of Behavioural Medicine, 28,* 114–118.

Wolfson, N. (n. d). *Yoga journal, incorporating yoga: In boardrooms from Manhattan to Silicon Valley, the mantra "let's do lunch" is being replaced by "let's do yoga."* Retrieved May 17, 2010 from http://www.yogajournal.com/lifestyle/294.

Yadav, R. K., & Das, S. (2001). Effect of yogic practice on pulmonary functions in young females. *Indian Journal of Physiology and Pharmacology, 45*(4), 493-6.

Zohar, D. (2005). Spiritually intelligent people. *Leader to leader journal, 38,* 45-51.

Permissions

The contributors of this book come from diverse backgrounds, making this book a truly international effort. This book will bring forth new frontiers with its revolutionizing research information and detailed analysis of the nascent developments around the world.

We would like to thank Dr. Ganesh R. Naik, for lending his expertise to make the book truly unique. He has played a crucial role in the development of this book. Without his invaluable contribution this book wouldn't have been possible. He has made vital efforts to compile up to date information on the varied aspects of this subject to make this book a valuable addition to the collection of many professionals and students.

This book was conceptualized with the vision of imparting up-to-date information and advanced data in this field. To ensure the same, a matchless editorial board was set up. Every individual on the board went through rigorous rounds of assessment to prove their worth. After which they invested a large part of their time researching and compiling the most relevant data for our readers. Conferences and sessions were held from time to time between the editorial board and the contributing authors to present the data in the most comprehensible form. The editorial team has worked tirelessly to provide valuable and valid information to help people across the globe.

Every chapter published in this book has been scrutinized by our experts. Their significance has been extensively debated. The topics covered herein carry significant findings which will fuel the growth of the discipline. They may even be implemented as practical applications or may be referred to as a beginning point for another development. Chapters in this book were first published by InTech; hereby published with permission under the Creative Commons Attribution License or equivalent.

The editorial board has been involved in producing this book since its inception. They have spent rigorous hours researching and exploring the diverse topics which have resulted in the successful publishing of this book. They have passed on their knowledge of decades through this book. To expedite this challenging task, the publisher supported the team at every step. A small team of assistant editors was also appointed to further simplify the editing procedure and attain best results for the readers.

Our editorial team has been hand-picked from every corner of the world. Their multi-ethnicity adds dynamic inputs to the discussions which result in innovative outcomes. These outcomes are then further discussed with the researchers and contributors who give their valuable feedback and opinion regarding the same. The feedback is then collaborated with the researches and they are edited in a comprehensive manner to aid the understanding of the subject.

Apart from the editorial board, the designing team has also invested a significant amount of their time in understanding the subject and creating the most relevant covers. They scrutinized every image to scout for the most suitable representation of the subject and create an appropriate cover for the book.

The publishing team has been involved in this book since its early stages. They were actively engaged in every process, be it collecting the data, connecting with the contributors or procuring relevant information. The team has been an ardent support to the editorial, designing and production team. Their endless efforts to recruit the best for this project, has resulted in the accomplishment of this book. They are a veteran in the field of academics and their pool of knowledge is as vast as their experience in printing. Their expertise and guidance has proved useful at every step. Their uncompromising quality standards have made this book an exceptional effort. Their encouragement from time to time has been an inspiration for everyone.

The publisher and the editorial board hope that this book will prove to be a valuable piece of knowledge for researchers, students, practitioners and scholars across the globe.

List of Contributors

Ricardo Dias and Rui Lima
Polytechnic Institute of Bragança, ESTiG/IPB, C. Sta. Apolonia, Bragança, Portugal. CEFT, Faculty of Engineering of the University of Porto (FEUP), R. Dr. Roberto Frias, Porto, Portugal.

Valdemar Garcia
Polytechnic Institute of Bragança, ESTiG/IPB, C. Sta. Apolonia, Bragança, Portugal.

Andrei Doncescu and Sebastien Regis
University of Toulouse, LAAS CNRS UPR 8001, France.

Katsumi Inoue
National Institute of Informatics, Japan.

Nathalie Goma
IPBS CNRS, France.

Airton Ramos
State University of Santa Catarina, Brazil.

Andrea Lima Schneider
University of Joinville, Brazil.

Qiyi Tang
Department of Microbiology/AIDS Research Program, Ponce School of Medicine, Ponce, PR, USA.

Benjamin Silver and Hua Zhu
Department of Microbiology and Molecular Genetics, New Jersey Medical School, Newark, NJ, USA

S. Lakshmana Prabu
Anna University of Technology, Tiruchirappalli, Tamil Nadu, India.

T. N. K. Suriyaprakash
Periyar College of Pharmaceutical Sciences, The Tamil Nadu Dr. M.G.R. Medical University, Chennai, Tamil Nadu, India

José María De la Roca Chiapas
Universidad de Guanajuato, División de Salud, Departamento de Psicología, Mexico. Organización Filosófica Nueva Acrópolis México, México.

Yuelin Zhang
Tokyo University of Agriculture and Technology, Japan.

Shigeru Aomura and Hiromichi Nakadate
Tokyo Metropolitan University, Japan.

Satoshi Fujiwara
Yokohama City University, Japan.

Damir Kralj
Ministry of the Interior, PD Karlovac, Croatia

Nevena Ackovska
University Sts Cyril and Methodius, Institute of Informatics, Macedonia

Liljana Bozinovska and Stevo Bozinovski
South Carolina State University, USA.

Mahmud Hasan
Computer Science Program, Faculty of Science, University Brunei Darussalam, Brunei.

Estevam Barbosa de Las Casas, João Batista Novaes Jr., Elissa Talma, Willian Henrique Vasconcelos, Tulimar P. Machado Cornacchia, Iracema Maria Utsch Braga, Carlos Alberto Cimini Jr. and Rodrigo Guerra Peixoto
Universidade Federal de Minas Gerais, Brazil Universidade de Campinas, Brazil.

Rudra B. Bhandari
University of Patanjali, Haridwar, Uttarakhand, India.

Churna B. Bhandari
Department of Physics, Case Western University, Ohio, USA.

Balkrishna Acharya
University of Patanjali, Haridwar, Uttarakhand, India.

Pranav Pandya
Dev Sanskriti Vishwavidyalaya, Uttarakhand, India.

Kartar Singh
University of Patanjali, Haridwar, Uttarakhand, India.

Vinod K. Katiyar
Department of Mathematics, Indian Institute of Technology, Roorkee, Uttarakhand, India.

Ganesh D. Sharma
Department of Yogic Sciences, University of Patanjali, Haridwar, Uttarakhand, India.

Printed in the USA
CPSIA information can be obtained
at www.ICGtesting.com
JSHW011452221024
72173JS00005B/1040

9 781632 390998